Texts and Monographs in Physics

Series Editors:

R. Balian, Gif-sur-Yvette, France
W. Beiglböck, Heidelberg, Germany
H. Grosse, Wien, Austria
E. H. Lieb, Princeton, NJ, USA
N. Reshetikhin, Berkeley, CA, USA
H. Spohn, München, Germany
W. Thirring, Wien, Austria

Springer
Berlin
Heidelberg
New York
Hong Kong
London
Milan
Paris
Tokyo

Physics and Astronomy ONLINE LIBRARY
springeronline.com

L.N. Savushkin   H. Toki

The Atomic Nucleus as a Relativistic System

With 45 Figures and 35 Tables

 Springer

UNIVERSITY OF ROCHESTER

Professor Lev N. Savushkin
St. Petersburg University for Telecommunications
Department of Physics
nab.r.Moika, 61
191186 St. Petersburg, Russia
E-Mail: savushkin@peterlink.ru

Professor Hiroshi Toki
Osaka University, RCNP
Mihogaoka 10-1, Ibaraki
Osaka 567-0047, Japan
E-mail: toki@rcnp.osaka-u.ac.jp

Library of Congress Cataloging-in-Publication Data.

Savushkin, L.N. (Lev Nikolaevich), 1939–
The atomic nucleus as a relativistic system / L.N. Savushkin, H. Toki.
p.cm. – (Texts and monographs in physics, ISSN 0172-5998)
Includes bibliographical references and index.
ISBN 3-540-40492-9 (acid-paper)
1. Nuclear physics. 2. Relativity (Physics) I. Toki, H. II. Title. III. Series.
QC782.S38 2003   539.7–dc21   2003054429

ISSN 0172-5998
ISBN 3-540-40492-9 Springer-Verlag Berlin Heidelberg New York

This work is subject to copyright. All rights are reserved, whether the whole or part of the material is concerned, specifically the rights of translation, reprinting, reuse of illustrations, recitation, broadcasting, reproduction on microfilm or in any other way, and storage in data banks. Duplication of this publication or parts thereof is permitted only under the provisions of the German Copyright Law of September 9, 1965, in its current version, and permission for use must always be obtained from Springer-Verlag. Violations are liable for prosecution under the German Copyright Law.

Springer-Verlag is a part of Springer Science+Business Media

springeronline.com

© Springer-Verlag Berlin Heidelberg 2004
Printed in Germany

The use of general descriptive names, registered names, trademarks, etc. in this publication does not imply, even in the absence of a specific statement, that such names are exempt from the relevant protective laws and regulations and therefore free for general use.

Typesetting by the authors
Final layout by Frank Herweg, Leutershausen
Cover design by *design & production* GmbH, Heidelberg

Printed on acid-free paper   57/3141/tr - 5 4 3 2 1 0

Preface

Relativity plays an important role in atomic nuclei. This importance of relativity has been increasingly recognized in recent years; this can be seen from the recent rapid increase in the number of publications in the field of nuclear physics. Relativity is important because both the scalar and the vector interactions, which are the most important ingredients of the relativistic framework for describing nuclei, have been realized to be very large in comparison with the magnitude of the nucleon mass. As a consequence, strong spin–orbit splittings naturally arise; these have a central role in providing the correct nuclear shell structure. Overwhelming success has been demonstrated in the description of the spin observables in proton scattering from nuclei within the relativistic version of the impulse approximation.

Another important role of relativity has been found in the description of nuclear matter. The relativistic Brueckner–Hartree–Fock theory is able to bring the saturation properties of nuclear matter close to the experimental values, which was not possible in the corresponding nonrelativistic approach. This is again due to the strong scalar and vector interactions, which cancel each other out in the central potential in the nonrelativistic description, but provide very large density-dependent repulsion. Pseudospin symmetry can be generated naturally and treated as a relativistic symmetry in the Dirac phenomenology involving attractive scalar and repulsive vector potentials.

Because of these essential features of the relativistic description, we can describe many observables of nuclei using relativistic many-body theory. It seems that all of the observables will be reformulated in terms of the relativistic theory in the near future. Hence, we felt it was a good time to write this book in order to collect together important material on this subject. This book is aimed at advanced students and researchers. We have tried to write about all of the ingredients of the relativistic theory and to give a detailed comparison with experimental results. We have also tried to go through the literature and to summarize its essential content. Hence, a large number of references is provided in the book for the further use of readers.

We have profited from fruitful collaborations on this subject and discussions with many researchers in the field. We are particularly indebted to V.N. Fomenko, B.L. Birbrair, S. Marcos, Y. Sugahara, K. Sumiyoshi, I. Tanihata, D. Hirata, B. Carlson, H. Shen, K. Oyamatsu, and R. Brockmann.

We are very thankful also to Adelheid Duhm, Elke Sauer, and Petra Treiber for the kind help and fruitful cooperation in preparing our manuscript for publication.

Osaka,
August 2003

Lev Savushkin
Hiroshi Toki

Contents

1	**Introduction**		1
2	**Relativistic Quantum Mechanics and Fields**		7
	2.1	The Dirac Equation	7
	2.2	Solutions for Free Particles	8
	2.3	Relativistic Free-Field Theories	11
		2.3.1 Real Scalar Field	12
		2.3.2 Vector Field	13
		2.3.3 Electromagnetic Field	13
		2.3.4 Spinor Dirac Field	14
		2.3.5 Real Scalar Field with Quartic Self-Interaction	14
	2.4	The Dirac Equation in a Central Potential	20
3	**Basic Features of the Meson Theory of Nucleon–Nucleon Interactions**		25
	3.1	One-Boson Exchange Potentials in Configuration Space	25
	3.2	One-Boson Exchange Potentials in Momentum Space	32
4	**The Relativistic Mean-Field Approximation for Nuclear Structure**		39
	4.1	General Characteristics of the Relativistic Framework	39
		4.1.1 Large Scalar and Vector Fields	39
		4.1.2 Spin–Orbit Force	40
		4.1.3 Saturation	40
	4.2	Relativistic Mean-Field Approximation for Finite Nuclei	48
	4.3	Relation of the RMF Model to the Skyrme–Hartree–Fock Approach	52
	4.4	Renormalization of the Kinetic Energy to Obtain Saturation in Nuclear Matter	62
	4.5	Relativistic Mean-Field Approximation for Deformed Nuclei	64
	4.6	Optimal Parameter Sets for the Relativistic Mean-Field Model	67

5 Electromagnetic Interactions of Nucleons in the Relativistic Framework 75
5.1 The Vector Dominance Model and the Nuclear Coulomb Potential 75
5.1.1 Results of Calculations 80
5.2 Nuclear Magnetic Moments in the Relativistic Approach ... 82

6 The Relativistic Approach to Nucleon–Nucleus Scattering 89
6.1 Energy Dependence of the Real Part of the Optical Potential 89
6.2 Coulomb–Nuclear Interference Effects in Nucleon–Nucleus Scattering 92
6.3 Relativistic Impulse Approximation 94
6.4 $\bar{p}$–Nucleus Scattering 96

7 Pion Dynamics and Chiral Symmetry 99
7.1 Pionic Excitations in Nuclear Matter 99
7.2 Equations of Motion for a Pion Field in a Nuclear Medium 101
7.3 Pionic Polarization in Infinite Nuclear Matter 104
7.4 Contribution of the Δ_{33} Resonance to the Pion Polarization Operator 107
7.5 Basic Equations of the Linear σ and σ–ω Models 113
7.6 Chiral σ–ω Model for Finite Nuclei 122
7.7 Effective Gauge-Invariant Nuclear Lagrangian 125
7.8 Mean-Field Results for Nuclear Matter and Finite Nuclei ... 130

8 The Relativistic Hartree–Fock Approach 137
8.1 The Relativistic Hartree–Fock Lagrangian 137
8.2 The Relativistic Hartree–Fock Approach for Symmetric Nuclear Matter 141
8.3 The Relativistic Hartree–Fock Approach for Finite Nuclei .. 146
8.4 Determination of Parameters and Numerical Results 148
8.5 The Relativistic Hartree–Fock Approach with Meson Self-Coupling Terms 154
8.6 Spin–Orbit Interaction 160
8.7 Pseudospin as a Relativistic Symmetry 165

9 Brueckner–Hartree–Fock Methods for Nuclear Matter and Finite Nuclei 173
9.1 The Brueckner–Hartree–Fock Approach 173
9.2 The Brueckner–Bethe–Goldstone Method for Internucleon Correlations 175
9.3 Δ-Isobar for Nuclear Interactions and Nuclear Structure ... 182

	9.4	Relativistic Extension of the BHF Theory for Nuclear Matter	187
	9.5	Finite Nuclei	194
10	**Excited Nuclear States in the Relativistic RPA Method**		201
	10.1	RRPA Method for Nuclear Matter and Finite Nuclei	201
	10.2	RRPA with Nonlinear Interactions for Giant Resonances	203
	10.3	Construction of the Meson Propagators	206
	10.4	Results for Collective States and Giant Resonances	208
	10.5	Cranked Relativistic Mean-Field Theory	216
11	**The Equation of State of Nuclear Matter for Supernovas and Neutron Stars**		219
	11.1	Thomas–Fermi Method for Nonuniform Matter	219
	11.2	Equation of State of Nuclear Matter	223
	11.3	Neutron Star Matter and Neutron Star Profiles	232
12	**Alternative Relativistic Models**		239
	12.1	Quark–Meson Coupling Models	239
	12.2	The Relativistic Point-Coupling Model	246
	12.3	Scalar Derivative Coupling Models	253
13	**Some Recent Applications of Relativistic Nuclear Theory**		261
	13.1	Mean Fields in Colliding Nuclear Matter	261
	13.2	Hartree–Fock–Bogoliubov Approximation	263
	13.3	Spin–Orbit Splitting for Single-Particle and Single-Hole Energies	266
	13.4	The Anomalous Kink in the Isotope Shifts of Pb Nuclei	267
	13.5	Electroweak Interactions in Nuclei in the Relativistic Framework	268
	13.6	Hypernuclei in the Relativistic Framework	270
	13.7	Theoretical Analysis of $A(e,e'p)B$ Reactions	271
	13.8	Exclusive Pion Production in Nucleon–Nucleus Scattering	272
	13.9	The Role of Relativity in Few-Body Systems	274
	13.10	Systematic Study of Even–Even Nuclei up to the Drip Lines	276
	13.11	Exotic Nuclei and Superheavy Nuclei	279
	13.12	Dilepton Production by Bremsstrahlung of Meson Fields in Nuclear Collisions	281
	13.13	(p, n) Spin Experiments and Relativity in Nuclear Physics	282
	13.14	Role of Currents (ω and ρ Fields)	284
	13.15	Fission Barriers	285
	13.16	Chiral Dynamics and Saturation of Nuclear Structure	286

14 Summary and Outlook ... 289

A Appendices ... 293
- A.1 Four-Dimensional Notation and the Dirac Matrices ... 293
- A.2 Properties of the Ground State of Nuclear Matter in the Walecka Model ... 296
- A.3 General Form of Local Dirac Equation ... 297
- A.4 Equivalent Local Dirac Nuclear Models ... 299
- A.5 Nucleon Effective Mass in the Nuclear Medium ... 300
- A.6 Radial Equations for the Upper and Lower Components $G(r)$ and $F(r)$ ... 302
- A.7 Boundary Conditions for Wave Functions and Meson Potentials ... 304
- A.8 Generalized Weinberg Transformation ... 305
- A.9 Expansions of the Vertex Functions for Various Mesons ... 306
- A.10 Globally Chirally Invariant Lagrangian for the Model with an Axial Meson ... 307
- A.11 Direct and Exchange Matrix Elements for the Two-Body Spin–Orbit Force ... 309
- A.12 Hartree–Fock Procedure for the Point-Coupling Model ... 318

References ... 323

Index ... 347

1 Introduction

Since the beginning of the 1970s, there has been a broad development in the direction, where the nucleus is treated as a relativistic system. Many new, important results have been obtained in this field during the last 30 years. In particular, it has been established that the anomalously large spin–orbit coupling in nuclei, which is treated purely phenomenologically in the nonrelativistic framework, can be explained naturally in the relativistic approach. Also, pseudospin symmetry has been the subject of many investigations, for both spherical and deformed nuclei. Recently it has been understood that the notion of pseudospin has deep roots in the relativistic theory.

One of the advantages of the relativistic (Dirac) description, compared with the Schrödinger description, is that in the framework of the relativistic theory a natural explanation is provided for the energy dependence of the real part of the optical potential at intermediate energies. The optical potential has the characteristic shape of a "wine-bottle bottom" when expressed in the Schrödinger form. This dependence has been confirmed experimentally by investigations of elastic scattering of protons by nuclei at intermediate energies.

The relativistic form of the Kerman–McManus–Thaler theory has been widely developed. This approach does not contain free parameters. Relativistic calculations have been performed in the framework of this theory of the total cross section, analyzing power, and spin rotation function for elastic scattering of polarized protons with energies ~ 500 MeV by ^{40}Ca and ^{208}Pb. It has been demonstrated that the transition from the nonrelativistic description to the relativistic one leads to a dramatic improvement in the agreement with experiment for polarization observables.

Even in the Hartree approximation with the inclusion of nonlinear self-interactions of mesonic fields, the relativistic treatment provides good descriptions of the experimental data for the total binding energies of nuclei, the charge distribution radii, and the separation energies. The binding energies and the charge radii are reproduced to an accuracy of less than a percent. The density distributions of spherical doubly magic nuclei obtained by the relativistic approach are in good agreement with the electron scattering data. It has been shown also that the saturation property, a fundamental property of nuclei, is a relativistic kinematical effect. Self-consistent relativistic calculations have been performed for both spherical and deformed nuclei.

The kink effect in the behavior of the r.m.s. charge radii in the Pb isotopic chain appears to be an inherent feature of the standard relativistic mean-field theory.

The relativistic mean-field theory has been extended into a rotating frame as a cranked relativistic mean-field (CRMF) theory and applied to a systematic investigation of all the experimentally known superdeformed rotational bands in the $A \sim 140$–150 region. It is remarkable that the CRMF theory (with only six or seven parameters) is able to generate a microscopic description of rotational bands in superdeformed nuclei.

The further success of the relativistic theory has been connected with the development of the relativistic Hartree–Fock method which demonstrates a very important role of pions, especially in describing spin–orbit splittings. This development is particularly important for incorporating chiral symmetry into nuclear physics.

Combination of the vector dominance model and the relativistic mean-field theory has shown that there is a natural way to describe the electromagnetic structure of nucleons, i.e. this theory can be successfully applied to the description of composite particles. This point that a nucleon is a composite particle has also been demonstrated by the quark–meson coupling model, in which the mesons couple not to point-like nucleons but to confined quarks.

It is a challenging aim to describe nucleon–nucleon (NN) forces and nuclear properties on the basis of QCD, an aim that has not yet been achieved. One of the possible ways to solve this problem at the present stage is to impose the QCD symmetries in the context of nuclear structure. For example, chiral symmetry is a symmetry of QCD (in the limit of vanishing quark masses) and should be considered as an approximate symmetry of the strong interactions. Important contributions have also been made to this field. Models with both linear and nonlinear realizations of chiral symmetry have been developed and successfully applied to the description of the ground-state properties of finite nuclei and nuclear matter.

The relativistic mean-field (RMF) theory has been successfully extended to unstable nuclei. The results obtained in this field compare extremely well with the existing data. A large amount of experimental data on unstable nuclei is expected from forthcoming accelerators in the near future.

The success of the relativistic Hartree and Hartree–Fock approximations has ensured a good basis for the study of nuclear excitations in relativistic models, the relativistic random-phase approximation (RPA) method providing a reasonable way to construct collective states. Relativistic RPA calculations have been extended to include nonlinear terms (corresponding to self-interactions of the meson fields), which enables us to use experimental data on excited states for further investigations of the relativistic models. An alternative method to describe nuclear excitations is the time-dependent relativistic mean-field theory, whose development is also extremely important.

Relativistic effects may be most important for fields of anomalous parity (for which the multipole order determines the spatial parity through the factor $(-)^{L+1}$), i.e. for fields that couple the upper and lower components of the nucleon wave function. Such effects can be very important in the investigation of the stability of nuclear ground states in external fields; fields whose symmetry differs from the symmetry of the ground state are of great interest.

Relativistic effects in nuclei can also be important in studies of electromagnetic and β transitions and of processes determined by the Coulomb interaction. In addition, a relativistic approach, in particular the use of relativistic wave functions, is needed to describe characteristic phenomena in the region of intermediate energies, for example (p, π^+) reactions.

The RMF theory has been used also to construct a table of the equation of state (EOS) covering a wide variety of densities, chemical compositions and temperatures for use in supernova simulations. This table has been applied to calculate the properties of neutron stars, and large-scale numerical simulations of the birth of neutron stars and supernova explosions are being prepared.

An advantage of the relativistic approach to the description of nuclear properties considered in the present book is that in this case it is possible to give a unified description not only of the interaction of nucleons with nuclei but also of the interaction of nuclei with other elementary particles, namely antiprotons, pions, hyperons, and kaons.

The development of the relativistic description of nuclei is inseparably connected with the success of the meson theory, in particular the meson theory of nucleon–nucleon interactions. During the last two decades the meson approach to the description of nuclear systems has developed very rapidly, and extensive reviews of the meson theory of nuclear forces have been published. Important successes have been achieved in the description of nuclear matter and finite nuclei with allowance for meson degrees of freedom, the $\Delta(1236)$ resonance, three-body forces, etc.

The relativistic theory of nuclear structure is being developed in two basic directions. In the first case, one investigates effective Lagrangians with a set of mesons whose parameters are fitted to reproduce the observed saturation properties of nuclear matter and finite nuclei, this being done either in the Dirac–Hartree or in the Dirac–Hartree–Fock approximation. This theory, involving nonlinear self-interactions of the meson fields, has achieved a high degree of sophistication and can describe quantitatively many properties of nuclear ground states.

The second direction is based on the Dirac–Brueckner–Hartree–Fock theory of nuclear structure. The point of departure in this case is provided by the NN forces in free space (of the type known as one-boson exchange potentials, OBEPs), which are fitted to the NN interaction in free space. The

Brueckner G-matrix is then completely determined; it does not contain any more free parameters. In such an approach, the saturation properties are not parameterized but must be obtained directly on the basis of the NN forces. A great achievement of the relativistic approach is that, in contrast to the nonrelativistic Brueckner theory, it leads to a saturation point quite close to the empirical value (and not on the Coester line).[1] Theories of this type have been developed for both infinite nuclear matter and finite nuclei.

It is a long-standing problem in nuclear-structure physics as to whether the hadrons (nucleons and mesons) in a nuclear medium have the same properties as bare particles. In a nonrelativistic theory, there is no other choice than to consider the intrinsic properties of nucleons as fixed and hence density-independent. The relativistic mean-field approximation is a type of theory in which the properties of a nucleon in a medium (the nucleon mass, charge radius, etc.) become interaction-dependent quantities. The nucleon mass, for example, is greatly reduced in the relativistic theory in nuclear matter in comparison with the free nucleon mass, its value being uniquely determined by the spin–orbit coupling. The properties of mesons (coupling constants and masses) also appear to be density-dependent.

In recent years there has been a strong growth of interest in the relativistic treatment of the properties of nuclear matter and finite nuclei on the same footing. The number of studies in this direction is increasing.

The systematics of the properties of nuclear matter, finite nuclei (both stable nuclei and exotic nuclei, far from the line of β stability), nuclear dynamics, different aspects of astro-nuclear physics, etc. have also been considered in the framework of relativistic theory. All of the above-mentioned subjects are considered in this book.

At this point, we would like to emphasize that this book aims at a presentation of the most recent developments in the field of relativistic many-body theory in nuclear physics. There are good books that explain the fundamentals of relativistic field theory [1–20]. Whenever difficulties arise, we urge readers to refer to those books. We focus, rather on a discussion of the most recent material in the relativistic description of nuclei. However, for the convenience of the reader we have provided a chapter of pedagogical character Chap. 2, containing the elements of relativistic quantum mechanics and fields. We consider Chap. 2 to be the theoretical background of the present consideration.

We have arranged this book as follows. In Chap. 3 we present the basic features of the meson theory of nucleon–nucleon interactions. We introduce

---

[1] Conventional nonrelativistic Brueckner–Hartree–Fock calculations usually give a saturation point at too large a density and/or with too little binding energy with respect to the empirical value. This problem is solved in the nonrelativistic framework by introducing three-nucleon forces. Relativistic Brueckner–Hartree–Fock calculations generate the empirical saturation properties of nuclear matter without introducing a three-nucleon force.

the one-boson exchange potentials, which are the basic ingredient of relativistic many-body theory. We then discuss in Chap. 4 the relativistic mean-field approximation for nuclear matter and finite nuclei. Here we demonstrate the success of the relativistic mean-field approach in reproducing the experimental data. We show how the relativistic theory compares with the popular nonrelativistic method.

In Chap. 5 we discuss the electromagnetic properties of nuclei in the relativistic description. In that chapter, we describe how to take into account the compositeness of the nucleons in terms of the vector dominance model. We discuss also the nuclear magnetic moments in the relativistic framework.

In Chap. 6 we present the relativistic description of nucleon–nucleus scattering. When expressed in terms of the nonrelativistic language, the energy dependence of the optical potential is very nicely reproduced. We discuss also the spin observables, which essentially arise from the relativistic spinors. By introducing the G-parity transformation, we can describe also the scattering of antiprotons by nuclei.

Pions are the Goldstone bosons of the chiral-symmetry breaking. However, they are not included in the relativistic mean-field approximation owing to the spin saturation in the nuclear ground state. In Chap. 7 we describe first the pionic polarizations in nuclei. We introduce also the delta isobars, which couple strongly with nucleons through pions. We take the linear σ model as a realization of the chiral symmetry in a nuclear many-body system. In Chap. 7 we stay at the mean-field level to see the effects of the chiral symmetry on the nuclear ground state.

We then discuss in Chap. 8 the relativistic Hartree–Fock approach to nuclear matter and finite nuclei. We find several interesting effects arising from the Fock terms. Particularly interesting is the role of pions in the spin–orbit splittings for spin-unsaturated nuclei. It is very important to describe nuclei in the Hartree–Fock framework using a chirally symmetric Lagrangian.

The strong omega-meson exchange and also the scalar-meson exchange potentials need special attention when one is solving the nucleon–nucleon system to all orders. This is achieved by using the Brueckner–Bethe–Goldstone equation. We demonstrate in Chap. 9 the success of the Brueckner–Hartree–Fock approach in describing nuclear matter and finite nuclei. This approach provides the microscopic support for the relativistic mean-field and the Hartree–Fock approaches to nuclear many-body systems.

A consistent framework for describing nuclear excited states is provided by the relativistic random-phase approximation. We describe the relativistic RPA in Chap. 10 in the framework of the mean-field apprximation. We compare the RPA results with the giant resonances of various nuclei.

In Chap. 11, we discuss the application to astrophysics of some results from nuclear physics. The chapter is devoted to the description of the equation of state of nuclear matter in the relativistic mean-field approximation. The relativistic EOS is presented in detail, and its application to neutron stars is described.

In order to complement the discussion of the relativistic theory, we describe alternative relativistic models in Chap. 12. It should be interesting to make comparisons of these various models in order to determine the possible direction of further research and eventually to obtain the correct picture.

There are many projects that can be carried out in the field of relativistic many-body theory. We can present only a fraction of the relativistic descriptions of the various observables. We use Chap. 13 to present some recent topics in relativistic nuclear physics. There are many directions in which one could apply relativistic many-body techniques to nuclear physics. This chapter should help readers to find the appropriate references for further developments.

Chapter 14 is devoted to a summary of the relativistic many-body approaches and considers the outlook. We have added several appendices to help readers to understand relativistic many-body theory.

2 Relativistic Quantum Mechanics and Fields

2.1 The Dirac Equation

The Dirac equation is a relativistic equation for describing spin-1/2 particles. Consider the Dirac equation for a free particle with mass M. The Schrödinger-like form is an equation of first order in both space and time, given by (we use natural units where $\hbar = c = 1$)

$$i\frac{\partial \psi}{\partial t} = H_D \psi . \tag{2.1}$$

The Hamiltonian H_D is given by

$$H_D = \boldsymbol{\alpha} \cdot \boldsymbol{p} + \beta M , \tag{2.2}$$

containing the Hermitian matrices $\boldsymbol{\alpha}$ and β; $\boldsymbol{p}$ is the momentum operator, given by $\boldsymbol{p} = -i\nabla$. The following 4×4 (the smallest possible) representation for the Dirac matrices is chosen here (see Appendix A1):

$$\beta = \begin{pmatrix} 1 & 0 \\ 0 & -1 \end{pmatrix}, \quad \alpha_i = \begin{pmatrix} 0 & \sigma_i \\ \sigma_i & 0 \end{pmatrix}, \tag{2.3}$$

where $i = 1, 2, 3$. The matrices are given in 2×2 block form and σ_i are Pauli matrices. In this case the fermion wave function ψ has four components also

$$\psi = \begin{pmatrix} \psi_1 \\ \psi_2 \\ \psi_3 \\ \psi_4 \end{pmatrix} . \tag{2.4}$$

If the three-vector current density $\boldsymbol{j}$ and the probability density ρ are defined as

$$\boldsymbol{j} = \psi^\dagger \boldsymbol{\alpha} \psi , \tag{2.5}$$

$$\rho = \psi^\dagger \psi = \overline{\psi} \gamma_0 \psi , \tag{2.6}$$

where $\psi^\dagger = (\psi_1^*, \psi_2^*, \psi_3^*, \psi_4^*)$, $\gamma^0 = \gamma_0 = \beta$, and the Dirac adjoint wave function $\overline{\psi}$ is defined by

$$\overline{\psi} = \psi^\dagger \gamma^0 , \tag{2.7}$$

one obtains the continuity equation

$$\partial \rho / \partial t + \nabla \boldsymbol{j} = 0 . \tag{2.8}$$

To satisfy the relativistic relation between the energy E_p and the momentum $\boldsymbol{p}$ of the particle,

$$E_p^2 = \boldsymbol{p}^2 + M^2 , \tag{2.9}$$

each component of the wave function ψ should obey the Klein–Gordon equation. If we introduce the four-vector current density $j^\mu = (\rho, \boldsymbol{j})$, where $\mu = 0, 1, 2, 3$, the continuity equation (2.8) can be written also in the invariant form

$$\frac{\partial}{\partial x^\mu} j^\mu = \partial_\mu j^\mu = 0 . \tag{2.10}$$

The relationship between the covariant and contravariant components of the relativistic four-vectors is explained in more detail in Appendix A1. The four-vector current density may be presented also in the following way:

$$j^\mu = \bar{\psi}\gamma^\mu\psi ; \tag{2.11}$$

the matrices γ^μ ($\mu = 0, 1, 2, 3$) are defined in Appendix A1. The Dirac equation can be also written down in the covariant representation:

$$(i\gamma^\mu \partial_\mu - M)\psi = 0 . \tag{2.12}$$

2.2 Solutions for Free Particles

The solutions of the free-particle Dirac equation (2.1), (2.2) have the following form [10, 14],

$$\psi^{(\pm)}(\boldsymbol{x}, t) = N_p e^{i(\boldsymbol{p}\cdot\boldsymbol{x} \mp E_p t)} \begin{pmatrix} \chi \\ \eta \end{pmatrix} , \tag{2.13}$$

where the plus and minus signs correspond to the positive- and negative-energy solutions, rescpectively; $E_p = (\boldsymbol{p}^2 + M^2)^{1/2}$ is considered to be always positive. M is the nucleon mass; χ and η each have two components. First, consider the positive-energy solutions. In this case, from the Dirac equation (2.1), one obtains:

$$\boldsymbol{\sigma} \cdot \boldsymbol{p}\eta + M\chi = E_p \chi , \tag{2.14}$$
$$\boldsymbol{\sigma} \cdot \boldsymbol{p}\chi - M\eta = E_p \eta . \tag{2.15}$$

Expressing η in terms of χ,

$$\eta = \left(\frac{\boldsymbol{\sigma} \cdot \boldsymbol{p}}{E_p + M}\right) \chi , \tag{2.16}$$

one obtains the value of N_p in (2.13) from the normalization condition

$$N_{\rm p} = \sqrt{\frac{E_{\rm p} + M}{2E_{\rm p}}} \ . \tag{2.17}$$

Usually the positive-energy solution is written in terms of the nucleon Dirac spinor $u(p, s)$ defined by the following equation:

$$u(p, s) = \sqrt{\frac{E_{\rm p} + M}{2E_{\rm p}}} \begin{pmatrix} 1 \\ \boldsymbol{\sigma} \cdot \boldsymbol{p}/(E_{\rm p} + M) \end{pmatrix} \chi^{(s)} \ , \tag{2.18}$$

where $\chi^{(s)}$ is the two-component Pauli spinor. We obtain two independent positive-energy solutions, taking

$$\chi^{(1/2)} = \begin{pmatrix} 1 \\ 0 \end{pmatrix} \ , \qquad \chi^{(-1/2)} = \begin{pmatrix} 0 \\ 1 \end{pmatrix} \ . \tag{2.19}$$

The positive-energy solutions are then expressed in the following form:

$$\psi^{(+)}(\boldsymbol{x}, t) = e^{i(\boldsymbol{p}\cdot\boldsymbol{x} - E_{\rm p}t)} u(p, s) \ . \tag{2.20}$$

For the negative-energy case, one obtains

$$\boldsymbol{\sigma} \cdot \boldsymbol{p}\, \eta + M\chi = -E_{\rm p}\chi \ , \tag{2.21}$$
$$\boldsymbol{\sigma} \cdot \boldsymbol{p}\, \chi - M\eta = -E_{\rm p}\eta \ . \tag{2.22}$$

Using the first equation, χ may be easily expressed in terms of η, i.e.

$$\chi = -\left(\frac{\boldsymbol{\sigma} \cdot \boldsymbol{p}}{E_{\rm p} + M}\right) \eta \ , \tag{2.23}$$

and the negative-energy solution can be expressed in the following form:

$$\psi^{(-)}(\boldsymbol{x}, t) = e^{i(\boldsymbol{p}\cdot\boldsymbol{x} + E_{\rm p}t)} v(p, s) \ , \tag{2.24}$$

where $v(p, s)$ is given by the equation

$$v(p, s) = \sqrt{\frac{E_{\rm p} + M}{2E_{\rm p}}} \begin{pmatrix} \boldsymbol{\sigma} \cdot \boldsymbol{p}/(E_{\rm p} + M) \\ 1 \end{pmatrix} \chi^{(s)} \ . \tag{2.25}$$

The Feynman notation for the scalar product of any four-vector p with the γ-matrices will often be used: $\not{p} = \gamma_\mu p^\mu$, where $p^\mu = (E_{\rm p}, \boldsymbol{p}) \equiv (p^0, \boldsymbol{p})$. The Dirac equation is written as

$$(\not{p} - M)u(p, s) = 0 \ , \tag{2.26}$$
$$(\not{p} + M)v(p, s) = 0 \ , \tag{2.27}$$

and for the Dirac conjugate wave functions we have

$$\bar{u}(p,s)(\not{p} - M) = 0 \,, \tag{2.28}$$
$$\bar{v}(p,s)(\not{p} + M) = 0 \,. \tag{2.29}$$

The normalization of u and v can be chosen in the following form, for example:

$$\bar{u}(p,s)u(p,s') = \delta_{ss'} \frac{M}{E_p} = \delta_{ss'} \frac{M}{\sqrt{p^2 + M^2}} \,, \tag{2.30}$$
$$\bar{u}(p,s)v(p,s') = \bar{v}(p,s)u(p,s') = 0 \,, \tag{2.31}$$
$$\bar{v}(p,s)v(p,s') = -\delta_{ss'} \frac{M}{E_p} = -\delta_{ss'} \frac{M}{\sqrt{p^2 + M^2}} \,. \tag{2.32}$$

One should pay attention to the negative sign in the latter expression, this sign manifests itself also in the completeness condition

$$\frac{E_p}{M} \left\{ \sum_s u(p,s)\bar{u}(p,s) - \sum_s v(p,s)\bar{v}(p,s) \right\} = 1 \,. \tag{2.33}$$

Notice that the normalization conditions are chosen differently by different authors. Here we utilize the conditions introduced by Serot and Walecka [10].

Let us define two projection operators:

$$\Lambda_+(p) = \frac{E_p}{M} \sum_s u(p,s)\bar{u}(p,s) = \frac{\not{p} + M}{2M} \,, \tag{2.34}$$
$$\Lambda_-(p) = -\frac{E_p}{M} \sum_s v(p,s)\bar{v}(p,s) = \frac{-\not{p} + M}{2M} \,. \tag{2.35}$$

For particles on the mass shell, i.e. for $p^2 = M^2$, we have

$$\Lambda_+^2 = \Lambda_+ \,, \quad \Lambda_-^2 = \Lambda_- \,, \quad \Lambda_+ \cdot \Lambda_- = \Lambda_- \cdot \Lambda_+ = 0 \,, \quad \Lambda_+ + \Lambda_- = 1 \,. \tag{2.36}$$

All these results follow directly from (2.30)–(2.32). From (2.26) and (2.27) it can be seen also that

$$\Lambda_+ u = u \,, \quad \Lambda_- u = 0 \,, \tag{2.37}$$
$$\Lambda_+ v = 0 \,, \quad \Lambda_- v = v \,, \tag{2.38}$$
$$\bar{u}\Lambda_- = \bar{v}\Lambda_+ = 0 \,. \tag{2.39}$$

If any state ψ is expanded in terms of the Dirac spinors

$$\psi = \sum_s \{a_s u(p,s) + b_s v(p,s)\} \,, \tag{2.40}$$

the operators $\Lambda_\pm$ will project out the positive- and negative-energy subspaces [14]:

$$\Lambda_+\psi = \sum_s a_s u(p,s)\,, \tag{2.41}$$

$$\Lambda_-\psi = \sum_s b_s v(p,s)\,. \tag{2.42}$$

For this reason the operators $\Lambda_{(\pm)}$ are referred to as energy projection operators. For particles, these operators are defined by (2.34) and (2.35).

In its initial version of a one-body relativistic equation, the Dirac theory met with certain problems related to the existence of the negative-energy solutions. In this case one needs to prevent particles with positive energy from falling into negative-energy states via coupling to the radiation field, for instance. The solution to this problem was also suggested by Dirac. In accordance with Dirac's idea, we assume that the physical vacuum is the state in which all states with negative energy are occupied. In this case the problem of stability of the states with positive energy does not arise, because transitions to states with negative energy would be blocked by the completely occupied Dirac sea of states with negative energies (because of the Pauli principle).

It is easy to understand that M is the lowest possible energy of a single-particle state with positive energy (it is not equal $-\infty$ as one would expect if the Dirac-sea states were not completely filled). It is possible also to consider the ground-state energy to be zero, and in this case one obtains the Dirac picture for the vacuum, referred to as a hole theory. In the framework of this theory an antiparticle is treated as a "hole" (as a vacant state) in the otherwise completely occupied Dirac sea. Treated in this way (as an absence of a particle), the antiparticle has a positive energy with respect to the vacuum. Hole theory established a connection between the antiparticle states and the states with negative energy, this connection may be expressed in terms of charge conjugation symmetry (see [14] for more detail). However, at present, hole theory can be considered only as an intermediate step in developing modern field theory.

2.3 Relativistic Free-Field Theories

The field equations may be obtained using Hamilton's principle. A relativistically invariant Lagrangian density is defined by

$$\mathcal{L} = \mathcal{L}\left(\psi_\alpha(x)\,,\partial_\mu\psi_\alpha(x)\right), \tag{2.43}$$

i.e. it is a function of the field variables $\psi_\alpha(x)$ and their first derivatives $\partial_\mu\psi_\alpha(x)$ (the subscript α labels different components of the corresponding field). In principle, higher derivatives may be included also.

The respective action integral is defined as follows:

$$S = \int \mathcal{L}(\psi_\alpha(x),\partial_\mu\psi_\alpha(x))\,\mathrm{d}^4x\,, \tag{2.44}$$

The integration is performed over the whole space. Also, one can introduce a proper Lagrangian $L(t)$ by means of the following equation:

$$L(t) = \int \mathcal{L}(\psi_\alpha(x), \partial_\mu \psi_\alpha(x)) \, d^3x \, . \tag{2.45}$$

In the latter equation the integration is carried out over the three-volume V. In this case one obtains

$$S = \int_{t_1}^{t_2} L(t) \, dt \, . \tag{2.46}$$

The action principle postulates that the variation of the action vanishes if one performs an arbitrary variation of the field $\delta\psi_\alpha$ with the constraint of vanishing $\delta\psi_\alpha$ on the boundaries:

$$\delta S = 0 \, , \qquad \psi_\alpha \to \psi_\alpha + \delta\psi_\alpha \, , \tag{2.47}$$

$$\delta\psi_\alpha|_{t_1} = \delta\psi_\alpha|_{t_2} = 0 \, , \qquad \delta\psi_\alpha(t)|_{S_V} = 0 \, , \tag{2.48}$$

where S_V denotes the surface corresponding to the boundary.

In this case, one immediately obtains the set of the Euler–Lagrange equations

$$\frac{\partial \mathcal{L}}{\partial \psi_\alpha} - \partial_\mu \frac{\partial \mathcal{L}}{\partial \partial_\mu \psi_\alpha} = 0 \, , \qquad \alpha = 1, 2, 3, \ldots, N \, , \tag{2.49}$$

that determine the motion of the field ψ_α. We should mention that all the dynamical observables can be derived from the Lagrangian, as well as the conservation laws. It appears that every invariance of the field equations provides a conservation law (see Noether's theorem [1, 3, 14]).

Let us introduce the notation

$$\frac{\partial \mathcal{L}}{\partial \partial_\mu \psi_\alpha} \equiv \pi_\alpha^\mu(x) \, , \tag{2.50}$$

relating a four-vector $\pi_\alpha^\mu(x)$ to every field component $\psi_\alpha(x)$. Consider the time component of the four-vector introduced above,

$$\pi^\alpha(x) \equiv \pi_\alpha^0(x) = \frac{\partial \mathcal{L}}{\partial \partial_0 \psi_\alpha} = \frac{\partial \mathcal{L}}{\partial \dot{\psi}_\alpha} \, ; \tag{2.51}$$

this time component plays a significant role and will be referred to as the canonical conjugate of the field $\psi_\alpha(x)$ in what follows.

We now show examples of the Lagrangian and the corresponding field equations.

2.3.1 Real Scalar Field

The Lagrangian for a scalar field is given by

$$\mathcal{L} = \frac{1}{2}\left(\partial_\mu \varphi \cdot \partial^\mu \varphi - m^2 \varphi^2\right), \tag{2.52}$$

and the respective Lorentz-invariant equation is the scalar Klein–Gordon equation,

$$(\Box + m^2)\varphi(x) = 0. \tag{2.53}$$

2.3.2 Vector Field

A vector field $V_\mu(x)$ ($\mu = 0, 1, 2, 3$) satisfies the vectorial Klein–Gordon equation

$$(\Box + m^2)V_\mu(x) = 0 \tag{2.54}$$

and is subject to the covariant subsidiary condition

$$\partial^\mu V_\mu(x) = 0. \tag{2.55}$$

Consider the vector-field antisymmetric tensor

$$G_{\mu\nu}(x) = \partial_\mu V_\nu(x) - \partial_\nu V_\mu(x), \tag{2.56}$$

which satisfies the Proca equations

$$\partial^\mu G_{\mu\nu}(x) + m^2 V_\nu(x) = 0. \tag{2.57}$$

The corresponding Lagrangian density is given by

$$\mathcal{L} = \frac{1}{4} G_{\mu\nu} G^{\mu\nu} - m^2 V_\nu V^\nu. \tag{2.58}$$

2.3.3 Electromagnetic Field

The special case $m = 0$ of a vector field is the electromagnetic field, determined by the vector potential $A_\mu(x)$ satisfying the Maxwell equations

$$\Box A_\mu(x) = 0 \tag{2.59}$$

and the Lorentz condition

$$\partial_\mu A^\mu(x) = 0. \tag{2.60}$$

If one introduces the electromagnetic field strength tensor by the following equation,

$$F_{\mu\nu}(x) = \partial_\mu A_\nu - \partial_\nu A_\mu, \tag{2.61}$$

one can derive the equations mentioned above from the following Lagrangian

$$\mathcal{L} = \frac{1}{4} F_{\mu\nu} F^{\mu\nu}. \tag{2.62}$$

2.3.4 Spinor Dirac Field

In this case the Lagrangian has the following form

$$\mathcal{L} = \overline{\psi}(i\gamma^\mu \partial_\mu - M)\psi . \tag{2.63}$$

Using this Lagrangian and performing a variation with respect to $\overline{\psi}$, it is easy to obtain the Dirac equation for ψ. This equation is given by (2.12), while for $\overline{\psi}$ one has

$$\overline{\psi}(i\gamma^\mu \partial_\mu + M) = 0 . \tag{2.64}$$

2.3.5 Real Scalar Field with Quartic Self-Interaction

Consider the Lagrangian

$$\mathcal{L} = \frac{1}{2}\partial^\mu \varphi \cdot \partial_\mu \varphi - U(\varphi) . \tag{2.65}$$

From the Euler–Lagrange equation (2.49), one obtains

$$\Box \varphi + F(\varphi) = 0 , \tag{2.66}$$

where $F(\varphi)$ may be either a polynomial or a series in φ. In the first case one obtains

$$\Box \varphi + \sum_{n \geq 1}^{N} \lambda_n \varphi^n = 0 . \tag{2.67}$$

Taking

$$\lambda_1 = m^2 , \qquad \lambda_n = 0 \tag{2.68}$$

for $n > 1$, one obtains the linear Klein–Gordon equation (2.53).

If we use

$$\lambda_1 = m^2 , \quad \lambda_2 = 0 , \quad \lambda_3 = \frac{\lambda}{6} , \quad \lambda_n = 0 \tag{2.69}$$

for $n > 3$, we arrive at the well-known equation

$$(\Box + m^2)\varphi + \frac{\lambda}{6}\varphi^3 = 0 \tag{2.70}$$

of what is called the φ^4 theory. It plays an essential role in gauge theories (see below).

We have defined above the dynamical field variables $\psi_\alpha(x)$ and $\pi^\alpha(x)$. Now, in accordance with the second-quantization procedure [3], let us impose the well-known equal-time canonical commutation and anticommutation relations for Bose and Fermi fields, respectively:

$$[\pi_\alpha(x), \psi_\beta(x')] \,\delta(t - t') = -i\delta_{\alpha\beta}\delta(x - x') , \tag{2.71}$$

$$[\psi_\alpha(x), \psi_\beta(x')] \,\delta(t - t') = [\pi_\alpha(x), \pi_\beta(x')] \,\delta(t - t') = 0 \tag{2.72}$$

and

$$\{\pi_\alpha(x), \psi_\beta(x')\} \delta(t-t') = \mathrm{i}\delta_{\alpha\beta}\delta(x-x') , \qquad (2.73)$$
$$\{\psi_\alpha(x), \psi_\beta(x')\} \delta(t-t') = \{\pi_\alpha(x), \pi_\beta(x')\} \delta(t-t') = 0 . \qquad (2.74)$$

These equations can be easily generalized to arbitrary spacelike $[(x-x')^2 < 0]$ intervals also [3]. The introduction of (2.71)–(2.74) corresponds to replacing the dynamical variables $\psi_\alpha(x)$ and $\pi^\alpha(x)$ by the respective Hermitian operators.

Consider the tensor

$$T^{\mu\nu}(x) \equiv -g^{\mu\nu}\mathcal{L} + \pi^\mu_\alpha(x)\partial^\nu \psi_\alpha(x) , \qquad (2.75)$$

which will be referred to as the canonical energy–momentum tensor. The Euler–Lagrange equations (2.49) ensure the conservation of this quantity:

$$\partial_\mu T^{\mu\nu}(x) = 0 , \qquad \nu = 0, 1, 2, 3 . \qquad (2.76)$$

Let us mention that the conservation law given by (2.76) is related, in accordance with Noether's theorem, to the invariance of the action principle $\delta S = 0$ under transformations of the following type:

$$x^\mu \to x'^\mu = x^\mu + \varepsilon^\mu , \qquad \text{where } \varepsilon^\mu = \text{const}, \qquad (2.77)$$

i.e. under space–time translations. In particular, let us consider

$$T_{00} = T^{00} = \pi_\alpha(x)\partial_0\psi_\alpha(x) - \mathcal{L}(x) = \mathcal{H}(x) , \qquad (2.78)$$

so that

$$\int T_{00}\,\mathrm{d}^3x = \int \mathcal{H}(x)\,\mathrm{d}^3x = H \qquad (2.79)$$

determines H, i.e. the Hamiltonian of the field (the total field energy).

Let us introduce also the four-momentum

$$P^\mu = (H, \boldsymbol{P}) = g^{\mu\nu}P_\nu \qquad (2.80)$$

defined by

$$P_\nu = \int T_{0\nu}\,\mathrm{d}^3x \qquad (2.81)$$

This four-vector P_ν appears to be a constant of the motion.

Consider the real scalar field as an example. In this case the field is described by a single-component wave function $\varphi(x)$, the Lagrangian of the system is given by (2.52) and the respective Klein–Gordon equation is given by (2.53). From (2.50) one obtains

$$\pi^\mu(x) = \partial^\mu \varphi(x) \qquad (2.82)$$

and, in particular, the momentum conjugate to $\varphi(x)$ is given by

$$\pi(x) \equiv \pi^0(x) = \dot{\varphi}(x) \,. \tag{2.83}$$

On quantization, the real field φ becomes a Hermitian operator $\varphi^\dagger = \varphi$, satisfying the equal-time commutation relations:

$$[\varphi(\boldsymbol{x},t), \dot{\varphi}(\boldsymbol{x}',t)] = \mathrm{i}\delta(\boldsymbol{x}-\boldsymbol{x}') \,, \tag{2.84}$$

$$[\varphi(\boldsymbol{x},t), \varphi(\boldsymbol{x}',t)] = [\dot{\varphi}(\boldsymbol{x},t), \dot{\varphi}(\boldsymbol{x}',t)] = 0 \,. \tag{2.85}$$

The Hamiltonian for the neutral scalar field has the following form:

$$H = \int \mathcal{H}(\dot{\varphi},\varphi)\, \mathrm{d}^3 x = \frac{1}{2} \int \left\{ \dot{\varphi}^2(x) + [\nabla\varphi(x)]^2 + m^2 \varphi^2(x) \right\} \mathrm{d}^3 x \,. \tag{2.86}$$

The three-dimensional field momentum is obtained from

$$\boldsymbol{P} = -\int \dot{\varphi} \nabla \varphi \, \mathrm{d}^3 x \,. \tag{2.87}$$

The procedure described below is given in more detail elsewhere (see [3] for example). Let us expand $\varphi(x)$ in a complete set of solutions of the Klein–Gordon equation. This complete set includes positive- and negative-energy solutions (owing to the relativistic energy–momentum relation) as in the Dirac case discussed above in detail.

Taking this into account, we obtain

$$\varphi(x) = \sum_{\boldsymbol{k}} \left(\frac{1}{2L^3 \omega_k} \right)^{1/2} \left\{ a_{\boldsymbol{k}} \mathrm{e}^{-\mathrm{i}kx} + a_{\boldsymbol{k}}^\dagger \mathrm{e}^{\mathrm{i}kx} \right\} \,. \tag{2.88}$$

Each operator $a_{\boldsymbol{k}}$ enters (2.88) paired with its adjoint $a_{\boldsymbol{k}}^\dagger$. This fact ensures that $\varphi(x)$ is Hermitian. We must mention that in (2.88) a box normalization is used, i.e. we consider the system to be placed in a box (cube) of volume $V = L^3$, in this case we impose periodic boundary conditions and the momenta become discrete:

$$k_i = \frac{2\pi n_i}{L}\,, \quad i = x, y, z\,, \quad n_i = 0, \pm 1, \pm 2, \pm \ldots \,. \tag{2.89}$$

Notice also that in (2.88) we have $kx = k_\mu x^\mu = k^0 t - \boldsymbol{k} \cdot \boldsymbol{x}$, whereas $k^0 = +\omega_k = (\boldsymbol{k}^2 + m^2)^{1/2}$. From (2.88) and the commutation relations (2.84) and (2.85), it is easy to derive the commutation relations for the operators $a_{\boldsymbol{k}}$ and $a_{\boldsymbol{k}}^\dagger$:

$$\left[a_{\boldsymbol{k}}, a_{\boldsymbol{k}'}^\dagger \right] = \delta_{\boldsymbol{k}\boldsymbol{k}'} \,, \tag{2.90}$$

$$[a_{\boldsymbol{k}}, a_{\boldsymbol{k}'}] = \left[a_{\boldsymbol{k}}^\dagger, a_{\boldsymbol{k}'}^\dagger \right] = 0 \,. \tag{2.91}$$

For this reason we may treat $a_{\bm{k}}$ and $a_{\bm{k}}^{\dagger}$ as the annihilation and creation operators of particles (of the quanta of the field) with momentum $\bm{k}$ and energy $\omega_{\bm{k}}$, while the operators

$$N_{\bm{k}} = a_{\bm{k}}^{\dagger} a_{\bm{k}} \tag{2.92}$$

can be considered as operators of the number of particles, with eigenvalues given by $n_{\bm{k}} = 0, 1, 2, \ldots$. Since for the real scalar field $\varphi^{\dagger} = \varphi$ the particles associated with φ are identical to their antiparticles, these particles must be neutral. Using (2.86) and (2.87) and the expansion given by (2.88), we obtain

$$H = \sum_{\bm{k}} \omega_{\bm{k}} \left(a_{\bm{k}}^{\dagger} a_{\bm{k}} + \frac{1}{2} \right), \tag{2.93}$$

$$\bm{P} = \sum_{\bm{k}} \bm{k} a_{\bm{k}}^{\dagger} a_{\bm{k}}. \tag{2.94}$$

It is possible to define the ground state of the neutral scalar field, i.e. the vacuum state $|0\rangle$, by the following equation:

$$a_{\bm{k}} |0\rangle = \langle 0| a_{\bm{k}}^{\dagger} = 0 \tag{2.95}$$

for all $\bm{k}$.

There is a problem connected with (2.93). The problem is that the energy related to the vacuum state, which is due to the presence of the term $1/2$ in (2.93), becomes formally infinite. This problem may be solved by dropping this infinite term because infinite energy shifts cannot be measured. Let us consider the vacuum expectation value of the time-ordered product of two scalar field operators,

$$\langle 0 | T(\varphi(x)\varphi(x')) | 0 \rangle$$
$$= \langle 0 | \varphi(x)\varphi(x') | 0 \rangle \theta(t-t') + \langle 0 | \varphi(x')\varphi(x) | 0 \rangle \theta(t'-t) \tag{2.96}$$
$$\equiv i\Delta_{\text{F}}(x - x'),$$

where θ is the Heaviside step function

$$\theta(t) = \begin{cases} 1 & \text{if } t > 0 \\ 0 & \text{if } t < 0 \end{cases}. \tag{2.97}$$

The value of Δ_{F} defined by this equation will be referred to as the Feynman propagator for the free scalar field. It can be calculated in a straightforward way, and one obtains

$$i\Delta_{\text{F}}(x - x') = i \int \frac{d^4 k}{(2\pi)^4} \Delta_{\text{F}}(k) e^{-ik(x-x')}, \tag{2.98}$$

where the Fourier image of the propagator is defined as

$$\Delta_{\mathrm{F}}(k) = \frac{1}{k^2 - m^2 + \mathrm{i}\eta} \,, \tag{2.99}$$

η being an infinitely small positive value. Here, $k^2 = k_\mu \cdot k^\mu = k_0^2 - \boldsymbol{k}^2$. The same quantization procedure can be carried out for the vector field $V_\mu(x)$. We shall not describe that procedure in detail here. The respective propagator is given by

$$\langle 0 | \, T\left(V_\mu(x) V_\nu(x')\right) | 0 \rangle = \mathrm{i}\left(g_{\mu\nu} + \frac{1}{m^2}\partial_\mu\partial_\nu\right) \Delta_{\mathrm{F}}(x - x') \,, \tag{2.100}$$

and the Fourier image has the following form:

$$\mathcal{D}^{\mathrm{F}}_{\mu\nu}(k) = \left[-g_{\mu\nu} + \frac{k_\mu k_\nu}{m^2}\right] \frac{1}{k^2 - m^2 + \mathrm{i}\eta} \,. \tag{2.101}$$

The Dirac Lagrangian is given by (2.63), and the respective field equations are given by (2.12), (2.64). Also, from (2.50) one obtains

$$\pi^\mu \equiv \mathrm{i}\overline{\psi}\gamma^\mu \tag{2.102}$$

and

$$\pi \equiv \pi^0 \equiv \mathrm{i}\overline{\psi}\gamma_0 = \mathrm{i}\psi^\dagger \tag{2.103}$$

The energy-momentum tensor defined by (2.75) has the following form:

$$T^{\mu\nu} = \mathrm{i}\overline{\psi}\gamma^\mu\partial^\nu\psi \,. \tag{2.104}$$

We also obtain for the total field energy

$$H = P_0 = \mathrm{i}\int \overline{\psi}\gamma_0 \dot{\psi}\, \mathrm{d}^3 x \,, \tag{2.105}$$

and for the field momentum

$$\boldsymbol{P} = \int \psi^\dagger(-\mathrm{i}\nabla)\psi\, \mathrm{d}^3 x \,, \tag{2.106}$$

and the baryon number is given by

$$B = \int \psi^\dagger \psi\, \mathrm{d}^3 x \,. \tag{2.107}$$

The values given above by (2.105)–(2.107) appear to be constants of the motion. Since we are now dealing with fermions, we should use the equal-time anticommutation relations set out in (2.74), which also can be written in the following form:

$$\{\psi(\boldsymbol{x}), \overline{\psi}(\boldsymbol{x}')\} = \gamma^0 \delta(\boldsymbol{x} - \boldsymbol{x}') \,, \tag{2.108}$$

$$\{\psi(\boldsymbol{x}), \psi(\boldsymbol{x}')\} = \{\overline{\psi}(\boldsymbol{x}), \overline{\psi}(\boldsymbol{x}')\} = 0 \,. \tag{2.109}$$

In the equations given above, we have suppressed the Dirac spinor indices labeling the spinor components.

For the basis of the Fourier expansion of the Dirac spinors, we consider the complete set of positive-energy solutions $u(\boldsymbol{k}, s)\mathrm{e}^{-ikx}$ and negative-energy solutions $v(\boldsymbol{k}, s)\mathrm{e}^{ikx}$. So we have

$$\psi(x) = \frac{1}{V^{1/2}} \sum_{\boldsymbol{k},s} \left[a_{\boldsymbol{k},s} u(\boldsymbol{k}, s)\mathrm{e}^{-ikx} + b^{\dagger}_{\boldsymbol{k},s} v(\boldsymbol{k}, s)\mathrm{e}^{ikx} \right] , \qquad (2.110)$$

$$\overline{\psi}(x) = \frac{1}{V^{1/2}} \sum_{\boldsymbol{k},s} \left[a^{\dagger}_{\boldsymbol{k},s} \overline{u}(\boldsymbol{k}, s)\mathrm{e}^{ikx} + b_{\boldsymbol{k},s} \overline{v}(\boldsymbol{k}, s)\mathrm{e}^{-ikx} \right] . \qquad (2.111)$$

We utilize here the normalization conditions for u and v given by (2.32). Imposing now the anticommutation relations (2.109 and 2.110) for $\psi(\boldsymbol{x})$ and $\overline{\psi}(\boldsymbol{x})$, we can work out the anticommutation relations in the momentum space:

$$\begin{aligned} \left\{ a_{\boldsymbol{k},s},\, a^{\dagger}_{\boldsymbol{k}',s'} \right\} &= \delta_{ss'} \delta_{\boldsymbol{k}\boldsymbol{k}'} , \\ \left\{ b_{\boldsymbol{k},s},\, b^{\dagger}_{\boldsymbol{k}',s'} \right\} &= \delta_{ss'} \delta_{\boldsymbol{k}\boldsymbol{k}'} , \end{aligned} \qquad (2.112)$$

and all the other anticommutators are equal to zero:

$$\{a_{\boldsymbol{k},s},\, a_{\boldsymbol{k}',s'}\} = \{b_{\boldsymbol{k},s},\, b_{\boldsymbol{k}',s'}\} = \{a_{\boldsymbol{k},s},\, b_{\boldsymbol{k}',s'}\} = \cdots = 0 . \qquad (2.113)$$

In this case we may treat $a_{\boldsymbol{k},s}$ and $b_{\boldsymbol{k},s}$ as the annihilation operators of the fermion and of its antiparticle respectively, and also we may treat $a^{\dagger}_{\boldsymbol{k},s}$ and $b^{\dagger}_{\boldsymbol{k},s}$ as the creation operators of the fermion and of its antiparticle respectively. Also, it can be seen that ψ can either annihilate a fermion or create an antifermion; at the same time, $\overline{\psi}$ can either create a fermion or annihilate an antifermion.

Let us define the vacuum state $|0\rangle$ by the following equation:

$$a_{\boldsymbol{k},s} |0\rangle = b_{\boldsymbol{k},s} |0\rangle = 0 . \qquad (2.114)$$

The single-fermion and single-antifermion states are described by

$$a^{\dagger}_{\boldsymbol{k},s} |0\rangle , \quad b^{\dagger}_{\boldsymbol{k},s} |0\rangle . \qquad (2.115)$$

Making use of (2.105)–(2.107) and also the Fourier expansions (2.110) and (2.111), we obtain

$$H = P_0 = \sum_{\boldsymbol{k},s} E_k \left(a^{\dagger}_{\boldsymbol{k},s} \cdot a_{\boldsymbol{k},s} + b^{\dagger}_{\boldsymbol{k},s} \cdot b_{\boldsymbol{k},s} - 1 \right) , \qquad (2.116)$$

$$\boldsymbol{P} = \sum_{\boldsymbol{k},s} \boldsymbol{k} \left(a^{\dagger}_{\boldsymbol{k},s} \cdot a_{\boldsymbol{k},s} + b^{\dagger}_{\boldsymbol{k},s} \cdot b_{\boldsymbol{k},s} \right) . \qquad (2.117)$$

$$B = \sum_{k,s} \left(a^\dagger_{k,s} \cdot a_{k,s} - b^\dagger_{k,s} \cdot b_{k,s} + 1 \right) . \tag{2.118}$$

The contribution of the c-numbers in (2.116)–(2.118) needs to be dropped, in the same way as was done in the example of the neutral scalar field. This can be achieved by introducing normal ordering of operators [14]. We say in summary, therefore, that the use of anticommutation relations and normal ordering of the operators solves the problem of negative-energy states correctly. The Feynman baryon propagator in the vacuum is defined by the following equation:

$$\langle 0 | \, T\left(\psi(x) \overline{\psi}(x') \right) | 0 \rangle = \mathrm{i} S_\mathrm{F}(x - x') , \tag{2.119}$$

where S_F can be written as

$$S_\mathrm{F}(x) = \frac{1}{(2\pi)^4} \int \mathrm{d}^4 k\, \mathrm{e}^{-\mathrm{i} k x} \frac{\slashed{k} + M}{k^2 - M^2 + \mathrm{i}\eta} , \quad -\infty < k_0 < \infty . \tag{2.120}$$

It is also possible to define the propagator for a system of free fermions at finite density. This procedure has been performed, for example, in the review paper by Serot and Walecka [10].

2.4 The Dirac Equation in a Central Potential

The most general form of the Dirac equation compatible with the requirement of good parity P and good total angular momentum $\boldsymbol{J}$ is given by (see Appendix A3)

$$H_\mathrm{D} \psi(\boldsymbol{r}) = \{ \boldsymbol{\alpha} \cdot \boldsymbol{p} + \beta[M + U(r)] \} \psi(\boldsymbol{r}) = E \psi(\boldsymbol{r}) , \tag{2.121}$$

so that

$$U(r) = S(r) + \beta V(r) - \gamma^r U^r_\mathrm{V}(r) - \beta\gamma^r U^r_\mathrm{T}(r) , \tag{2.122}$$

where $S(r)$ is a scalar field, $V(r)$ is a time component of the vector field, $U^r_\mathrm{V}(r)$ comes from the space components of the vector field, and $U^r_\mathrm{T}(r)$ is a contribution from the tensor potential. It is easy to verify that

$$[H_\mathrm{D}, \boldsymbol{J}] = 0 , \quad [H_\mathrm{D}, \boldsymbol{J}^2] = 0 , \quad [H_\mathrm{D}, P] = 0 , \tag{2.123}$$

where P stands for the relativistic parity operator and $\boldsymbol{J}$ is the relativistic total-angular momentum operator, both operators being defined in Appendix A1. For example,

$$\boldsymbol{J} = \boldsymbol{L} + \boldsymbol{S} = \boldsymbol{L} + \frac{\hbar}{2} \boldsymbol{\Sigma} , \tag{2.124}$$

where

2.4 The Dirac Equation in a Central Potential

$$\boldsymbol{L} = \boldsymbol{r} \times \boldsymbol{p} \begin{pmatrix} 1 & 0 \\ 0 & 1 \end{pmatrix}, \qquad \boldsymbol{S} = \frac{\hbar}{2} \begin{pmatrix} \boldsymbol{\sigma} & 0 \\ 0 & \boldsymbol{\sigma} \end{pmatrix}. \qquad (2.125)$$

We have also

$$[\boldsymbol{J}^2, J_z] = 0, \qquad [H_D, J_z] = 0, \qquad (2.126)$$

and the eigenfunction $\psi(\boldsymbol{r})$ in (2.121) can be labeled by the quantum numbers E, $j(j+1)$, and m.

Let us define the operator

$$K = \beta(\boldsymbol{\Sigma} \cdot \boldsymbol{L} + 1); \qquad (2.127)$$

we can verify easily that this operator commutes with the Dirac Hamiltonian, i.e.

$$[H_D, K] = 0. \qquad (2.128)$$

The operator K can be represented in the following form:

$$K = \beta \left(\boldsymbol{J}^2 - \boldsymbol{L}^2 + \frac{1}{4} \right). \qquad (2.129)$$

Let us write the four-component Dirac spinor ψ in terms of two-component spinors:

$$\psi = \begin{pmatrix} \psi_A \\ \psi_B \end{pmatrix}, \qquad (2.130)$$

where ψ_A and ψ_B correspond to the upper and lower components, respectively, of ψ. From what has been said above, one obtains [19]

$$H_D \begin{pmatrix} \psi_A \\ \psi_B \end{pmatrix} = E \begin{pmatrix} \psi_A \\ \psi_B \end{pmatrix}, \qquad P \begin{pmatrix} \psi_A \\ \psi_B \end{pmatrix} = \pm \begin{pmatrix} \psi_A \\ \psi_B \end{pmatrix}, \qquad (2.131)$$

$$\boldsymbol{J}^2 \begin{pmatrix} \psi_A \\ \psi_B \end{pmatrix} = j(j+1) \begin{pmatrix} \psi_A \\ \psi_B \end{pmatrix}, \qquad J_z \begin{pmatrix} \psi_A \\ \psi_B \end{pmatrix} = m \begin{pmatrix} \psi_A \\ \psi_B \end{pmatrix}, \qquad (2.132)$$

$$K \begin{pmatrix} \psi_A \\ \psi_B \end{pmatrix} = -\kappa \begin{pmatrix} \psi_A \\ \psi_B \end{pmatrix}. \qquad (2.133)$$

It is useful also to notice that although $[H_D, \boldsymbol{L}^2] \neq 0$, we have $[H_D, \boldsymbol{S}^2] = 0$ and also

$$\boldsymbol{L}^2 \psi_A = l_A(l_A + 1)\psi_A, \qquad (2.134)$$

$$\boldsymbol{L}^2 \psi_B = l_B(l_B + 1)\psi_B, \qquad (2.135)$$

where l_A and l_B can be treated as the orbital angular-momentum quantum numbers of the upper and lower components, respectively, of the wave function. From quantum mechanics, it is known that the eigenfunctions of the operators $\boldsymbol{L}^2$, $\boldsymbol{S}^2$, $\boldsymbol{J}^2$, and J_z are the spin–angle functions $\Omega_{jlm}(\theta, \varphi, s)$, defined as the tensor product of the orbital and spin functions

$$\Omega_{jlm}(\theta,\varphi,s) = \sum_{m_s,m_l} \left\langle \frac{1}{2}m_s l m_l \middle| jm \right\rangle Y_{lm_l}(\theta,\varphi)\chi_{m_s/2} . \tag{2.136}$$

Here, $Y_{lm_l}(\theta,\varphi)$ are the spherical harmonics with $l = 0, 1, 2, \ldots$ and $m_l = -l, -l+1, \ldots, +l$; $\chi_{m_s/2}$ are the spin functions (simultaneous eigenfunctions of the operators $\boldsymbol{S}^2$ and S_z (with $m_s = -s, -s+1, \ldots, +s$); and $\langle (1/2)m_s l m_l | jm \rangle$ are the Clebsch–Gordan coefficients. The spin–angle functions Ω_{jlm} have the parity $\pi = (-)^l$. So it is easily seen that for $s = 1/2$ and $l = j \pm 1/2$ with $j = 1/2, 3/2, \ldots$, the upper component ψ_A is proportional to Ω_{jlm}. In the same way one can obtain the result that the lower component ψ_B is determined by $\Omega_{jl'm}$ with a parity equal to $(-)^{l'}$. Bearing in mind that l' can only be equal to $j \pm 1/2$, one obtains the result that $l' = l \pm 1$, and

$$l' = l + 1 = j + 1/2 \quad \text{if} \quad l = j - 1/2 , \tag{2.137}$$
$$l' = l - 1 = j - 1/2 \quad \text{if} \quad l = j + 1/2 . \tag{2.138}$$

From (2.129) and (2.133) one also obtains

$$\left(\boldsymbol{J}^2 - \boldsymbol{L}^2 + \frac{1}{4}\right)\psi_A = -\kappa\psi_A , \tag{2.139}$$

$$\left(\boldsymbol{J}^2 - \boldsymbol{L}^2 + \frac{1}{4}\right)\psi_B = \kappa\psi_B , \tag{2.140}$$

where κ, the eigenvalue of the operator $\boldsymbol{J}^2 - \boldsymbol{L}^2 + 1/4$, is given by [19]

$$\kappa = j(j+1) - l'(l'+1) + \frac{1}{4} . \tag{2.141}$$

Also,

$$\kappa = \begin{cases} -(l+1), & j = l+1/2 , \\ l, & j = l-1/2 . \end{cases} \tag{2.142}$$

It is clear that $\kappa \neq 0$, and that $\kappa = \pm 1, \pm 2, \pm 3, \ldots$, the values of j and l being determined by κ:

$$j = |\kappa| - \frac{1}{2} , \tag{2.143}$$

$$l = \begin{cases} \kappa, & \kappa > 0 , \\ -\kappa - 1, & \kappa < 0 . \end{cases} \tag{2.144}$$

Finally, the solution of (2.121), (2.122) can be written in the following form:

$$\psi_\kappa(r) = \begin{pmatrix} g_\kappa(r)\Omega_{jlm}(\theta,\varphi,s) \\ if_\kappa(r)\Omega_{jl'm}(\theta,\varphi,s) \end{pmatrix} , \tag{2.145}$$

where $g_\kappa(r)$ and $f_\kappa(r)$ are the radial amplitudes of the upper and lower components, respectively, of the Dirac spinor ψ_κ, and depend on the specific

2.4 The Dirac Equation in a Central Potential

choice of the potentials in (2.122). It is easy to verify also the following identity:

$$(\boldsymbol{\sigma} \cdot \boldsymbol{n})\Omega_{jlm} = -\Omega_{jl'm} , \qquad (2.146)$$

where $\boldsymbol{n}$ is a unit vector along the direction $\boldsymbol{r}$. The factor i in front of $f_\kappa(r)$ in (2.145) has been introduced to have real bound-state functions $g_\kappa(r)$ and $f_\kappa(r)$ in (2.121).

Notice that the orbital angular momenta l and l' are determined by j and the parity π:

$$l = \begin{cases} j + \frac{1}{2}, & \text{for } \pi = (-)^{j+1/2} , \\ j - \frac{1}{2}, & \text{for } \pi = (-)^{j-1/2} , \end{cases} \qquad (2.147)$$

and

$$l' = \begin{cases} j - \frac{1}{2}, & \text{for } \pi = (-)^{j+1/2} , \\ j + \frac{1}{2}, & \text{for } \pi = (-)^{j-1/2} . \end{cases} \qquad (2.148)$$

We must mention also that the potentials in (2.122) may include isovector components. To take these components into account, the Dirac spinor ψ_κ should be multiplied by the isospin wave function χ_τ.

3 Basic Features of the Meson Theory of Nucleon–Nucleon Interactions

3.1 One-Boson Exchange Potentials in Configuration Space

Quantum chromodynamics (QCD) is considered to be the underlying theory for strong interactions. This means that the nucleon–nucleon interaction can be calculated, in principle, on the basis of QCD. However, in practice to calculate the NN force for nuclear-structure purposes, one needs to know this interaction at low energies and momenta, i.e. in the region where QCD cannot be treated perturbatively. Lattice QCD methods cannot be used in this case either, the system of two nucleons being still too complicated a system for this type of calculation. There is another way to construct the NN interaction within the Bethe–Salpeter approach with separable kernel, which was developed in [21], but this method will not be described here.

However, it is possible to develop models for the NN force that include explicitly the QCD degrees of freedom, i.e. quarks and gluons. Examples are the nonrelativistic constituent quark models (CQMs) [22] and the quark cluster models based on CQMs [23–27]. Constituent quark cluster models are efficient for describing the basic features of the nucleon–nucleon interaction. However, these models have not yet been used widely for nuclear-structure calculations (see, however, Sect. 12.1). In [28, 29] it is shown that in the low energy region the relevant degrees of freedom of QCD can be taken into account in the framework of the meson theory (see Fig. 3.1), where the NN force is presented in terms of the quark- and meson-exchange pictures [30]. Following this point of view, it is reasonable to treat the NN force in terms of mesons that realize the interaction between nucleons.

The attempt to describe the interaction between nucleons and the properties of nuclei in the framework of a unified approach on the basis of a meson exchange mechanism is an old but attractive idea. In recent years, interest in this problem has grown in connection with the construction of a model of nucleon–nucleon interaction known as one-boson exchange potentials (OBEPs), which take into account the exchange of mesons and resonances with different space–time transformation properties. This interest is fully understandable since the use of meson potentials puts the theory of nuclear structure on a new footing, eliminating phenomenology in the choice

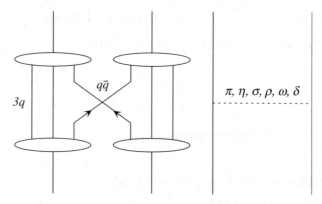

Fig. 3.1. Diagrammatic representation of the NN force in terms of QCD and meson degrees of freedom

of the nucleon–nucleon interaction and relegating it to the description of elementary particles, so that in each of those two fields one can obtain new results by using regularities established in the other field.

The pioneering attempt to construct a meson theory of nuclear forces was made in 1935 by Yukawa, who predicted the existence of a new fundamental particle which acted as a quantum for interactions between nucleons. At that time no mesons had been observed experimentally. After the discovery of the pion, for a long time the pion was considered to be responsible for the total internucleon interaction. However, starting from 1961, the number of mesons observed increased tremendously. From investigations of their interactions with nucleons, the essential role of many of them (first of all, the vector ρ and ω mesons) in generating the NN forces became evident.

In nuclear physics, in contrast to electrodynamics, there are several types of quanta of the nuclear field, differing by their quantum numbers and masses. For this reason, the NN potential is a superposition of components with different space–time transformation properties and different radii of action. In calculations of the NN potential, the range of the potential is divided into three parts: the external part $r > 2.14$ fm, the intermediate part 0.71 fm $< r < 2.14$ fm, and the internal part $r < 0.71$ fm. These three regions are determined by mesons and resonances[1] that make contributions to the NN potential (from the uncertainty principle, it follows that the radius of action of a nuclear force is determined by the Compton wavelength of the respective quantum).

---

[1] In this chapter we shall restrict ourselves to considering only the basic features of the meson theory of nucleon–nucleon forces because the current status of the relativistic theory of nuclear structures is essentially based on the same philosophy. However, it should be emphasized that there are now indications that nuclear forces can be understood almost quantitatively at a deeper microscopic level, i.e. on the basis of the fundamental principles of QCD [30–32].

3.1 One-Boson Exchange Potentials in Configuration Space

Table 3.1. Properties of light mesons involved in the NN interaction [33]

Meson	Mass (MeV)	I^G, J^P	Type of meson–nucleon coupling
π	139	$1^-, 0^-$	$f_\pi/m_\pi \overline{\psi}\gamma_5\gamma^\mu \partial_\mu \boldsymbol{\pi} \cdot \boldsymbol{\tau}\psi$ or $ig_\pi \overline{\psi}\gamma_5 \boldsymbol{\pi} \cdot \boldsymbol{\tau}\psi$
σ	400–1200	$0^+, 0^+$	$g_\sigma \overline{\psi}\psi\varphi$ *
ω	782	$0^-, 1^-$	$g_\omega \overline{\psi}\gamma^\mu\psi\omega_\mu + f_\omega/(2M)\overline{\psi}\sigma^{\mu\nu}\psi\partial_\mu\omega_\nu$
ρ	770	$1^+, 1^-$	$g_\rho \overline{\psi}\gamma^\mu \boldsymbol{\rho}_\mu \cdot \boldsymbol{\tau}\psi + f_\rho/(2M)\overline{\psi}\sigma^{\mu\nu}\partial_\mu\boldsymbol{\rho}_\nu \cdot \boldsymbol{\tau}\psi$
a_0	980	$1^-, 0^+$	$g_{a_0}\overline{\psi} \cdot \boldsymbol{a_0} \cdot \boldsymbol{\tau}\psi$
η	548	$0^+, 0^-$	$ig_\eta \overline{\psi}\gamma_5\eta\psi$
a_1	1260	$1^-, 1^+$	$g_a \overline{\psi}\gamma_5\gamma^\mu \boldsymbol{a}_\mu \cdot \boldsymbol{\tau}\psi$

* The scalar–isoscalar meson is denoted to σ in what follows, while φ stands for the corresponding field. It should be emphasized that the scalar–isoscalar field can generate only an attractive interaction

At present, it can be considered as an established fact that the external and intermediate parts of the NN potential in meson theory are determined by one- and two-pion exchanges, as well as by one-meson exchanges of ω, ρ, a_0, a_1, and some other mesons. For the internal part, for internucleon distances less than 0.71 fm, the picture of a single-boson exchange is not acceptable, because many processes make contributions to the NN force in this region, and for this reason this part of the interaction is described phenomenologically. We note here that extensive efforts to extract the short-range part of the NN force from of the quark–gluon dynamics have been made by many scientists (see [30–32] and references therein).

Table 3.1 represents a set of mesons expected to make a contribution to the NN interaction. The table gives some quantum numbers of these mesons: J^P, the spin and parity; I^G, the isotopic spin and G-parity; their masses; and the form of their coupling to nucleons. Most of these mesons have been observed in experiments [33, 34]; the only exception is the σ (see the comments on this subject in [35]), and all are involved in the theoretical framework of the one-boson exchange scheme.

The following notation is used in Table 3.1: γ_μ, γ_5 are the Dirac matrices, and $\sigma^{\mu\nu} = i/2[\gamma^\mu, \gamma^\nu]$, where the standard notations are used [2]; M, m_π denote the bare masses of the nucleon and the pion; ψ is the nucleon field; $\boldsymbol{\pi}$, φ, ω_μ, $\boldsymbol{\rho}_\mu$, $\boldsymbol{a_0}$, $\boldsymbol{a}_\mu$ are the respective meson fields (note that $\boldsymbol{\rho}_\mu$, $\boldsymbol{\pi}$, $\boldsymbol{a_0}$ and $\boldsymbol{a}_\mu$ are vectors in isospin space); g and f are coupling constants; and $\boldsymbol{\tau}$ are the Pauli isospin matrices. One can obtain an idea about the values of the coupling constants from experimental data on interactions of elementary particles; see [36, 37], for example. Note that four-dimensional relativistic notation is explained in more detail in Appendix A1; the expressions for the

Dirac matrices are also given there, as well as some useful formulae containing these matrices.

The tail of the NN force, i.e. its external region, is completely determined by the one-pion exchange. The pion may be coupled to the nucleon field by either pseudovector or pseudoscalar coupling. The pseudovector coupling is more widely used in the relativistic OBEP framework. However, the pseudoscalar interaction is also acceptable; a mixture of these two forms of πN coupling is also utilized sometimes [38, 39].

The intermediate-range attraction, the important part of the NN force, requires a proper treatment of the two-pion contribution to describe it. A two-pion exchange potential was introduced by groups in Bonn [34], Paris [40], and Stony Brook [41, 42]. However, in many NN models the attraction produced by the 2π exchange is simulated by a scalar–isoscalar σ meson with mass around 500–600 MeV. The scalar–isoscalar field φ is an essential component of the theory in this case. A discussion of the problem of scalar–isoscalar mesons can be found in [35] (see also "Note on scalar mesons" in the Particle Listings under $f_0(1370)$ in [33]). There are also other aspects of the problem connected with scalar–isoscalar mesons in the context of nuclear structure. An object of this type is needed to provide a chiral partner of the pion in the linear σ model. It may also play the role of the "Higgs" field for the nucleon (generating, partly or entirely, the nucleon mass). One more position remains for a Higgs boson for the ω field, the latter field being treated as a gauge field. All these features of the scalar–isoscalar object will be discussed in more detail in Chap. 7.

The isoscalar–vector ω meson is responsible for the repulsive part of the NN interaction. The ω meson may be coupled to nucleons either by vector interaction (g_ω) or by tensor interaction (f_ω). The coupling constant f_ω is small in the NN channel and for this reason the tensor interaction is usually omitted. However, this interaction is included, together with the tensor interaction of the ρ meson, in the vector meson dominance model to determine the proper values of the anomalous nucleon magnetic moments. The tensor interaction of the ω meson also becomes more important in the case of the Λ and Σ particles.

The ρ–meson–nucleon interaction also contains two components. The relative role of the tensor interaction (tensor to vector) of the ρ meson is more strong than for the ω meson (in the nuclear structure context). The tensor part of the ρ interaction is important in the NN system. Its role is also very much strengthened in the relativistic Hartree–Fock approximation for finite nuclei in comparison with the relativistic Hartree approximation.

The scalar–isovector a_0 meson, which has been named $\delta(980)$, has an isotopic spin equal to unity and has very specific properties for a scalar: its mass (982 MeV) is low, compared with its $I = 1$ partners (1260 MeV, 1320 MeV, 1235 MeV); its width (54 ± 10) MeV is also small compared with these partners (≥ 100 MeV). The a_0 meson is needed for a consistent description of both

3.1 One-Boson Exchange Potentials in Configuration Space

of the S-wave phase shifts. However, it gives a very small contribution. For this reason, in many cases it is omitted in the description of NN scattering. The a_0 meson is important in relativistic Brueckner–Hartree–Fock (RBHF) calculations for finite nuclei if one is to obtain different values of the effective masses for protons and neutrons.

The pseudoscalar mesons $\eta(548)$ and $\eta'(958)$ are usually missing in the framework of the theory considered here, because their masses are larger and their coupling constants are smaller [37] than the respective values for the pion.

The ϕ meson (≈ 1020 MeV), which has been observed in experiments, has the same space–time transformation properties (quantum numbers) as the ω meson, but because of large mass, it is also not considered in the theory.

The axial meson a_1 (1260) is observed experimentally; its field is denoted by $\boldsymbol{a}^\mu$ in what follows. If the chiral symmetry is adopted, the field $\boldsymbol{a}^\mu$ is treated as the chiral partner of the ρ meson field. The contribution of a_1 exchange to the nucleon–nucleon potential has been studied in [43, 44]. Different aspects of the role of the axial meson in nuclear physics have been investigated in a number of publications [45–49] and will also be discussed below.

Let us mention that the process of obtaining the one-boson exchange potential is accompanied by a regularization procedure, and that the mesonic nucleon form factors $F(k^2)$ are introduced at each meson–nucleon vertex, similarly to the electromagnetic nucleon form factors. The reason for introducing these values is to make the potential obtained in the meson theory finite at $r = 0$.

The introduction of the nucleon form factor $F(k^2)$ corresponds to stretching of the point-like nucleons over a spatial region of finite dimensions. It is a phenomenological method to take into account the finite size of nucleons. There is a certain arbitrariness in choosing $F(k^2)$. In any case, the introduction of $F(k^2)$ is manifested by replacing the conventional Yukawa potential by a regularized combination of Yukawa functions. For example, this combination may be taken in the following form [50, 51]:

$$J(r) = -\frac{g^2}{4\pi}\left(\frac{\Lambda^2}{\Lambda^2 - m^2}\right)^2 \left[\frac{e^{-mr}}{r} - \frac{e^{-\Lambda r}}{r}\left(1 + \frac{\Lambda^2 - m^2}{2\Lambda}r\right)\right], \quad (3.1)$$

where m is a meson mass and Λ is called a regulator mass. We should mention that the function $J(r)$ used here differs in sign from that in [50, 51].

There are various theoretical methods (that of Fock functionals, for example) which can be used to obtain NN potentials on the basis of the interaction Lagrangians given in Table 3.1. To get an idea of what kind of forces are generated by different types of mesons, we present the total interaction potential between two nucleons in the following relativistic form [50, 51].

If the interaction between nucleons is realized via the exchange of scalar meson, we have

$$V_S^D = (\boldsymbol{\tau}_1 \cdot \boldsymbol{\tau}_2)^{I_S} \cdot \gamma_1^0 \gamma_2^0 J^S(r). \quad (3.2)$$

For the vector mesons, we obtain

$$V_V^D = -(\boldsymbol{\tau}_1 \cdot \boldsymbol{\tau}_2)^{I_V} \cdot \gamma_1^0 \gamma_2^0 \gamma_1^\mu \gamma_{2\mu} J^V(r) \ . \tag{3.3}$$

Finally, for the pseudoscalar mesons, we obtain

$$V_{PS}^D = -(\boldsymbol{\tau}_1 \cdot \boldsymbol{\tau}_2)^{I_{PS}} \cdot \gamma_1^0 \gamma_2^0 \gamma_1^5 \gamma_2^5 J^{PS}(r) \ , \tag{3.4}$$

where $\gamma_i^0, \gamma_i^5, \gamma_i^\mu$ ($i = 1, 2$) are the Dirac matrices [50, 51], while the functions $J^S(r)$, $J^V(r)$, and $J^{PS}(r)$ are given by (3.1) with the corresponding masses, regulator masses, and coupling constants (the indices 1 and 2 relate to the interacting nucleons); the factor $(\boldsymbol{\tau}_1 \cdot \boldsymbol{\tau}_2)^I$ determines the isotopic dependence of the interaction for isovector ($I = 1$) and isoscalar ($I = 0$) mesons.

The procedure for deriving the OBEPs in the coordinate representation (3.2)–(3.4) includes two approximations: (a) the adiabatic approximation is used and (b) retardation effects are ignored. Equations (3.2)–(3.4) are also approximated to within an accuracy of v^2/c^2. In this case the total potential is given by

$$V_{\text{tot}} = V_c(r) + V_\sigma(r)(\boldsymbol{\sigma}_1 \cdot \boldsymbol{\sigma}_2) + V_{LS}(r)\boldsymbol{l} \cdot \boldsymbol{S}$$
$$+ V_\Delta(r)\nabla^2 + V_T(r)S_{12} + V_\nabla(r)(\boldsymbol{r} \cdot \boldsymbol{\nabla}) \ , \tag{3.5}$$

where $\boldsymbol{r} = \boldsymbol{r}_{12} = \boldsymbol{r}_1 - \boldsymbol{r}_2$ is the vector connecting the two nucleons, $\boldsymbol{p} = \boldsymbol{p}_{12} = (\boldsymbol{p}_1 - \boldsymbol{p}_2)/2$ is the relative momentum of the nucleons, $\boldsymbol{l} = \boldsymbol{r} \times \boldsymbol{p}$ is the angular momentum of the relative motion, $\boldsymbol{S} = (1/2)(\boldsymbol{\sigma}_1 + \boldsymbol{\sigma}_2)$ is the operator of the total spin and, $S_{12} = [(3/r^2)(\boldsymbol{\sigma}_1 \cdot \boldsymbol{r})(\boldsymbol{\sigma}_2 \cdot \boldsymbol{r}) - (\boldsymbol{\sigma}_1 \cdot \boldsymbol{\sigma}_2)]$ is the tensor operator for two nucleons. The structure of the functions $V_c(r)$, $V_\sigma(r)$, $V_{LS}(r)$, $V_T(r)$, $V_\Delta(r)$, $V_\nabla(r)$ is determined by the space–time transformation properties of the mesons exchanged. The contributions of the various mesons to each component of V_{tot} are given below.

Scalar mesons:

$$V_c(r) = J^S(r) + \frac{a^2 \left(\nabla^2 J^S(r)\right)}{4} \ , \tag{3.6}$$

$$V_\sigma(r) = 0 \ , \quad V_T(r) = 0 \ , \tag{3.7}$$

$$V_{LS}(r) = -\frac{1}{2}a^2 \frac{1}{r}\frac{dJ^S}{dr} \ , \tag{3.8}$$

$$V_\Delta(r) = a^2 J^S(r) \ , \tag{3.9}$$

$$V_\nabla(r) = a^2 \frac{1}{r}\frac{d}{dr} J^S(r) \ . \tag{3.10}$$

Vector mesons:

$$V_c(r) = -J^V(r) - \frac{1}{2}a^2 \frac{f}{g} \left\{\nabla^2 J^V(r)\right\} \ , \tag{3.11}$$

3.1 One-Boson Exchange Potentials in Configuration Space

$$V_\sigma(r) = -\frac{1}{6}a^2\left(1+\frac{f}{g}\right)^2\{\nabla^2 J^V(r)\}, \tag{3.12}$$

$$V_{LS}(r) = -\frac{3}{2}a^2\left(1+\frac{4}{3}\frac{f}{g}\right)\frac{1}{r}\frac{dJ^V(r)}{dr}, \tag{3.13}$$

$$V_T(r) = \frac{1}{2}a^2\left(1+\frac{f}{g}\right)^2 r\frac{d}{dr}\left(\frac{1}{r}\frac{dJ^V(r)}{dr}\right), \tag{3.14}$$

$$V_\Delta(r) = a^2 J^V(r), \tag{3.15}$$

$$V_\nabla(r) = a^2\frac{1}{r}\frac{dJ^V(r)}{dr}. \tag{3.16}$$

Pseudoscalar mesons:

$$V_c(r) = V_{LS}(r) = V_\Delta(r) = V_\nabla(r) = 0, \tag{3.17}$$

$$V_\sigma(r) = -\frac{1}{12}a^2\{\nabla^2 J^{PS}(r)\}, \tag{3.18}$$

$$V_T(r) = -\frac{1}{12}a^2 r^2 \frac{1}{r}\frac{d}{dr}\left(\frac{1}{r}\frac{dJ^{PS}(r)}{dr}\right). \tag{3.19}$$

In the above, $a^2 = 1/M^2$; the brackets { } indicate that ∇^2 operates only on the functions inside these brackets. If the exchange is realized via an isovector meson, each of the operators V_i ($i = \sigma$, LS, ∇, Δ, T) is multiplied by $\tau_1 \cdot \tau_2$.[2]

We should mention that in the static limit ($v/c \to 0$), the interaction potential is reduced to a very simple form

$$V_S = J^S(r), \quad V_V = -J^V(r), \quad V_{PS} = 0, \tag{3.20}$$

while the operator in (3.5) is given by a sum of the static part (3.20) (velocity-independent) and relativistic corrections of the order v^2/c^2. The operator (3.5) has such a structure that, for each type of meson and for given values of g, f, m, and Λ, all relativistic corrections (spin–orbit and tensor forces, in particular) are uniquely determined by the static limit of the respective OBEP and do not need additional fitting parameters when this operator is used in calculations of nucleon–nucleon scattering or a many-body problem.

All PVS models (i.e. models with exchanges of pseudoscalar, vector, and scalar mesons) contain combinations of scalar and vector potentials with a near cancellation of a very strong attractive (generated by a scalar meson) and a slightly less strong, repulsive (produced by a vector meson ω) static term [50, 51], so that the resulting (attractive) static potential is relatively weak. This combination leads also to an important increase of the role of relativistic effects in the nucleon–nucleon interaction even at low energies (relativistic

---

[2] We do not show here the contributions of the axial and antisymmetric tensor interactions. These contributions can be found in [50], for example.

corrections of the order of v^2/c^2, are connected with exchanges of scalar and vector mesons, are determined by the very strong static parts of the respective potentials, and, in contrast to the static parts, have additive contributions). Such a manifestation of relativistic effects is usually referred to as "maximal relativity", while "minimal relativity" refers to considering the relativistic kinematics properly. At present there exist a number of models of one-boson exchange potentials in coordinate and momentum space, which differ in the masses and coupling constants of some of the mesons taken into consideration, in the regularization procedure, and in the quality of reproduction of the experimental data on NN scattering, deuteron properties and nuclear matter.

One-boson exchange (OBE) models contain a small number (in comparison with phenomenological potentials) of fitting parameters (5–10), and all of them have a definite physical meaning. The fitting parameters include the regulator cut-off masses Λ, and the coupling constants and masses of the σ, ρ, ω mesons. One can find a complete set of the mesons used and the respective values of the parameters for several different OBE models in [34, 50, 51]. The fitting parameters, as a rule, are determined from NN scattering over a whole range of energies (0–450 MeV) and from a description of deuteron properties (binding energy, quadrupole moment, etc.), the quality of the description being not inferior to that obtained with one of the best phenomenological potentials, the Reid potential with a soft core. The most up-to-date version of the NN potential is given in [52], in both configuration space and momentum space.

3.2 One-Boson Exchange Potentials in Momentum Space

In this section we consider the general properties of the OBEP developed by the Bonn group in momentum space. One can find a much more detailed discussion of this problem in the excellent review papers [34]. The nucleon interaction potential in momentum space may be obtained as a sum of amplitudes of Feynman diagrams (see Fig. 3.2), describing the process of exchange of scalar, vector, and pseudoscalar mesons between two nucleons:

$$\langle \lambda_1' \lambda_2' \boldsymbol{q}' | V | \lambda_1 \lambda_2 \boldsymbol{q} \rangle = \sum_\alpha V_\alpha(\boldsymbol{q}', \boldsymbol{q}) , \qquad (3.21)$$

where $\boldsymbol{q}$ and $\boldsymbol{q}'$ are the relative nucleon momentum before and after the interaction. The amplitude for meson α has the form

$$V_\alpha(\lambda_1' \lambda_2' \boldsymbol{q}'; \lambda_1 \lambda_2 \boldsymbol{q}) = \bar{u}_{\lambda_2'}(-\boldsymbol{q}') \Gamma_\alpha^{(2)} u_{\lambda_2}(-\boldsymbol{q}) P_\alpha \bar{u}_{\lambda_1'}(\boldsymbol{q}') \Gamma_\alpha^{(1)} u_{\lambda_1}(\boldsymbol{q}) , \qquad (3.22)$$

where $u(\boldsymbol{q})$ is the Dirac spinor for a free nucleon with momentum $\boldsymbol{q}$, Γ_α is the vertex operator of the meson–nucleon interaction, and P_α is the meson propagator.

3.2 One-Boson Exchange Potentials in Momentum Space

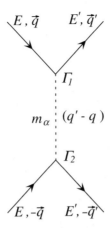

Fig. 3.2. One-boson exchange diagram. The *solid lines* correspond to the nucleons and the *dashed lines* to mesons

For scalar mesons, the values of Γ_α and P_α are given by

$$\Gamma_\sigma = g_\sigma \, , \qquad P_\sigma = \frac{1}{\Delta^2 - m_\sigma^2} \, . \tag{3.23}$$

For pseudoscalar mesons,

$$\Gamma_{\rm PS} = g_\pi i\gamma_5 \, , \qquad P_{\rm PS} = \frac{1}{\Delta^2 - m_\pi^2} \, . \tag{3.24}$$

For vector mesons,

$$\Gamma_{\rm V} = g_{\rm V}\gamma_\mu \, , \quad \Gamma_{\rm T} = -\frac{f_{\rm V}}{2M}(q' + q)_\mu \, , \quad P_{\rm V} = -\frac{g_{\mu\nu}}{\Delta^2 - m_{\rm V}^2} \, , \tag{3.25}$$

where $\Gamma_{\rm V}$ is related to the vector part of the interaction and $\Gamma_{\rm T}$ to the tensor part, $\Delta^2 = (q_0' - q_0)^2 - (\boldsymbol{q}' - \boldsymbol{q})^2$, and $g_{\mu\nu}$ is the metric tensor ($g_{00} = 1$, $g_{11} = g_{22} = g_{33} = -1$, $g_{\mu\neq\nu} = 0$).

It is convenient to describe the nucleon spin state by the quantum number of helicity. In this case the Dirac spinor has the following form:

$$u_\lambda(\boldsymbol{q}) = \left(\frac{E_q + M}{2M}\right)^{1/2} \begin{pmatrix} 1 \\ 2\lambda q/(E_q + M) \end{pmatrix} |\lambda\rangle \, . \tag{3.26}$$

Here $|\lambda\rangle$ is an eigenstate of the chirality operator. By inserting (3.23)–(3.25) into (3.21), one can obtain expressions for the OBEP in terms of the relative momenta $\boldsymbol{q}$ and $\boldsymbol{q}'$ and the helicities $\lambda_1, \lambda_2, \lambda_1', \lambda_2'$.

However, in many applications it is more convenient to use the angular-momentum representation. One can obtain this starting from the relation

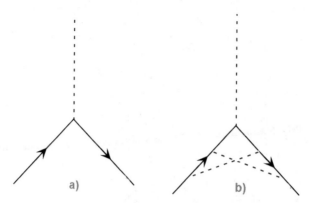

Fig. 3.3. Vertex parts of the Feynman diagrams: **a)** simplest vertex; **b)** higher order diagram

$$\langle \lambda_1' \lambda_2' \boldsymbol{q}' | V | \lambda_1 \lambda_2 \boldsymbol{q} \rangle = \frac{1}{4\pi}$$
$$\times \sum_J (2J+1) d^J_{\lambda \lambda'}(\theta) \langle \lambda_1' \lambda_2' | V^J(q',q) | \lambda_1 \lambda_2 \rangle \,, \quad (3.27)$$

where $\lambda = \lambda_1 - \lambda_2$, $\lambda' = \lambda_1' - \lambda_2'$, θ is the scattering angle (the angle between the two vectors $\boldsymbol{q}'$ and $\boldsymbol{q}$), $d^J_{\lambda\lambda'}(\theta)$ is the rotation matrix, and $\langle \lambda_1' \lambda_2' | V^J(q',q) | \lambda_1 \lambda_2 \rangle$ is the OBEP in the J-representation.

Using the property of orthogonality of the d-functions, it is easy to work out a relation inverse to (3.27):

$$\langle \lambda_1' \lambda_2' | V^J(q',q) | \lambda_1 \lambda_2 \rangle = 2\pi \int_0^\pi \sin\theta \, d\theta \, d^J_{\lambda\lambda'}(\theta) \langle \lambda_1' \lambda_2' \boldsymbol{q}' | V | \lambda_1 \lambda_2 \boldsymbol{q} \rangle \,. \quad (3.28)$$

By inserting the expressions for $\langle \lambda_1' \lambda_2' \boldsymbol{q}' | V | \lambda_1 \lambda_2 \boldsymbol{q} \rangle$ into this equation, one can obtain the potentials in the J-representation after sophisticated calculations. The results are given in [34].

However, OBEPs derived in this way still cannot be used directly in the problem of NN scattering and in calculations of many-nucleon systems because of the divergences of the expressions obtained for the OBEPs in the limit of high relative momenta $\boldsymbol{q}$, $\boldsymbol{q}'$. The physical reason for these divergences is connected with the picture of point-like nucleons used in the calculations of the OBEPs. The nucleon–meson interaction, that enters (3.21) is obtained when the simplest vertex Feynman diagram (see Fig. 3.3a) is taken into account.

Considering the diagrams of higher order (see Fig. 3.3b) leads to the appearance of an additional factor $F_\alpha(\boldsymbol{q}, \boldsymbol{q}')$ in the vertex part, which goes to zero sufficiently rapidly as the relative nucleon momentum increases, so the divergence of the OBEP at high momenta is compensated. Regrettably, however, the consideration of the diagrams of higher order is a complicated problem and has been solved only approximately.

In many cases the meson–nucleon form factors are introduced phenomenologically. When OBEPs are considered in momentum space, two main types of form factors are utilized:

$$F_\alpha(\boldsymbol{q}',\boldsymbol{q}) = \left[\frac{\Lambda_\alpha^2 - m_\alpha^2}{\Lambda_\alpha^2 - \Delta^2}\right]^n, \qquad (3.29)$$

where $n = 1$ or $1/2$, and

$$F_\alpha(\boldsymbol{q}',\boldsymbol{q}) = \frac{\Lambda_\alpha^2}{\Lambda_\alpha^2 - \Delta^2}\left(\frac{\Lambda_{\alpha,V}^2 - m_{\alpha,V}^2}{\Lambda_{\alpha,V}^2 - \Delta^2}\right)^{1/2}, \qquad (3.30)$$

where Λ_α, $\Lambda_{\alpha,V}$ are free parameters determined from experimental measurements of processes of meson radiation by nucleons, nucleon–nucleon scattering, electromagnetic processes involving nucleons, etc. Even if the form factors are well determined by experiment (this can be said now only about pions), this is still not sufficient for a complete determination of the OBE interaction, since the off-shell behavior of the form factors needs to be known in the case of many-nucleon systems. In momentum space in the OBE models with phenomenological form factors, one usually uses the form factors in (3.29) for scalar and pseudoscalar mesons, while one usually uses the form factors in (3.30) for vector mesons. When the form factors have been taken into account, the OBEP can be given in the following form:

$$\langle \lambda_1'\lambda_2'\boldsymbol{q}' | \widetilde{V} | \lambda_1\lambda_2\boldsymbol{q}\rangle = \sum_\alpha F_\alpha^2(\boldsymbol{q}',\boldsymbol{q})V_\alpha(\boldsymbol{q}',\boldsymbol{q}). \qquad (3.31)$$

The introduction of form factors, in the general case, makes it impossible to write out an expression for the OBEP in the representation of helicity and of angular momentum $\langle \lambda_1'\lambda_2' | V^J(q',q) | \lambda_1\lambda_2\rangle$. However, for the form factors given by (3.29) and (3.30), this is possible.

The form factors considered above are introduced into the OBE framework in a phenomenological and, to a certain extent, arbitrary way. At present, along with the form factors mentioned above, form factors derived on the basis of accounting for multimeson exchanges and obtained as a result of summing certain classes of diagrams are widely used. These form factors, calculated in the eikonal approximation (known from electrodynamics), enable one to obtain the amplitude of a process in which a certain interaction, conditionally treated as a "hard" one, is accompanied by exchange of an arbitrary number of "soft" mesons (in electrodynamics, the exchange of only photons is taken into account).

The introduction of the eikonal form factors into the theory of nuclear forces should, apparently, be considered as an important development of the theory, since they have greatly improved the results of calculations of few-

Table 3.2. Relativistic OBEPs obtained using the Thompson equation and the pseudovector coupling for π and η [34]. The meson parameters are given. In all cases here, $f_\rho/g_\rho = 6.1$, $f_\omega/g_\omega = 0.0$ and $n_\alpha = 1$

		Potential				
	BM-A		BM-B		BM-C	
m_α (MeV)	$g_\alpha^2/4\pi$	Λ_α (GeV)	$g_\alpha^2/4\pi$	Λ_α (GeV)	$g_\alpha^2/4\pi$	Λ_α (GeV)
π 138.03	14.9	1.05	14.6	1.2	14.6	1.3
η 548.8	7	1.5	5	1.5	3	1.5
ρ 769	0.99	1.3	0.95	1.3	0.95	1.3
ω 782.6	20	1.5	20	1.5	20	1.5
δ 983	0.7709	2.0	3.1155	1.5	5.0742	1.5
σ 550	8.3141	2.0	8.0769	2.0	8.0279	1.8

and many-body (nucleon) systems, and have eliminated, to a certain extent, the arbitrariness in the choice of the OBEPs.

The momentum-space one-boson exchange potentials, considered in this section [34] are among the best ones on the world market. There is still one more reason to discuss this type of NN force in this introduction. Recently, it has been demonstrated by Brockmann and Machleidt [53] that the relativistic Brueckner–Hartree–Fock theory is capable of describing nuclear-matter saturation starting from the nucleon–nucleon interaction determined by scattering experiments. These calculations were based on relativistic OBEPs developed by the Bonn group (BM-A, BM-B, and BM-C); the parameters of these potentials are given in Table 3.2. The potentials BM-A, BM-B, and BM-C differ mainly in the strength of the tensor force, increasing from A to C.

All OBE models of the free-space NN interaction utilize the current experimental values for the meson masses. However, the free nucleon–meson coupling constants are not determined uniquely, and there is a certain diversity in the values obtained by different groups.

All recent models of the NN forces contain a one-pion exchange term, with the same πNN coupling constant (in particular, they describe the difference between the masses of the charged $\pi_\pm$ and neutral π_0). This long-range part of the NN interaction is considered to be well understood. However, even this part is treated in a different way in different models. Recently, a model has been developed by the Bonn group [54] in which the NN potential is derived in the framework of the relativistic meson field theory by applying ps coupling. The potential is calculated in momentum space and contains nonlocal terms in the short-range and long-range parts, including the pion-exchange contribution. It has been shown that the nonlocalities included in the relativistic description of this Bonn potential lead to smaller D-state probabilities in the deuteron [55], the potential is referred to as the charge-dependent (CD) Bonn potential.

4 The Relativistic Mean-Field Approximation for Nuclear Structure

4.1 General Characteristics of the Relativistic Framework

In the previous chapter we considered briefly the general properties of the mesonic PVS models for NN interaction. This was done because of two reasons. First of all, the relativistic theory of nuclear structure, which is the main subject of our present considerations, has many features in common with the PVS models of NN forces. Second, the early publications in this field [51, 56–65] were based on the relativistic OBEPs in coordinate space introduced in Chap. 3; these papers contain many important features of the relativistic theory in its current status.

4.1.1 Large Scalar and Vector Fields

The presence of two strong fields (nearly equal in magnitude and opposite in sign) with the transformation properties of a relativistic scalar $S(r)$ (attractive) and of the time component of a relativistic vector $V(r)$ (repulsive) is assumed here. These fields are formed by corresponding mesons (the σ meson in the first case and the ω meson in the second case).

The presence of these two fields in the nucleus is confirmed by relativistic theories with interacting nucleons and mesons [59], and by theories with nucleons interacting with Skyrme-type interactions [66]. It follows also from QCD sum rules [67] and may be considered as a firmly established fact.[1]

Using expressions for the nucleon effective mass (see below) and central potential in terms of mesonic fields, and also the fact that the depth of the shell potential ≈ -50 MeV, one obtains the following values for the depth of the scalar field S and the value of the vector meson field V inside the nucleus: $S \approx -420$ MeV, $V \approx +330$ MeV. To obtain this estimate, it is sufficient to use only phenomenological values, which are used in the nonrelativistic theory and are well known from experiment, and one needs to make no assumptions concerning the nature of the scalar and vector fields.

---

[1] In this connection, one should also mention [68–73]. However, those papers were written at a time when no mesons except the pion had been observed in experiments. So one might say that those papers were published "ahead of time".

4.1.2 Spin–Orbit Force

The anomalously large spin–orbit coupling is a direct indication of the important role of relativistic effects in the nucleus. To clarify this statement, let us compare the situation in an atom with that in a nucleus. In the atomic case the spin–orbit potential is determined by the Thomas formula

$$U_{\text{LS}} = \frac{1}{4M_e^2} \frac{1}{r} \frac{dU}{dr} l \cdot \sigma , \tag{4.1}$$

where U is the central potential (Coulomb field plus self-consistent field). The use of (4.1) in the nucleus leads to (a) the wrong size of the effect (the central potential in nuclear case is a potential well with a depth equal to $\sim -50\,\text{MeV}$ and with a radius of the order of magnitude of the nuclear radius, so it has a positive derivative, while the experimental sign of the spin–orbit force is negative); and (b) an absolute value of the effect approximately 30 times smaller than the observed value. For this reason, a negative factor (-30) is introduced in (4.1) in the nuclear case; this factor characterizes the scale of the strengthening of the relativistic effects in the nucleus. A strong, parameter-free spin–orbit force is a peculiar feature of the relativistic theory [51, 59, 60].

4.1.3 Saturation

Saturation [56] is a very important feature of the current relativistic approaches connected with taking the lower component of the nucleon wave function into account (for details, see below). In particular, the effect of saturation in the relativistic theory is related to a reduction in the expectation values of the relativistic analogue of the kinetic-energy operator [56]. We should mention also that relativistic Brueckner–Hartree–Fock calculations, which will be discussed below, are based on free NN forces of the type introduced in Chap. 3.

In 1974 Walecka [74, 75] suggested an elegant way to develop a relativistic approach (quantum hadrodynamics, QHD), based on a Lagrangian (field-theoretic) description of nuclear systems rather than on NN potentials. His model incorporates general features of the relativistic approach mentioned above and contains new effects which will be discussed below. In the framework of this unique approach, the model describes the following set of experimental facts: (a) the saturation property, (b) the ultrarelativistic limit for the equation of state in the high-density approximation, (c) the spin–orbit interaction in the nucleus, and (d) the energy dependence of the real part of the optical potential. The basic features of the Walecka model for the ground state of nuclear matter are considered in Appendix A2.

Generalized versions of the Walecka model are widely used now. The main reason for the success of the relativistic theory is its simplicity and ability to reproduce a large amount of experimental data [51, 76–85]. The

4.1 General Characteristics of the Relativistic Framework

description of the properties of nuclear structure on the basis of the Walecka Lagrangian puts nuclear theory on a qualitatively new level, at which it can be regarded as a relativistic quantum field theory. Of course, the exact relativistic quantum-field description corresponding to the Walecka Lagrangian is very complicated.[2] However, it turns out that in the framework of this model, it is possible to obtain many important results by using a mean-field approximation in which the fields that occur in the Walecka approach are treated as classical fields. Generally speaking, this approximation requires a theoretical justification, which has been provided by Walecka only in the limit of high densities. The use of the model in the mean-field approximation at the observed nuclear densities is justified at present only by the large number of experimental results that are reproduced in the framework of the Walecka model in this approximation.[3] In what follows, it is this approximation that is mostly considered. It deals with such properties of nuclei as masses, radii and single-particle energies, as well as the basic nuclear bulk properties [74–76].

It should be emphasized that even the reproduction of the saturation property (the correct binding energy per nucleon $E/A = -15.75$ MeV at the "empirically observed" density $\rho = 0.17$ fm^{-3} is obtained for infinite nuclear matter) is an important result; the simultaneous description of a large body of experimental data is an undoubted success of the relativistic mean-field theory.

In this theory, the strong interaction is produced by the exchange of mesons between nucleons through the Yukawa couplings (see Fig. 3.2), i.e. it is generated in the same way as in the PVS models of the NN interaction.

We start with a relativistic covariant Lagrangian which is composed of the meson and nucleon degrees of freedom and the electromagnetic field:

---

[2] The Walecka model is a renormalizable theory [76]. However, one meets with serious problems in QHD when using perturbation theory because of the large values of the coupling constants obtained. For example, it has been shown that loop expansions do not converge [86].

[3] In most applications of the relativistic mean-field theory to finite nuclei, the negative-energy eigenfunctions of the single-particle potentials are excluded from consideration, this approximation being referred to as the no-Fermi-sea approximation (it neglects the vacuum polarization). Different aspects of the no-sea approximation for nuclear matter and finite nuclei have been considered in [81, 86–90]. The mean-field and no-Fermi-sea approximations are very strong, they ignore essential many-body effects (for example, exchange terms and short-range correlations) and important quantum field effects (nucleonic and mesonic vacuum polarization, coupling, and mass renormalization). So the Lagrangian of the relativistic mean-field theory may be considered as an effective Lagrangian for relativistic calculations in the same sense as the Skyrme potential is an effective force for classical nonrelativistic mean-field calculations.

4 The Relativistic Mean-Field Approximation for Nuclear Structure

$$\begin{aligned}
\mathcal{L}_{\text{RMF}} = {} & \bar{\psi}\Big(i\gamma_\mu\partial^\mu - M - g_\sigma\varphi - g_\omega\gamma_\mu\omega^\mu - g_\rho\gamma_\mu\boldsymbol{\tau}\cdot\boldsymbol{\rho}^\mu - ig_\pi\gamma_5\boldsymbol{\tau}\boldsymbol{\pi} \\
& - e\gamma_\mu A^\mu \frac{1+\tau_3}{2}\Big)\psi + \frac{1}{2}\partial_\mu\varphi\partial^\mu\varphi - \frac{1}{2}m_\sigma^2\varphi^2 - \frac{1}{3}g_2\varphi^3 - \frac{1}{4}g_3\varphi^4 \\
& + \frac{1}{2}(\partial_\mu\boldsymbol{\pi}\partial^\mu\boldsymbol{\pi} - m_\pi^2\boldsymbol{\pi}\cdot\boldsymbol{\pi}) - \frac{1}{4}\omega_{\mu\nu}\cdot\omega^{\mu\nu} + \frac{1}{2}m_\omega^2\omega_\mu\omega^\mu + \frac{1}{4}c_3(\omega_\mu\omega^\mu)^2 \\
& - \frac{1}{4}\boldsymbol{\rho}_{\mu\nu}\cdot\boldsymbol{\rho}^{\mu\nu} + \frac{1}{2}m_\rho^2\boldsymbol{\rho}_\mu\cdot\boldsymbol{\rho}^\mu - \frac{1}{4}F_{\mu\nu}\cdot F^{\mu\nu} \,,
\end{aligned} \quad (4.2)$$

where the standard notation is used for γ-matrices [2, 91]. Here ψ denotes an $SU(2)$ baryon field of mass M (proton and neutron); φ, ω_μ and ρ_μ^a (where the Greek indices may take four values, $\mu, \nu = 0, 1, 2, 3$, while the Latin indices may take three values, $a = 1, 2, 3$) are σ, ω, and ρ meson fields with masses m_σ, m_ω, and m_ρ, respectively [76, 92].[4] $\omega_{\mu\nu}$, $\rho_{\mu\nu}^a$ and $F_{\mu\nu}$ are the antisymmetric field tensors for the ω_μ, ρ_μ^a, A_μ fields (A_μ being the electromagnetic field), given by

$$\begin{cases}
\omega_{\mu\nu} = \partial_\mu\omega_\nu - \partial_\nu\omega_\mu \,, \\
\boldsymbol{\rho}_{\mu\nu} = \partial_\mu\boldsymbol{\rho}_\nu - \partial_\nu\boldsymbol{\rho}_\mu \,, \\
F_{\mu\nu} = \partial_\mu A_\nu - \partial_\nu A_\mu \,,
\end{cases} \quad (4.3)$$

where the bold variables stand for vectors in isotopic space.

The constants g_σ, g_ω, g_ρ, g_π are coupling constants for the corresponding meson–nucleon interactions,[5] and g_2, g_3, and c_3 are the nonlinear parameters for the self-interactions of the scalar and vector fields (the cubic self-interaction $\sim \omega^3$ is not introduced because of its parity).[6]

For all fields Φ considered here $(\psi, \varphi, \omega_\mu, \boldsymbol{\rho}_\mu, \boldsymbol{\pi}, A_\mu)$, the Euler–Lagrange equations of motion have the following form:

$$\frac{\partial\mathcal{L}}{\partial\Phi} - \partial^\mu\frac{\partial\mathcal{L}}{\partial[\partial^\mu\Phi]} = 0 \,. \quad (4.4)$$

For the Lagrangian (4.2), one obtains the Dirac equation for the nucleon wave functions,

---

[4] In (4.2) the pion–nucleon coupling is represented in a pseudoscalar form; it may be written down also in a pseudovector form. A more detailed discussion of the pion–nucleon interaction will be given in Chap. 7.

[5] The tensor coupling of the ρ mesons and nucleons is usually not considered in the relativistic mean-field approximatin (RMFA). This coupling mostly affects the value of the spin–orbit interaction in finite nuclei (see Chap. 8). The tensor ωN coupling is very weak (see Chap. 5).

[6] The simplest general form for the functional containing the φ and ω fields may include terms of the type $\varphi\omega^2$ and $\varphi^2\omega^2$ also. The role of these terms will be discussed in Chap. 8.

4.1 General Characteristics of the Relativistic Framework

$$\left\{ \gamma_\mu \left(i\partial^\mu + g_\omega \omega^\mu + g_\rho \boldsymbol{\tau} \cdot \boldsymbol{\rho}^\mu + e\frac{1+\tau_3}{2} A^\mu \right) \right.$$

$$\left. + (M + g_\sigma \varphi) + ig_\pi \gamma_5 \boldsymbol{\tau} \cdot \boldsymbol{\pi} \right\} \psi = 0 . \quad (4.5)$$

For the meson fields, one obtains the Klein–Gordon equations for the scalar and pseudoscalar fields,

$$(\Box + m_\sigma^2)\varphi = -g_\sigma \bar{\psi}\psi - g_2 \varphi^2 - g_3 \varphi^3 , \quad (4.6)$$

$$(\Box + m_\pi^2)\boldsymbol{\pi} = -g_\pi \cdot \bar{\psi}\boldsymbol{\tau}\gamma_5 \psi , \quad (4.7)$$

and the Proca equation with a source term, for example, for the ω meson field:

$$\partial_\nu \omega^{\nu\mu} + m_\omega^2 \omega^\mu = g_\omega \bar{\psi}\gamma^\mu \psi - c_3 (\omega_\mu \cdot \omega^\mu) \omega^\mu . \quad (4.8)$$

However, since the nucleon current satisfies the continuity equation

$$\partial^\mu [\bar{\psi}\gamma_\mu \psi] = 0 , \quad (4.9)$$

the Proca equation reduces to a Klein–Gordon equation

$$(\Box + m_\omega^2)\omega^\mu = g_\omega \bar{\psi}\gamma^\mu \psi - c_3 (\omega_\mu \cdot \omega^\mu) \omega^\mu . \quad (4.10)$$

The same equations are obtained for the ρ meson and for the electromagnetic field:

$$(\Box + m_\rho^2)\boldsymbol{\rho}^\mu = g_\rho \bar{\psi}\boldsymbol{\tau}\gamma^\mu \psi , \quad (4.11)$$

$$\Box A^\mu = e\bar{\psi}\frac{1}{2}(1+\tau_3)\gamma^\mu \psi , \quad (4.12)$$

where $\tau_3 = \pm 1$ (the positive sign corresponds to protons and the negative sign to neutrons), and the following densities and currents are introduced:

$$\rho_s(x) = \bar{\psi}(x)\psi(x) , \quad \text{scalar density,} \quad (4.13)$$

$$\boldsymbol{\rho}_{\text{ps}}(x) = \bar{\psi}(x)\gamma^5 \boldsymbol{\tau}\psi(x) , \quad \text{pseudoscalar density,} \quad (4.14)$$

$$j^\mu(x) = \bar{\psi}(x)\gamma^\mu \psi(x) , \quad \text{baryon current,} \quad (4.15)$$

$$\boldsymbol{j}^\mu(x) = \bar{\psi}(x)\gamma^\mu \boldsymbol{\tau}\psi(x) , \quad \text{isovector current,} \quad (4.16)$$

$$j_c^\mu(x) = \bar{\psi}(x)\gamma^\mu \frac{1}{2}(1+\tau_3)\psi(x) , \quad \text{electromagnetic current,} \quad (4.17)$$

where $x = (t, \boldsymbol{r})$ is the space–time coordinate.

In the mean-field approximation [74, 76] the meson fields are treated as classical fields, taking the expectation values of the ground states as

$$\varphi \longrightarrow \langle \varphi \rangle , \quad (4.18)$$

$$\omega^\mu \longrightarrow \langle \omega^\mu \rangle \,, \tag{4.19}$$

$$\rho^\mu \longrightarrow \langle \rho^\mu \rangle \,. \tag{4.20}$$

Interactions with pions are included in the equations given above. However, it should be mentioned that in the mean-field approximation adopted in this chapter, the expectation value of the pion field in the ground state equals zero,[7]

$$\langle \pi \rangle = 0 \,, \tag{4.21}$$

and in the case of the ground state 0^+, only scalar and vector fields need to be considered.

The pion contribution to the ground-state properties of nuclear matter and finite nuclei are revealed in the Hartree–Fock approach. It is a common feature of the relativistic mean-field theory that the attractive isoscalar–scalar σ meson and the repulsive isoscalar–vector ω meson provide together the saturation property of symmetric nuclear matter.

If one considers static infinite matter, the meson fields are assumed to be time-independent, while the time dependence of the nucleon spinors is given by a factor $\exp(-iE_s t)$:

$$i\partial_t \psi_{is} = E_{is} \psi_{is} \,, \tag{4.22}$$

$$\partial_t^2 \sigma = 0 \,, \tag{4.23}$$

$$\partial_t^2 \omega_\mu = 0 \,, \tag{4.24}$$

$$\omega_i = 0 \,, \quad i = 1, 2, 3 \,, \tag{4.25}$$

$$A_\mu = 0 \quad \text{(for nuclear matter)} \,, \tag{4.26}$$

$$\rho^\mu = 0 \quad \text{(for symmetric nuclear matter)} \,. \tag{4.27}$$

In this case simplified equations are obtained, where the derivative terms in the Klein–Gordon equations for mesons vanish automatically (owing to the translational invariance of infinite nuclear matter), the Coulomb interaction being switched off completely in this case. For finite nuclei, the derivative terms remain and provide the ranges of the interaction. The spatial components of the vector meson fields vanish under rotational symmetry. Only the third isospin component of the meson field ρ_μ has a nonvanishing value, because of charge conservation. Hence the equations for meson fields are reduced to [94]

$$\varphi_0 \equiv \langle \varphi \rangle = -\frac{g_\sigma}{m_\sigma^2} \langle \bar\psi \psi \rangle - \frac{1}{m_\sigma^2}(g_2 \varphi_0^2 + g_3 \varphi_0^3) \,, \tag{4.28}$$

$$\omega_0 \equiv \langle \omega^0 \rangle = \frac{g_\omega}{m_\omega^2} \langle \bar\psi \gamma^0 \psi \rangle - \frac{c_3}{m_\omega^2} \omega_0^3 \,, \tag{4.29}$$

$$\rho_0 \equiv \langle \rho_3^0 \rangle = \frac{g_\rho}{m_\rho^2} \langle \bar\psi \tau_3 \gamma^0 \psi \rangle \,, \tag{4.30}$$

---

[7] See, however, [93] in this connection.

4.1 General Characteristics of the Relativistic Framework

$$\langle \boldsymbol{\pi} \rangle = 0 \,, \tag{4.31}$$

and so the pseudoscalar density ρ_{ps} is equal to zero in the RMFA. The source terms in the above equations are the expectation values of the nucleon field where nucleons occupy single-particle orbits from the bottom to the Fermi surface with an occupation probability $f_{is} = 1$ at zero temperature:

$$\langle \bar{\psi}\psi \rangle = \sum_{is} f_{is} \bar{\psi}_{is} \psi_{is} \,, \tag{4.32}$$

$$\langle \bar{\psi}\gamma^0\psi \rangle = \sum_{is} f_{is} \bar{\psi}_{is} \gamma^0 \psi_{is} \,, \tag{4.33}$$

$$\langle \bar{\psi}\tau_3\gamma^0\psi \rangle = \sum_{is} f_{is} \bar{\psi}_{is} \tau_3 \gamma^0 \psi_{is} \,, \tag{4.34}$$

where the index i denotes the isospin degree of freedom (proton and neutron) and s denotes the index of eigenstates of the nucleon. The nucleon field ψ_{is} is obtained by solving the Dirac equation, where the potentials are created by the mean meson fields calculated above:

$$(-i\boldsymbol{\alpha}\cdot\boldsymbol{\nabla} + \beta M^* + g_\omega \omega_0 + g_\rho \tau_3 \rho_3^0)\psi_{is} = E_{is}\cdot\psi_{is} \,. \tag{4.35}$$

Here $M^* \equiv M + g_\sigma \varphi_0$ denotes the effective mass of a nucleon and E_{is} denotes the single-particle energy of the nucleons, while $\boldsymbol{\alpha}$ and β are the conventional Dirac matrices [2, 91]. The equations (4.28)–(4.35) are solved self-consistently. For infinite matter at zero temperature, we have only one equation for the effective mass, as a self-consistency equation

$$M^* = M - \frac{g_\sigma^2}{m_\sigma^2}\sum_i \frac{\gamma}{2\pi^2}\int_0^{k_{Fi}} dk\, k^2 \frac{M^*}{\sqrt{k^2 + M^{*2}}} - \frac{g_\sigma}{m_\sigma^2}(g_2\varphi_0^2 + g_3\varphi_0^3) \,. \tag{4.36}$$

Here the degeneracy factor γ is 4 for symmetric ($N=Z$) matter and 2 for pure neutron matter ($Z=0$), and k_{Fi} is the Fermi momentum for a nucleon.

When one applies the present framework to the equation of state at finite temperature for hot neutron stars, one can incorporate temperature [94] in a straightforward way [74, 76]. We replace the occupation probability in the summation over states in (4.32)–(4.34),

$$f_{is} = \theta(E_{Fi} - E_{is}) \,, \tag{4.37}$$

by a Fermi–Dirac distribution with a temperature T and chemical potential μ_i, namely

$$f_{is} = \frac{1}{1 + \exp\left[(\sqrt{k^2 + M^{*2}} - \nu_i)/T\right]} \,, \tag{4.38}$$

$$f_{\bar{i}s} = \frac{1}{1 + \exp\left[(\sqrt{k^2 + M^{*2}} + \nu_i)/T\right]} \,, \tag{4.39}$$

for nucleons and antinucleons, respectively. The chemical potential μ_i is related to the nucleon number density $\rho_{B,i} = \left[\gamma/(2\pi)^3\right] \int_0^\infty \mathrm{d}^3 k\, (f_{is} - f_{\bar{i}s})$ (where i denotes the nucleon species, i.e. proton or neutron) via the equations

$$\mu_p = \nu_p + \frac{g_\omega^2}{m_\omega^2}\rho_B + \frac{g_\rho^2}{m_\rho^2}(\rho_p - \rho_n)\,, \quad \mu_n = \nu_n + \frac{g_\omega^2}{m_\omega^2}\rho_B - \frac{g_\rho^2}{m_\rho^2}(\rho_p - \rho_n)\,, \quad (4.40)$$

and ν_i is the kinetic part of the chemical potential. In the equations above and in what follows, the quantum number s is the momentum k of a nucleon in nuclear matter. Instead of (4.36) at zero temperature, one obtains an equation for an effective mass at finite temperature given by

$$M^* = M - \frac{g_\sigma^2}{m_\sigma^2}\sum_i \frac{\gamma}{2\pi^2}\int_0^\infty \mathrm{d}k\, k^2 \frac{M^*}{\sqrt{k^2 + M^{*2}}}(f_{ik} + f_{\bar{i}k})$$

$$- \frac{g_\sigma}{m_\sigma^2}(g_2\varphi_0^2 + g_3\varphi_0^3)\,. \quad (4.41)$$

The procedure for solving the self-consistency equation for the effective mass is described in [94] (see Appendix A2). Once the value of the effective mass has been obtained from the above equation, the energy density and the pressure can be obtained from the energy-momentum tensor [74] $T^{\mu\nu}$ given by

$$T^{\mu\nu} = -g^{\mu\nu}\mathcal{L} + \frac{\partial \mathcal{L}}{\partial(\partial_\mu \Phi)}\partial^\nu \Phi\,, \quad (4.42)$$

where $g^{\mu\nu}$ is the metric tensor [2].

Owing to (4.4), $T^{\mu\nu}$ satisfies the continuity equation

$$\partial_\mu T^{\mu\nu} = 0\,, \quad (4.43)$$

while the four-momentum P^ν,

$$P^\nu = \int \mathrm{d}^3 r\, T^{0\nu}\,, \quad (4.44)$$

is conserved. The total energy is its time component

$$P^0 = E = \int \mathrm{d}^3 r\, \mathcal{H}(r)\,, \quad (4.45)$$

where $\mathcal{H}(r)$ is the Hamiltonian density

$$\mathcal{H} = T^{00} = \frac{\partial \mathcal{L}}{\partial \dot{\Phi}}\dot{\Phi} - \mathcal{L}\,. \quad (4.46)$$

The energy density for the nuclear-matter ground state is given by

4.1 General Characteristics of the Relativistic Framework

$$\mathcal{E} = \sum_i \frac{\gamma}{2\pi^2} \int_0^\infty dk\, k^2 \sqrt{k^2 + M^{*2}} (f_{\bar{i}k} + f_{ik})$$
$$+ \frac{g_\omega^2}{2m_\omega^2}(\rho_\mathrm{p} + \rho_\mathrm{n})^2 + \frac{g_\rho^2}{2m_\rho^2}(\rho_\mathrm{p} - \rho_\mathrm{n})^2$$
$$+ \frac{1}{2} m_\sigma^2 \varphi_0^2 + \frac{1}{3} g_2 \varphi_0^3 + \frac{1}{4} g_3 \varphi_0^4 - \frac{1}{4} c_3 \omega_0^4, \qquad (4.47)$$

while the pressure can be represented in the following form:

$$p = \sum_i \frac{\gamma}{6\pi^2} \int_0^\infty dk\, k^2 \frac{k^2}{\sqrt{k^2 + M^{*2}}} (f_{ik} + f_{\bar{i}k})$$
$$+ \frac{g_\omega^2}{2m_\omega^2}(\rho_\mathrm{p} + \rho_\mathrm{n})^2 + \frac{g_\rho^2}{2m_\rho^2}(\rho_\mathrm{p} - \rho_\mathrm{n})^2$$
$$- \frac{1}{2} m_\sigma^2 \varphi_0^2 - \frac{1}{3} g_2 \varphi_0^3 - \frac{1}{4} g_3 \varphi_0^4 + \frac{1}{4} c_3 \omega_0^4. \qquad (4.48)$$

Further details of the notations can be found in [94], where formulae for the physical quantities in the table of the equation of state are summarized (see Appendix A1 in [94]).

The original form of the Walecka model [74] includes only σ and ω meson fields and contains two parameters determined by reproducing the ground-state properties of nuclear matter. The isovector–vector ρ meson (g_ρ being its coupling constant to nucleons[8]) is added to describe the properties of isotopes and asymmetric matter. However, the Walecka model, with the ρ meson added, gives only a qualitative description of the ground-state properties of finite nuclei [100, 101]. In [102] it is recognized that incompressibility and the surface properties of nuclei cannot be described in the framework of this simple version; nuclear deformations [84] also cannot be reproduced in this case. Boguta and Bodmer [102] have studied the role of the self-interactions of the scalar meson field in nuclear matter (g_2 and g_3 are the respective self-coupling constants); the most complete investigation for finite nuclei has been described in [103, 104]. Self-interactions of mesonic fields are related to many-body forces; their role both in nuclear matter and in finite nuclei was studied earlier in [105–112] also. In [112] it is shown that the incompressibility and surface properties of nuclei are reproduced well in a nonlinear extension of the Walecka model with φ^3 and φ^4 terms, while in [92, 95, 103, 112] it is concluded also that the binding energies and the radii of the nuclei cannot be well reproduced simultaneously without these terms. The scalar-meson

---
[8] Knowledge of the strength of the interaction between the ρ meson and nucleon fields is essential to reproduce the properties of nuclei far from the stability line [95, 96]. The strength of this interaction may also greatly affect the chemical composition inside a neutron star [94, 97–99].

self-coupling constants are determined from fitting the surface properties of finite nuclei. However, the best fitting is obtained for negative values of g_3. In this case the quantum field theory based on such a Lagrangian becomes unstable. If one utilizes the mean-field approximation, this is not a very severe restriction. It should be mentioned, however, that it causes trouble in obtaining the self-consistent solutions for several light nuclei, such as ^{12}C [113, 114]. The solution of the problem has been suggested by Bodmer [115], who introduced quartic self-interactions of the ω meson field (a cubic self-interaction $\sim \omega^3$ is not introduced, because of its parity) and studied the effect of the ω^4 term on the equation of state extensively. He established an important softening of the equation of state due to the ω^4 term, caused by weakening of the vector potential from a linear behavior at high densities. The problem of negative values of g_3 was also solved by introducing an ω^4 term into (4.2) [116]. It should be emphasized that the simplest general form for a functional containing φ and ω fields may include also terms of the type $\sim \varphi\omega^2$ and $\varphi^2\omega^2$. These terms have been investigated in the framework of the relativistic Hartree–Fock scheme [117].

So, finally the Lagrangian considered[9] contains six (seven) parameters. These are the meson–nucleon coupling constants g_σ, g_ω, g_ρ; the meson masses m_σ, m_ω, m_ρ; and the meson–meson self-coupling constants g_2, g_3, c_3, the masses of the vector mesons being taken equal to their experimental values while c_3 is introduced in the most recent models only. Different sets of parameters have been calculated and utilized for describing bulk properties of finite nuclei, both spherical and deformed. We consider below the sets of parameters developed by different groups.

4.2 Relativistic Mean-Field Approximation for Finite Nuclei

The usual shell-model assumptions of good parity and time-reversal invariance are adopted below; this is the case, for example, for the ground state of even–even nuclei. The relativistic mean-field equations are given by [125] (the most general form of the Dirac equation compatible with invariance with respect to rotation and reflection (in space and time) is considered in Appendices A3 and A4)

$$[-i\boldsymbol{\alpha}\cdot\boldsymbol{\nabla}+\beta M^*(\boldsymbol{r})+V(\boldsymbol{r})]\psi_\kappa=E_\kappa\psi_\kappa\,. \tag{4.49}$$

Here $M^*(\boldsymbol{r}) = M + g_\sigma\varphi(\boldsymbol{r})$ is the effective mass of the nucleon (see Appendix A5 for different possible definitions of the nucleon effective mass).

---

[9] Most of the current relativistic nuclear models are based on the Lagrangian formalism. Nuclear systems may also be described using a Hamiltonian framework that includes the nucleon kinetic energy and potentials with relativistic corrections. Such a formalism is developed, for example, in [118–124].

4.2 Relativistic Mean-Field Approximation for Finite Nuclei

In the relativistic mean-field framework, the effective mass is significantly smaller than the bare nucleon mass and is space-dependent. The potential $V(r)$ is given by

$$V(\boldsymbol{r}) = g_\omega \omega^0(\boldsymbol{r}) + g_\rho \tau_3 \rho_3^0(\boldsymbol{r}) + e\frac{1+\tau_3}{2} A^0(\boldsymbol{r}) \,. \tag{4.50}$$

The meson fields are determined by the following equations for mesonic fields:

$$\left[-\nabla^2 + m_\sigma^2\right] \varphi(\boldsymbol{r}) = -g_\sigma \rho_\mathrm{s}(\boldsymbol{r}) - g_2 \varphi^2(\boldsymbol{r}) - g_3 \varphi^3(\boldsymbol{r}) \,, \tag{4.51}$$

$$\left[-\nabla^2 + m_\omega^2\right] \omega^0(\boldsymbol{r}) = g_\omega \rho_\mathrm{V}(\boldsymbol{r}) - c_3 \omega^{03}(\boldsymbol{r}) \,, \tag{4.52}$$

$$\left[-\nabla^2 + m_\rho^2\right] \rho_3^0(\boldsymbol{r}) = g_\rho \rho_3(\boldsymbol{r}) \,, \tag{4.53}$$

$$-\nabla^2 A^0(\boldsymbol{r}) = e\rho_\mathrm{p}(\boldsymbol{r}) \,. \tag{4.54}$$

Here ω^0 is the time component of the ω meson field, $\rho_3^0(\boldsymbol{r})$ is the third (neutral) isotopic component of the time component of the ρ meson field, $A^0(\boldsymbol{r})$ is the Coulomb potential, and the densities are given by

$$\rho_\mathrm{s} = \sum_\kappa v_\kappa^2 \bar{\psi}_\kappa \psi_\kappa \,, \quad \text{the Lorentz-scalar density}\,, \tag{4.55}$$

$$\rho_\mathrm{V} = \sum_\kappa v_\kappa^2 \bar{\psi}_\kappa \gamma_0 \psi_\kappa \,, \quad \text{the vector density (baryonic density)}\,, \tag{4.56}$$

$$\rho_3 = \sum_\kappa v_\kappa^2 \bar{\psi}_\kappa \gamma_0 \tau_{3\kappa} \psi_\kappa \,, \quad \text{the isotopic vector density}\,, \tag{4.57}$$

$$\rho_\mathrm{p} = \sum_\kappa v_\kappa^2 \bar{\psi}_\kappa \gamma_0 \frac{1+\tau_{3\kappa}}{2} \psi_\kappa \,, \quad \text{the proton density}\,. \tag{4.58}$$

In a central potential, single-nucleon wave functions have the following form:

$$\psi_\kappa = \frac{1}{r} \begin{pmatrix} iG_\kappa(r) \Omega_{jlm} \\ F_\kappa(r)(\boldsymbol{\sigma n}) \Omega_{jlm} \end{pmatrix} = \begin{pmatrix} \varphi \\ \chi \end{pmatrix} \,, \tag{4.59}$$

where $G_\kappa(r)$ is the upper component of the wave function, $F_\kappa(r)$ is the lower component, Ω_{jlm} is the spin–angle function, and κ is a quantum number of the relativistic theory defined as: $\kappa = \mp(j+1/2)$ for $j = l \pm 1/2$. The eigenvalue E_κ in (4.49) is given by $E_\kappa = M + \varepsilon_\kappa$. The spinor in the case of spherical symmetry (4.59) is characterized by the single-particle angular-momentum quantum number, the parity, and the isospin. For spherical nuclei, the system of (4.49)–(4.59) is reduced to two first-order coupled differential equations for $G_\kappa(r)$ and $F_\kappa(r)$ (the second-order differential equations for these functions are given in Appendix A6, while the boundary conditions are considered in Appendix A7):

$$\frac{\mathrm{d}G_\kappa(r)}{\mathrm{d}r} + \frac{\kappa}{r} G_\kappa(r) - [M^*(r) + E_\kappa - V(r)] F_\kappa(r) = 0 \,, \tag{4.60}$$

$$\frac{dF_\kappa(r)}{dr} - \frac{\kappa}{r} F_\kappa(r) + [E_\kappa - M^*(r) - V(r)] G_\kappa(r) = 0 , \quad (4.61)$$

$V(r)$ is given by (4.50), and the equations satisfied by the boson fields are

$$\frac{d^2\varphi(r)}{dr^2} + \frac{2}{r}\frac{d\varphi(r)}{dr} - m_\sigma^{*2}(r)\varphi(r) = g_\sigma \rho_s(r) ,$$
$$m_\sigma^{*2}(r) = m_\sigma^2 + g_2 \varphi + g_3 \varphi^2 , \quad (4.62)$$

$$\frac{d^2\omega_0(r)}{dr^2} + \frac{2}{r}\frac{d\omega_0(r)}{dr} - m_\omega^{*2}(r)\omega_0(r) = -g_\omega \rho_V(r) ,$$
$$m_\omega^{*2}(r) = m_\omega^2 + c_3 \omega_0^2 , \quad (4.63)$$

$$\frac{d^2 \rho_3^0(r)}{dr^2} + \frac{2}{r}\frac{d\rho_3^0(r)}{dr} - m_\rho^2 \rho_3^0(r) = -g_\rho \rho_3(r) , \quad (4.64)$$

$$\frac{d^2 A_0(r)}{dr^2} + \frac{2}{r}\frac{d A_0(r)}{dr} = -e \rho_p(r) , \quad (4.65)$$

where m_σ^* and m_ω^* are the effective meson masses. The densities become

$$\rho_s(r) = \sum_\kappa v_\kappa^2 \frac{2j_\kappa + 1}{4\pi} \frac{G_\kappa^2 - F_\kappa^2}{r^2} , \quad (4.66)$$

$$\rho_V(r) = \sum_\kappa v_\kappa^2 \frac{2j_\kappa + 1}{4\pi} \frac{G_\kappa^2 + F_\kappa^2}{r^2} , \quad (4.67)$$

$$\rho_3(r) = \rho_{V,p}(r) - \rho_{V,n}(r) . \quad (4.68)$$

For magic nuclei, the occupation probabilities v_κ^2 are equal to 1 for occupied levels and to zero for unoccupied levels. For an open-shell nucleus, we introduce a fractional occupation probability v_κ^2 for each nucleon state denoted by the quantum number κ. We take pairing correlations into account using the BCS theory [84, 103, 116]. In the BCS scheme the occupation probability becomes (we use a schematic pairing model with a constant gap [126])

$$v_\kappa^2 = \frac{1}{2}\left(1 - \frac{E_\kappa - E_F}{\sqrt{(E_\kappa - E_F)^2 + \Delta^2}}\right) \quad (4.69)$$

for a quantum state with a single-particle energy E_κ. The "unoccupation" probability u_κ^2 is defined as $u_\kappa^2 = 1 - v_\kappa^2$. We take the value $\Delta = 11.2$ MeV $\cdot A^{-1/2}$ for the gap energy. The Fermi energy E_F is obtained from the condition

$$\sum_\kappa v_\kappa^2 = N = \frac{1}{2}\sum_\kappa \left(1 - \frac{E_\kappa - E_F}{\sqrt{(E_\kappa - E_F)^2 + \Delta^2}}\right) , \quad (4.70)$$

where N is the number of protons or neutrons. In the BCS calculations it is possible to consider single-particle states up to one more major shell above

4.2 Relativistic Mean-Field Approximation for Finite Nuclei

the Fermi energy [116]. For most nuclei, not too close to the drip line, there is no problem in using this method. However, when nuclei approach the drip line, single-particle states above the Fermi level become unbound. In this case, we take only bound-state contributions into account in (4.70) for practical reasons. We then add the pairing energy

$$E_{\text{pair}} = -\Delta \sum_\kappa u_\kappa v_\kappa \tag{4.71}$$

as a part of the total energy.

In addition, we have to make center-of-mass corrections to the total energy and to the nuclear radius. In the relativistic formalism, it is difficult to separate the center-of-mass corrections from the intrinsic ones. Hence, we use the ansatz of the nonrelativistic case,

$$E_{\text{ZPE}} = \frac{\langle F | \hat{P}^2_{\text{total}} | F \rangle}{2M_{\text{total}}}, \tag{4.72}$$

where

$$M_{\text{total}} = AM, \qquad \hat{P}^2_{\text{total}} = \sum_i \hat{p}_i^2. \tag{4.73}$$

For the ground state it is possible to choose from the two following cases. In the first case, $|F\rangle$ is a harmonic-oscillator wave function so we can express E_{ZPE} analytically as

$$E_{\text{ZPE}} = \frac{3}{4} 41\, A^{-1/3} \;(\text{MeV}). \tag{4.74}$$

This simple form is used in the parameter set TM1 (see below) for heavy nuclei, because the nuclear binding energies are large and the center-of-mass corrections are not large. For light nuclei, the corrections become essential, and it would be more precise to take $|F\rangle = |F\rangle_{\text{RMF}}$ explicitly. In this case the matrix elements (4.72) are calculated directly, including the exchange-term contributions. This procedure is more involved and is used in obtaining the parameter set TM2 (see below) for light nuclei. However, in many parameter sets the center-of-mass correction is taken in the form given by (4.74) in calculations for both light and heavy nuclei.[10]

Putting all contributions together, the total energy of the system described by the Lagrangian (4.2) may be given in the following form:

$$E_{\text{total}} = E_{\text{RMF}} + E_{\text{pair}} - E_{mathrm ZPE} - AM. \tag{4.75}$$

Here the last term is subtracted to obtain the binding energy, while the first term in (4.75) is given by

---

[10] The nuclear radii should also be corrected for the center-of-mass motion. This procedure is described in [116].

$$E_{\text{RMF}} = \sum_\alpha \int d^3 r v_\alpha^2 \psi_\alpha^\dagger \left\{ -i\boldsymbol{\alpha} \cdot \boldsymbol{\nabla} + \beta M^* + g_\omega \omega^0 + g_\rho \tau_3 \rho_3^0 \right.$$

$$\left. + e \frac{1 + \tau_{3\alpha}}{2} A^0 \right\} \psi_\alpha + \int d^3 r \left\{ \frac{1}{2} (\nabla \sigma)^2 + U(\sigma) \right\}$$

$$- \int d^3 r \left\{ \frac{1}{2} (\nabla \omega^0)^2 + U(\omega^0) \right\}$$

$$- \int d^3 r \left\{ \frac{1}{2} (\nabla \rho_3^0)^2 + \frac{1}{2} m_\rho^2 (\rho_3^0)^2 \right\} - \int d^3 r \frac{1}{2} \{ (\nabla A_0)^2 \} \,, \quad (4.76)$$

where

$$U(\sigma) = \frac{1}{2} m_\sigma^2 \varphi^2 + \frac{1}{3} g_2 \varphi^3 + \frac{1}{4} g_3 \varphi^4 \,, \quad (4.77)$$

$$U(\omega) = \frac{1}{2} m_\omega^2 (\omega^0)^2 + \frac{1}{4} c_3 (\omega^0)^4 \,. \quad (4.78)$$

4.3 Relation of the RMF Model to the Skyrme–Hartree–Fock Approach

In the previous section we have considered a relativistic self-consistent approach to the description of the properties of finite nuclei. In the framework of this approach, an atomic nucleus is treated as a relativistic system of nucleons moving in static scalar and vector mesonic fields (see (4.5)). The nature of these fields is directly related in this approach to the mesons that generate the NN interaction. Now we shall discuss another (more phenomenological) relativistic approach. In this approach, values of the fields acting on the nucleons inside the nucleus are obtained from the experimentally determined properties of atomic nuclei, for example the depth of the shell and the value of the spin–orbit potentials in the nucleus. This approach starts from the method presented in [127–131]. We shall refer to this approach as the method of Dirac (relativistic) phenomenology. So, in the case of relativistic phenomenology, the actual nature of the fields in which the nucleons move inside the nucleus is not discussed, only their Lorentz space–time transformation properties being utilized [132, 133].

The main point of the present monograph is that the atomic nucleus is a relativistic system. However, 40 years of the existence of nuclear physics have been connected with the development of nonrelativistic approaches to the theory of the atomic nucleus, some successes being achieved in this field. In particular, starting from 1972, the nonrelativistic self-consistent Hartree–Fock theory with effective (density-dependent or three-particle) forces, introduced by Skyrme [134, 135], became widely known the SHF theory. A large number of publications have been produced in this field and reasonable description of nuclear bulk properties has been obtained.

4.3 Relation of the RMF Model to the Skyrme–Hartree–Fock Approach

The deep-rooted conviction that the nucleus is a nonrelativistic system is based on the fact that the binding energy of a nucleon in the nucleus is much smaller than its rest mass. However, one has to take into account the following circumstances:

1. The saturation property is caused by the strong repulsion between the nucleons at short distances. The latter generates the high-momentum component of the nucleon wave functions, and, as Brown and Jackson note [42], we cannot count on achieving a correct description of this component by remaining within the framework of the nonrelativistic theory. We recall in this connection that the description of the properties of nuclei in the nonrelativistic theory is achieved by the introduction of "effective forces" which are not directly related to the interaction between the free nucleons.
   It is natural to assume that the transition from the real forces to the effective forces which takes place in the nonrelativistic theory is caused by the need of compensate the error in describing the nucleon wave functions. In the present section we show that this is indeed the situation.
2. A direct indication of the important role of relativistic effects in the nucleus is the anomalously large spin–orbit coupling. The value of the spin–orbit potential [59, 60] characterizes the amount of strengthening of the relativistic effects in finite nuclei.

These circumstances constitute the basis for the assumption that the description of nuclear properties using real nuclear forces must be carried out within the framework of relativistic theory.

In this respect, the following question arises: what are the conditions for the successes of the nonrelativistic theory? Are they brought about by the use of phenomenological parameters, or there is a physical cause which allows the corresponding phenomenology? We answer this question below [132, 133].

Consider the original Walecka model [74] (without nonlinear self-interactions of the scalar and vector fields, i.e. $g_2 = g_3 = c_3 = 0$). In this model, the nucleus is considered as a system of nucleons in the static scalar and vector fields generated by the nucleons. The wave functions and the nucleon bound-state energies satisfy the Dirac equation

$$E_\lambda \psi_\lambda = [\boldsymbol{\alpha}\boldsymbol{p} + V + \beta(M+S)]\psi_\lambda , \qquad (4.79)$$

and the vector and scalar fields V and S satisfy the Klein–Gordon equations

$$(\Delta - m_V^2)V = -g_V^2 \rho_V , \qquad (4.80)$$

$$\rho_V = \sum_\lambda v_\lambda^2 \bar{\psi}_\lambda \beta \psi_\lambda = \sum_\lambda v_\lambda^2 (|\varphi_\lambda|^2 + |\chi_\lambda|^2) , \qquad (4.81)$$

$$(\Delta - m_s^2)S = g_s^2 \rho_s , \qquad (4.82)$$

$$\rho_s = \sum_\lambda v_\lambda^2 \bar{\psi}_\lambda \psi_\lambda = \sum_\lambda v_\lambda^2 (|\varphi_\lambda|^2 - |\chi_\lambda|^2) , \qquad (4.83)$$

where φ_λ and χ_λ are the upper and lower components, respectively, of the bispinor:

$$\psi_\lambda = \begin{pmatrix} \varphi_\lambda \\ \chi_\lambda \end{pmatrix} . \qquad (4.84)$$

For simplicity, we do not take the isotopic nature of the fields V and S into account. Since two fields S and V introduced in this section are not related directly to the mesons that generate the NN interaction, we use here the subscripts V and S (rather than ω and σ, as was done in the previous sections) for the respective coupling constants and masses.

Measuring the energies of the single-particle states from the nucleon rest mass, i.e.

$$E = M + \varepsilon \qquad (4.85)$$

(where, for a bound state, ε has the meaning of the binding energy), we write (4.79) in the form of a system of two equations for φ_λ and χ_λ:

$$\varepsilon_\lambda \varphi_\lambda = \boldsymbol{\sigma} \boldsymbol{p} \chi_\lambda + (V+S)\varphi_\lambda , \qquad (4.86)$$

$$\varepsilon_\lambda \chi_\lambda = \boldsymbol{\sigma} \boldsymbol{p} \varphi_\lambda - (2M + S - V)\chi_\lambda . \qquad (4.87)$$

We now make use of the fact that for the bound states, $\varepsilon_\lambda \ll M$, and, therefore, in the second equation (4.87) one can neglect the term containing ε_λ. Then, within an accuracy of terms of order ε_λ/M, the system (4.86), (4.87) takes the form

$$\chi_\lambda = \frac{1}{2\mathcal{M}(r)} \boldsymbol{\sigma} \cdot \boldsymbol{p} \varphi_\lambda , \qquad (4.88)$$

$$\varepsilon_\lambda \varphi_\lambda = \left[\boldsymbol{p} \frac{1}{2\mathcal{M}(r)} \boldsymbol{p} + U(r) + \frac{1}{r} \frac{d}{dr}\left(\frac{1}{2\mathcal{M}(r)}\right) \boldsymbol{l} \cdot \boldsymbol{\sigma} \right] \varphi_\lambda , \qquad (4.89)$$

where the following notation has been introduced:[11]

$$U(r) = V(r) + S(r) , \quad 2\mathcal{M}(r) = 2M + S(r) - V(r) . \qquad (4.90)$$

In deriving (4.90), we have utilized the following identity (valid for an arbitrary function $f(r)$):

$$(\boldsymbol{\sigma} \cdot \boldsymbol{p}) f (\boldsymbol{\sigma} \cdot \boldsymbol{p}) = \sigma_i \sigma_k p_i f p_k = (\delta_{ik} + i\varepsilon_{ikl}\sigma_l) p_i f p_k$$

$$= \boldsymbol{p} f \boldsymbol{p} + i \boldsymbol{\sigma}[\boldsymbol{p} f \times \boldsymbol{p}] = \boldsymbol{p} f \boldsymbol{p} + \frac{1}{r} \frac{df}{dr} \boldsymbol{l} \cdot \boldsymbol{\sigma} , \qquad (4.91)$$

where ε_{ikl} is the unit antisymmetric tensor of the third rank.

---

[11] Note that in the RMF theory, two types of definitions of the effective nucleon mass $M^*(r) = M + S(r)$ are introduced. There is only a minor difference between $M^*(r)$ and $\mathcal{M}(r)$ at normal density. Different definitions of the nucleon effective mass in the nuclear medium are given in Appendix A5.

4.3 Relation of the RMF Model to the Skyrme–Hartree–Fock Approach

Owing to the spatial structure of the effective nucleon mass, the nonrelativistic Hamiltonian for a finite nucleus contains a nonlocal, i.e. momentum-dependent, potential produced by the operator of the kinetic energy

$$\delta U_p = \mathbf{p}\frac{1}{2\mathcal{M}(r)}\mathbf{p} - \frac{1}{\langle 2\mathcal{M}\rangle}\mathbf{p}^2 \ . \tag{4.92}$$

This means that even at the Hartree level, the relativistic framework with local potentials suggests a nonlocality if considered in the nonrelativistic limit. The nonlocal interaction, to a certain extent, is caused by the variation of $S(r)$ related to the scalar-meson mass.

Equation (4.89), which is valid within an accuracy of order ε/M, has the form of the Schrödinger equation for a particle with the effective mass $\mathcal{M}(r)$ moving in the central potential $U(r)$ and the spin–orbit potential. The HFS method [136, 137] leads to equations of exactly the same form. The difference manifests itself in the fact that in the HFS method the spin–orbit potential is introduced by hand, while in the relativistic theory it is of the same origin as the effective mass. From (4.89), one obtains [132]

$$U_{LS} = \frac{1}{r}\frac{(V-S)'}{(2\mathcal{M})^2}\mathbf{l}\cdot\boldsymbol{\sigma} \ . \tag{4.93}$$

In contrast to formula (4.1), the expression (4.93) has the correct sign and magnitude. The increase of the force U_{LS} in comparison with (4.1) arises from two reasons: first from the inequality $|V-S| \gg |U|$, and second, because of the inequality $\mathcal{M} < M$. For estimation, we use the fact that the spin–orbit potential is known from experiment. Usually it is chosen in the form

$$U_{LS} = \frac{\alpha}{r}\frac{d\rho}{dr}\mathbf{l}\cdot\boldsymbol{\sigma} \ , \tag{4.94}$$

where $\rho(r)$ is the density of the nucleon distribution in the nucleus[12] and α is a known constant. Comparing (4.93) and (4.94), we obtain

$$\frac{1}{2\mathcal{M}(r)} = \frac{1}{2M} + \alpha\rho(r) \ . \tag{4.95}$$

In deducing (4.95), we have taken into account that outside the nucleus, $\rho(r) = 0$ and $\mathcal{M}(r) = M$. Substituting $\alpha = 85.5\,\text{MeV}\,\text{fm}^5$ [138] into (4.95) and assuming that inside the nucleus $\rho = 0.17\,\text{fm}^{-3}$, we obtain

$$\mathcal{M}/M = 0.6 \ . \tag{4.96}$$

From this result and from (4.79), we find that inside the nucleus,

---

[12] In (4.94) $\rho(r)$ is the nonrelativistic nucleon density (see below), i.e. the nonrelativistic limit of the vector density $\rho_V(r)$.

$$V - S = 0.8M \approx 750 \text{ MeV} . \tag{4.97}$$

We can estimate the depth of the central potential on the basis of the fact that the depth of the well generated in the relativistic framework is determined by the expression $2\mathcal{M}UR^2$ (where R is the radius of the nucleus), and that the depth of the shell potential is of the order of magnitude of -50 MeV. Taking into account that for the shell model $\mathcal{M} = M$, we find

$$U = \frac{M}{\mathcal{M}} U_{\text{sh}} \approx -90 \text{ MeV} . \tag{4.98}$$

From this, we have

$$S \approx -420 \text{ MeV} , \qquad V \approx +330 \text{ MeV} , \tag{4.99}$$

in agreement with the estimates of [59, 127]. We note that this estimate is quite reliable because it does not depend on any specific model for S and V, and it is obtained by using only well-known experimental quantities. Thus, the specific feature of the nucleus is the fact that the fields acting on a nucleon inside the nucleus are not small in comparison with its rest mass. This is precisely the reason why the nucleus is a relativistic system in spite of the inequality $\varepsilon_\lambda/M \ll 1$. The existence of the small parameter ε_λ/M enables one to use the quasi-nonrelativistic equation (4.89) to calculate the upper component of the wave functions. However, as follows from (4.88) and (4.96), the lower components of the wave functions are significantly enhanced, and neglecting them can lead to wrong physical results. Later we show this by reproducing saturation in the theory considered here, as well as by other examples.

Obviously, (4.89) resembles very much the Hartree–Fock equation with the Skyrme interaction [134]. The HFS method is based on effective forces containing two-particle and three-particle components

$$v = \sum_{i<j} v_{ij}^{(2)} + \sum_{i<j<k} v_{ijk}^{(3)} . \tag{4.100}$$

the interactions $v_{12}^{(2)}$ and $v_{123}^{(3)}$ in the configuration space are taken to be in the form of potentials of zero range (depending on velocity):

$$v_{12}^{(2)} = t_0(1 + x_0 P_\sigma)\delta(\boldsymbol{r}_1 - \boldsymbol{r}_2) + \frac{1}{2}t_1[\delta(\boldsymbol{r}_1 - \boldsymbol{r}_2)k^2 + k'^2\delta(\boldsymbol{r}_1 - \boldsymbol{r}_2)]$$
$$+ t_2 \boldsymbol{k}'\delta(\boldsymbol{r}_1 - \boldsymbol{r}_2)\boldsymbol{k} + \mathrm{i}W_0(\boldsymbol{\sigma}_1 + \boldsymbol{\sigma}_2)\boldsymbol{k}' \times \delta(\boldsymbol{r}_1 - \boldsymbol{r}_2)\boldsymbol{k} , \tag{4.101}$$

and

$$v_{123}^{(3)} = t_3 \delta(\boldsymbol{r}_1 - \boldsymbol{r}_2)\delta(\boldsymbol{r}_2 - \boldsymbol{r}_3) , \tag{4.102}$$

where $\{t_0, t_1, t_2, t_3, x_0, W_0\}$ are parameters; P_σ is a spin-exchange operator; $\boldsymbol{k}$ stands for an operator $(\boldsymbol{\nabla}_1 - \boldsymbol{\nabla}_2)/(2i)$ acting to the right, while $\boldsymbol{k}'$ is

4.3 Relation of the RMF Model to the Skyrme–Hartree–Fock Approach

an operator $-(\boldsymbol{\nabla}_1 - \boldsymbol{\nabla}_2)/(2i)$ acting to the left; the last term in (4.101) corresponds to the two-body spin–orbit forces. In Hartree–Fock calculations of even–even nuclei the three-body forces (4.102) are equivalent to repulsive two-body forces with a linear density dependence (providing saturation in the HFS method). For the Skyrme interaction the energy density $\mathcal{H}(\boldsymbol{r})$ [136] is an algebraic function of the nucleon densities ρ_n and ρ_p, the kinetic-energy densities τ_n and τ_p and also spin densities $\boldsymbol{J}_n$ and $\boldsymbol{J}_p$; these densities are determined by the following relations:

$$\rho(\boldsymbol{r}) = \sum_{\lambda,\sigma} |\varphi_\lambda(\boldsymbol{r},\sigma)|^2 , \qquad (4.103)$$

$$\tau(\boldsymbol{r}) = \sum_{\lambda,\sigma} |\boldsymbol{\nabla}\varphi_\lambda,(\boldsymbol{r},\sigma)|^2 \qquad (4.104)$$

$$\boldsymbol{J}(\boldsymbol{r}) = -i \sum_{\lambda,\sigma\sigma'} \varphi_\lambda^*(\boldsymbol{r},\sigma)[\boldsymbol{\nabla}\varphi_\lambda(\boldsymbol{r},\sigma') \times \langle\sigma|\,\boldsymbol{\sigma}\,|\sigma'\rangle] . \qquad (4.105)$$

The sums in (4.103)–(4.105) are calculated over single-particle states, and $\sigma, \sigma' = \pm 1/2$. For nuclei with $N = Z$ and no Coulomb field, one obtains

$$\rho_n = \rho_p = \frac{1}{2}\rho , \quad \tau_n = \tau_p = \frac{1}{2}\tau , \quad \boldsymbol{J}_n = \boldsymbol{J}_p = \frac{1}{2}\boldsymbol{J} , \qquad (4.106)$$

and the expression for the energy density $\mathcal{H}(\boldsymbol{r})$ has the following form:

$$\mathcal{H}_{\text{HFS}}(\boldsymbol{r}) = \frac{\tau}{2M} + \frac{3}{8}t_0\rho^2 + \frac{1}{16}(3t_1 + 5t_2)\rho\tau + \frac{1}{16}t_3\rho^3$$
$$+ \frac{1}{64}(9t_1 - 5t_2)(\boldsymbol{\nabla}\rho)^2 - \frac{3}{4}W_0\rho\boldsymbol{\nabla}\boldsymbol{J} . \qquad (4.107)$$

It is easily seen that (4.89) is very similar to the equation which appears in the HFS method. Let us follow this similarity in more detail. The Klein–Gordon equations (4.80), (4.82) can be solved by a perturbation method, so one has (where two dimensionless parameters, $C_{\text{s}} = g_{\text{s}}/m_{\text{s}}M$ and $C_{\text{V}} = g_{\text{V}}/m_{\text{V}}M$, are introduced) [133]

$$S = -\frac{C_{\text{s}}^2}{M^2}\left(\rho_{\text{s}} + \frac{1}{m_{\text{s}}^2}\nabla^2\rho_{\text{s}} + \ldots\right) , \qquad (4.108)$$

$$V = \frac{C_{\text{V}}^2}{M^2}\left(\rho_{\text{V}} + \frac{1}{m_{\text{V}}^2}\nabla^2\rho_{\text{V}} + \ldots\right) , \qquad (4.109)$$

the convergence of these series being fast enough, owing to the large values of the meson masses. Using (4.86), (4.87), it is possible to write ρ_{s} and ρ_{V} in terms of the nonrelativistic density $\rho(\boldsymbol{r})$ (4.103) and the kinetic-energy density $\tau(\boldsymbol{r})$:

$$\rho_{\text{s}} = \rho - \frac{\tau}{(2M)^2} , \qquad (4.110)$$

$$\rho_V = \rho + \frac{\tau}{(2\mathcal{M})^2}. \qquad (4.111)$$

Using (4.90), (4.108), (4.108), one obtains for the effective mass

$$2\mathcal{M} = 2M\left[1 - 2M\alpha_1\rho - 2M\alpha_2\nabla^2\rho + 2M\alpha_3\tau\left(\frac{M}{\mathcal{M}}\right)^2\right], \qquad (4.112)$$

where the parameters are given by

$$\alpha_1 = \frac{1}{4M^4}(C_s^2 + C_V^2), \qquad \alpha_2 = \frac{1}{4M^4}\left(\frac{C_s^2}{m_s^2} + \frac{C_V^2}{m_V^2}\right), \qquad (4.113)$$

$$\alpha_3 = \frac{1}{16M^6}(C_s^2 - C_V^2); \qquad (4.114)$$

for $U(r)$ one obtains

$$U = -a_1\rho - a_2\nabla^2\rho + \alpha_1\tau\left(\frac{M}{\mathcal{M}}\right)^2 \qquad (4.115)$$

and

$$a_1 = \frac{1}{M^2}(C_s^2 - C_V^2), \qquad a_2 = \frac{1}{M^2}\left(\frac{C_s^2}{m_s^2} - \frac{C_V^2}{m_V^2}\right). \qquad (4.116)$$

The spin–orbit potential is determined by a more precise (than (4.95)) equation:

$$\frac{1}{2\mathcal{M}} = \frac{1}{2M} + \alpha_2\nabla^2\rho - \alpha_3\tau + \alpha_1\rho\frac{M}{\mathcal{M}}. \qquad (4.117)$$

So the ratio $M/\mathcal{M}$ entering (4.115), (4.117) is determined by

$$\frac{M}{\mathcal{M}} = 1 + 2M\alpha_1\rho + 4M^2\alpha_1^2\rho^2 + \ldots + 2M\alpha_2\nabla^2\rho - 2M\alpha_3\tau. \qquad (4.118)$$

Consequently, the Hamiltonian density is

$$\mathcal{H} = \frac{\tau}{2M} - \frac{1}{2}a_1\rho^2 + \frac{1}{2}a_2(\nabla\rho)^2 + \alpha_1\tau\rho\frac{M}{\mathcal{M}} - \alpha_1\rho\frac{M}{\mathcal{M}}(\nabla\cdot\boldsymbol{J}). \qquad (4.119)$$

The original Walecka model [74] corresponds to the values $C_s^2 = 266.9$, $C_V^2 = 195.7$, $M = 938.3$ MeV, and $M^*/M = 0.556$. The numerical values of the coefficients entering (4.115)–(4.119) are given in Table 4.1.

Equations (4.115)–(4.119) establish a remarkable correspondence between the HFS approximation and the Walecka model. However, some remarks should be made:

1. The main difference between these two approaches is that in the HFS method the spin–orbit potential is introduced "by hand", whereas in the relativistic theory it has the same origin as the effective mass.

4.3 Relation of the RMF Model to the Skyrme–Hartree–Fock Approach

Table 4.1. Coefficients determining the contributions of the various terms of the central potential (4.115) and of the effective mass (4.118) calculated for the Walecka model [74]. In the cases where it was necessary the masses of the two mesons were taken as $m_V = 783\,\text{MeV}$ and $m_s = 550\,\text{MeV}$ [133]

Coefficient	Value
α_1 (MeV fm^5)	44.6
α_2 (MeV fm^7)	4.5
α_3 (MeV fm^7)	0.07
a_1 (MeV fm^3)	621.3
a_2 (MeV fm^5)	191.3

2. The effective nucleon mass in the relativistic case is more complicated than in the HFS theory, where it depends on the density only (compare (4.117) and (4.95)).
3. There is no term quadratic in the density in the central potential U (see (4.115)), unlike the usual Skyrme-type potentials (with the exception of the force Skyrme V [136]), this term arising from three-body forces. One can add such a term phenomenologically [132] to obtain a minimum for the potential energy. Nevertheless, this is not obligatory, since a ρ^2 term is hidden in the expression for the effective nucleon mass (4.117). In fact, if one makes the replacement $\Phi = [(2M + \varepsilon)/(2M + S - V + \varepsilon)]^{1/2}\varphi$ in the Schrödinger equation, there appears a new central potential of the form $S + V + 1/(2M)(S^2 - V^2)$, where the quadratic terms in the fields S and V will generate a ρ^2-term.
4. It is possible to exclude the nonrelativistic density from consideration and utilize only scalar densities ρ_s and vector densities ρ_V related by the following equation:[13]

$$\rho_s = \rho_V - \frac{2\tau}{(2\mathcal{M})^2} \,. \tag{4.121}$$

In this case, one finds equations similar to the previous ones with the following replacements [133]:

$$\alpha_1 \quad \text{by} \quad \alpha'_1 = \frac{C_s^2}{2M^4}, \tag{4.122}$$

$$\alpha_3 \quad \text{by} \quad \alpha'_3 = \frac{C_s^2}{8M^6}. \tag{4.123}$$

---

[13] The most precise expression connecting ρ_s and ρ_V for spin-saturated systems is given by

$$\rho_s = \rho_V - \frac{1}{\mathcal{N}}\left[\frac{\tau}{2\mathcal{M}^2} - \frac{1}{2}\left(\frac{1}{\mathcal{M}}\right)\left(\frac{1}{\mathcal{M}}\right)' \rho'\right], \tag{4.120}$$

where the quantity $\mathcal{N}$ accounts for the small contribution from the lower component of the nucleon wave function; $\mathcal{N} = 1 + \tau/(4\mathcal{M}^2)$.

The nucleon density is now ρ_V instead of ρ. The respective energy density is equivalent to a low-density expansion of the energy density introduced by Walecka [74] if $\mathcal{M}(r)$ is replaced by a Dirac effective nucleon mass $M^*(r)$.

An essential point is that a nonrelativistic reduction of the relativistic mean-field approximation is similar to the result obtained using the Skyrme force only when the nucleon effective mass is close to its bare mass. However, in the Walecka model $M^*/M = 0.556$. For this reason alone, the resulting Hamiltonian has a more general type and is more complicated than the Skyrme Hamiltonian, mainly because in the Hamiltonian (4.119) there appear terms in $\tau\rho^2$ and $\tau\rho^3$ and higher-order terms in ρ. There are no terms of this type in (4.107). Terms containing τ are of relativistic origin, so these extra terms correspond to relativistic effects not included in the Skyrme Hamiltonian.

It is possible to derive the coefficients t_0 and W_0 from (4.119). It is more difficult to say something about $t_1, t_2, t_3, \ldots$. If one tries to obtain an energy density of the Skyrme type (4.107), this can be done in the framework of two very crude approximations.

1. Consider the expansion of $M/\mathcal{M}$ (see (4.118)). In this case one obtains terms such as $\rho^{8/3}$, $\rho^{11/3}$, ... in the energy per particle in nuclear matter. These terms can be considered to give roughly the same effect as the ρ^2 term we discussed above. However, these terms appear here as higher-order effects of two-body interactions. In the Skyrme interaction these terms correspond to three-body interactions. In the framework of this approximation the energy density is given by

$$\mathcal{H}_1 = \frac{\tau}{2M} - \frac{1}{2}a_1\rho^2 + \frac{1}{3}b_1\rho^3 + \frac{1}{2}a_2(\nabla\rho)^2 + \alpha_1\tau\rho - \alpha_1\rho\frac{M}{\mathcal{M}}(\boldsymbol{\nabla}\cdot\boldsymbol{J}) \,. \quad (4.124)$$

The value of the coefficient b_1 may be obtained, for example, by demanding saturation at $\rho = \rho_0$ (the nuclear-matter density in the Walecka model). Comparing the result with (4.107) gives the first set of parameters in Table 4.2. However, one should bear in mind that (4.124) is not consistent with (4.115) and (4.117) for the central potential and the effective mass, respectively.

2. Replace, when necessary, the expansion (4.118) by its nuclear-matter value

$$\left(\frac{M}{\mathcal{M}}\right)_{\text{nm}} \simeq 1 - \frac{(C_s^2 + C_V^2)\rho_0}{2M^3} \,. \quad (4.125)$$

In practice, the density entering (4.125) should not be precisely ρ_0, because of the choice made in (4.110), (4.111). However this point will not influence the result too much.

4.3 Relation of the RMF Model to the Skyrme–Hartree–Fock Approach

Table 4.2. Coefficients of the Skyrme-type energy functional for the two approximations considered in Sect. 4.3 [133]

t_0 MeV fm^3	t_1 MeV fm^5	t_2 MeV fm^5	t_3 MeV fm^6	W_0 MeV fm^5
−828.4	569.7	−199.0	554.8	101.9
−828.4	612.1	−122.7	0	101.9

In the second case, the energy density is given by

$$H_2 = \frac{\tau}{2M} - \frac{1}{2}a_1\rho^2 + \frac{1}{2}a_2(\nabla\rho)^2 + \beta_1\tau\rho - \beta_1\rho(\boldsymbol{\nabla}\cdot\boldsymbol{J}), \quad (4.126)$$

where $\beta_1 = \alpha_1(M/\mathcal{M})_{\text{nm}}$. Thus one arrives at the second set of parameters of Table 4.2. However, in this case also, (4.126) is not consistent with (4.115) and (4.117). The dispersion of the values given in Table 4.2 can be seen. This fact indicates that it is necessary to be cautious. One has to consider these values as approximate values aimed at allowing one to write the nonrelativistic limit of the RMF model in a Skyrme-type fashion.

In fact, the nonrelativistic reduction of the RMFA is more complicated than the two possible approximations considered above. So it is better to work with the energy density (4.119), not with its Skyrme-type analogue.

Further comparisons between the relativistic mean-field theory including nonlinear self-interactions and the HFS theory in relation to the properties of finite nuclei and nuclear matter have been made in [81, 101, 139–143].[14] The main conclusions are the following. The differences between the HFS theory and the results of the RMF approach may be noticeable. The most important is the difference between the nucleon effective masses: $M^*/M = 0.79$ in the Skyrme model, while this value is equal to 0.57 in the relativistic theory, i.e M^* is smaller in the RMF models. Another big difference is related to the $\nabla^2\rho_0$ term.

So a comparison of the HFS theory with the nonrelativistic limit of the RMFA shows similarities in general and discrepancies in detail [81]. This comparison enables one to establish possible deficiencies of the nonrelativistic theory. Studies of these two models taken together can help one to obtain more reasonable parameterizations.

An important ingredient of the relativistic model is antinucleon degrees of freedom (ANDF), they are necessary even for low-energy phenomena. However, in this case the ANDF in the relativistic model could be hidden in the Landau parameters. For example, the excitation energy of the giant monopole state is described in terms of the Landau parameters F_0 and F_1, as in nonrelativistic approach. A similar role of the antinucleon in the relativistic

---

[14] The comparison of the relativistic mean-field theory (with nonlinear self-interactions) with the nonrelativistic HFS method has been carried out both numerically and analytically.

model can be found in the excitation energies of other giant resonances (see Chap. 10) and also in the nuclear convection current or magnetic moment (see Chap. 5). In particular, the continuity equation is not fulfilled without nucleon-antinucleon excitations.

To obtain a difference between the relativistic and nonrelativistic models, high-momentum-transfer phenomena need to be taken into consideration also. For example, in [144] it is emphasized that the neutron spin density in the relativistic approximation reproduces very well experimental data on elastic electron scattering which have not been reproduced in the nonrelativistic framework. Moreover, it is predicted ([144] and references therein) that the value of the relativistic Coulomb sum is strongly quenched in comparison with the nonrelativistic value. This quenching is related to the ANDF, as a result of which the ω meson mass in the nuclear medium is reduced or, equivalently, the nucleon size becomes larger.

It will be necessary to obtain new experimental data around a momentum transfer of 1 GeV in order to establish any more fundamental differences between the relativistic and nonrelativistic models.

4.4 Renormalization of the Kinetic Energy to Obtain Saturation in Nuclear Matter

The success of the relativistic Hartree approximation in reproducing the binding energies and single-particle nuclear levels for nuclei with closed shells is connected, in particular, with the fact that the expectation value for the nuclear ground state of the relativistic kinetic-energy operator,

$$\hat{T} = \boldsymbol{\alpha} \cdot \boldsymbol{p} + \beta M \, , \tag{4.127}$$

differs greatly from that obtained in the nonrelativistic case. A calculation of the expectation value of the operator (4.127) with the relativistic single-particle wave functions (4.59) leads to the following expression [56]:

$$\langle \psi_\kappa | \, \hat{T} \, | \psi_\kappa \rangle = \int_0^\infty [E_\kappa - S(r) - V(r)] G_\kappa^2(r) \, dr$$

$$+ \int_0^\infty [S(r) - V(r) + E_\kappa] F_\kappa^2(r) \, dr \, . \tag{4.128}$$

The first term of (4.128) has the form of a nonrelativistic mean value of the kinetic energy, while the second term can be treated as a relativistic correction. The second term gives the major relativistic effect. In the old-fashioned relativistic picture this term would be small, because the smaller component $F(r)$ is of the order of $1/10$ of the larger component $G(r)$. However, this conclusion is invalid in the OBEP models and the Walecka model. As was

4.4 Renormalization of the Kinetic Energy

mentioned in Chap. 3, in the OBEP model the central potential results from the destructive cancellations of the contributions of scalar and vector mesons (this type of interference is necessary to fit NN data without hard cores). In the RMF formalism this feature of the NN potential results in a very large values for the single-particle potentials $S(r)$ and $V(r)$. The potential $S(r)$ is attractive and the vector potential $V(r)$ is repulsive, and their appearance with the same signs in (4.128) produces a nonrelativistic potential well of the conventional depth. In the second term of (4.128), $S(r)$ and $V(r)$ enter with opposite signs. So these potentials are additive and the integrand becomes significant even though the lower component $F(r)$ is small. The contribution of the second term in (4.128) is negative (5–10 MeV), so the ground-state expectation value is reduced in comparison with the nonrelativistic case [145].

Now we discuss the saturation mechanism in the RMFA [84, 132]. Note that if the scattering of two baryons in free space is calculated using the ladder approximation to the Bethe–Salpeter equation, then the interaction contained in the Walecka Lagrangian (4.2) (no self-interactions) can be replaced by an equivalent potential in momentum space [74]:

$$V(q)_\text{eq} = g_\text{V}^2 \frac{\gamma_\lambda^{(1)} \cdot \gamma_\lambda^{(2)}}{q^2 + m_\text{V}^2 - i\eta} - g_\text{s}^2 \frac{\mathbf{1}^{(1)} \cdot \mathbf{1}^{(2)}}{q^2 + m_\text{s}^2 - i\eta} \qquad (4.129)$$

where $q_\lambda = (-\mathbf{q}, q_0)$ is the four-dimensional momentum transfer, and $\gamma_\lambda^{(1,2)}$ and $\mathbf{1}^{(1,2)}$ are matrices (**1** is the unit matrix) referring to the first and second particles. If the baryons are assumed to be heavy and to move nonrelativistically, then we can use the approximations

$$\gamma_\lambda^{(1)} \cdot \gamma_\lambda^{(2)} \longrightarrow \mathbf{1}^{(1)} \cdot \mathbf{1}^{(2)} \quad \text{for} \quad |q_0| \ll |\mathbf{q}|, \qquad (4.130)$$

which lead to a potential that does not take into account retardation, does not depend on the spins, and in configuration space has the form

$$\begin{aligned} V_\text{eq}(|\mathbf{r}-\mathbf{r}'|) &= V_\text{V}(|\mathbf{r}-\mathbf{r}'|) + V_\text{s}(|\mathbf{r}-\mathbf{r}'|) \\ &= \frac{1}{4\pi}\left(g_\text{V}^2 \frac{e^{-m_\text{V}|\mathbf{r}-\mathbf{r}'|}}{|\mathbf{r}-\mathbf{r}'|} - g_\text{s}^2 \frac{e^{-m_\text{s}|\mathbf{r}-\mathbf{r}'|}}{|\mathbf{r}-\mathbf{r}'|}\right). \end{aligned} \qquad (4.131)$$

If $g_\text{V}^2 > g_\text{s}^2$, this potential is repulsive at short distances; if $m_\text{V} > m_\text{s}$, then the potential (4.129) is attractive at large distances. Thus, the interaction contained in (4.129) includes the main features of the nucleon–nucleon potentials that are responsible for saturation.

Notice that the Yukawa potentials entering (4.131) are just the Green's functions of the respective Klein–Gordon equations for the scalar and vector fields $S(r)$ and $V(r)$. So we have

$$S(\mathbf{r}) = \int \mathrm{d}^3 \mathbf{r}'\, V_\text{s}(|\mathbf{r}-\mathbf{r}'|)\, \rho_\text{s}(\mathbf{r}'), \qquad (4.132)$$

$$V(r) = \int \mathrm{d}^3 r' \, V_\mathrm{V}\!\left(|r - r'|\right) \rho_\mathrm{V}(r'). \tag{4.133}$$

If we ignore the difference between $\rho_s(r)$ and $\rho_V(r)$, it can be seen that the relativistic mean-field method is directly related to the Hartree method with a nucleon–nucleon potential of Yukawa type containing a σ-meson-induced strong attractive part and an ω-meson-generated strong repulsive part.

However, in the nonrelativistic case this simple model would not contain a saturation mechanism and one would obtain a collapse. In contrast, in the RMF approximation this phenomenon is prevented by a relativistic mechanism produced by the difference between the scalar and vector densities in (4.132), (4.133),

$$\rho_\mathrm{s} = \sum_\lambda v_\lambda^2 \left(|\varphi_\lambda|^2 - |\chi_\lambda|^2 \right), \tag{4.134}$$

$$\rho_\mathrm{V} = \sum_\lambda v_\lambda^2 \left(|\varphi_\lambda|^2 + |\chi_\lambda|^2 \right). \tag{4.135}$$

The vector density ρ_V is related to the normal baryon density; it is normalized to the total particle number and is determined by the sum of the squared values of the upper φ_λ and lower χ_λ components of the relativistic nucleon wave function. The scalar density ρ_s includes the difference between the same values, this difference decreasing whenever the lower components might appear to be important, in particular in the case of a possible collapse. As can be seen from (4.82), the scalar density is the source of the attractive potential S, which is the starting point of the possible collapse. So one has a relativistic mechanism which by itself stabilizes the system, reducing the attraction properly [83] at higher density.

Let us discuss this mechanism in more detail. To do this, we consider the equation connecting the scalar and vector densities (4.110), (4.111),

$$\rho_\mathrm{s} = \rho_\mathrm{V} - \frac{2\tau}{(2\mathcal{M})^2}, \tag{4.136}$$

τ being the kinetic-energy density.

For high densities, the kinetic energy increases even more than one might expect, because also the value of $\mathcal{M}$ decreases. This fact makes the scalar density much smaller, preventing the nuclear system from collapsing. Such a saturation mechanism is different from that operating in the nonrelativistic theory, where strongly repulsive density-dependent interactions are responsible for saturation.

4.5 Relativistic Mean-Field Approximation for Deformed Nuclei

In the case of deformed nuclei, the nucleon potentials do not possess spherical symmetry. However, we shall suppose [92] in this section that the potentials

4.5 Relativistic Mean-Field Approximation for Deformed Nuclei

and densities have axial symmetry since many deformed nuclei can be described by axially symmetric shapes. So, in this case, the source terms in the Dirac and Klein–Gordon equations are deformed. For this reason one has to modify the equations used above for spherical nuclei to incorporate the axial symmetry. Since there is no longer complete rotational symmetry, j (the total angular momentum) is no longer a good quantum number. However, the densities are invariant under rotation around the symmetry axis. One has to work with the cylindrical coordinates [92]

$$x = r_\perp \cos\varphi \,,$$
$$y = r_\perp \sin\varphi \,,$$
$$z = z \,. \tag{4.137}$$

The nucleon spinor ψ_i is now characterized by the quantum numbers

$$\Omega_i \,, \quad \text{parity, and isospin}\,, \tag{4.138}$$

where $\Omega_i = m_{l_i} + m_{s_i}$ is an eigenvalue of the symmetry operator J_z. Hence the Dirac spinor ψ_i can be written as

$$\psi_i(\mathbf{r},t) = \frac{1}{\sqrt{2\pi}} \begin{bmatrix} g_i(z,r_\perp)^+ \exp(\mathrm{i}(\Omega_i - \tfrac{1}{2})\varphi) \\ g_i(z,r_\perp)^- \exp(\mathrm{i}(\Omega_i + \tfrac{1}{2})\varphi) \\ \mathrm{i}f_i(z,r_\perp)^+ \exp(\mathrm{i}(\Omega_i - \tfrac{1}{2})\varphi) \\ \mathrm{i}f_i(z,r_\perp)^- \exp(\mathrm{i}(\Omega_i + \tfrac{1}{2})\varphi) \end{bmatrix} \chi_{t_i}(t) \,. \tag{4.139}$$

Substituting the value of ψ_i given by (4.139) to (4.49)–(4.54), we obtain

$$(M^* + V)g_i^+ + \partial_z f_i^+ + \left[\partial_{r_\perp} + \frac{\Omega_i + 1/2}{r_\perp}\right] f_i^- = E_i g_i^+ \,, \tag{4.140}$$

$$(M^* + V)g_i^- - \partial_z f_i^- + \left[\partial_{r_\perp} - \frac{\Omega_i - 1/2}{r_\perp}\right] f_i^+ = E_i g_i^- \,, \tag{4.141}$$

$$(M^* - V)f_i^+ + \partial_z g_i^+ + \left[\partial_{r_\perp} + \frac{\Omega_i + 1/2}{r_\perp}\right] g_i^- = -E_i f_i^+ \,, \tag{4.142}$$

$$(M^* - V)f_i^- - \partial_z g_i^- + \left[\partial_{r_\perp} - \frac{\Omega_i - 1/2}{r_\perp}\right] g_i^+ = -E_i f_i^- \,. \tag{4.143}$$

For each solution with positive Ω_i,

$$\psi_i \equiv \{g_i^+, g_i^-, f_i^+, f_i^-, \Omega_i\} \,, \tag{4.144}$$

one also has a time-reversed solution with the same energy,

$$\psi_{\bar{i}} = T\psi_i = \{-g_i^-, g_i^+, f_i^-, -f_i^+, -\Omega_i\} \,. \tag{4.145}$$

Here, T is the time reversal operator $T = i\sigma_y K$ (where K stands for the complex conjugation operator). For nuclei with time-reversal symmetry (for even–even nuclei), the contributions to the densities of the two time-reversed states are identical. So we obtain the densities

$$\rho_{S,V} = 2\sum_{i>0} n_i \left(\left(|g_i^+|^2 + |g_i^-|^2\right) \mp \left(|f_i^+|^2 + |f_i^-|^2\right)\right). \tag{4.146}$$

and, in the same way, the isovector density and the proton density. The sum in (4.146), $\sum_{i>0}$ is calculated only for the states with positive Ω_i values. All these densities determine the source terms for the boson fields $(\sigma, \omega^0, \rho_3^0, A^0)$, which satisfy a Klein–Gordon equation of the form (in cylindrical coordinates)

$$\left(-\frac{1}{r_\perp}\partial_{r_\perp} r_\perp \partial_{r_\perp} - \partial_z^2 + m_{\text{boson}}^2\right)\phi_{\text{boson}}(r_\perp, z) = \text{source terms}. \tag{4.147}$$

In the spherical case, the large and small components $g_i(r)$ and $f_i(r)$ of the relativistic nucleon wave function can be expanded separately in terms of the radial functions $R_{nl_i}(r)$ of a spherical harmonic-oscillator potential $V_{\text{osc}}(r) = (1/2)M\omega^2 r^2$, with an oscillator frequency $\hbar\omega_0$ and oscillator length $b_0 = \sqrt{\hbar/M\omega_0}$:

$$g_i(r) = \sum_{n=1}^{n_{\max}} g_n^i R_{nl_i}(r), \tag{4.148}$$

$$f_i(r) = \sum_{\tilde{n}=1}^{\tilde{n}_{\max}} f_{\tilde{n}}^i R_{\tilde{n}\tilde{l}_i}(r). \tag{4.149}$$

The upper limits $n_{\max}$ and $\tilde{n}_{\max}$ in (4.148), (4.149) are radial quantum numbers and are determined by the respective major shell quantum numbers $N_{\max} = 2(n_{\max} - 1) + l_i$ and $\tilde{N}_{\max} = 2(\tilde{n}_{\max} - 1) + \tilde{l}_i$.

In the case of a deformed nucleus with axial symmetry, the spinors $g_i^\pm$ and $f_i^\pm$ are expanded in terms of a deformed-oscillator-potential basis, the conservation of volume being taken into account. The frequencies $\hbar\omega_\perp$ and $\hbar\omega_z$ are expressed via the deformation parameter β_0 as follows:

$$\hbar\omega_z = \hbar\omega_0 \exp\left(-\sqrt{\frac{5}{4\pi}}\beta_0\right), \tag{4.150}$$

$$\hbar\omega_\perp = \hbar\omega_0 \exp\left(\frac{1}{2}\sqrt{\frac{5}{4\pi}}\beta_0\right). \tag{4.151}$$

The set consisting of (4.140)–(4.143) and (4.147) is solved self-consistently in cylindrical coordinates, and the wave functions are expanded in a deformed-oscillator-potential basis.

The input parameters needed to perform numerical calculations are

- the baryon and meson masses,
- the meson–nucleon coupling constants,
- the number of oscillator shells (the cutoff parameter up to which the fermion wave functions and meson fields, as well as the densities, are expanded; in many practical calculations [125] $N_{\max} = 8$ is taken for nucleons and $N_{\max} = 10$ for bosons),
- the parameters $\hbar\omega_0$ and β_0 in (4.150), (4.151) utilized to expand the relativistic spinors and fields.

From the converged solutions, one can calculate various physical quantities for deformed nuclei. The quadrupole moment Q and deformation parameter β are calculated as follows:

$$Q_{n,p} = \left\langle \sum_i 2r_i^2 P_2(\cos\theta_i) \right\rangle_{n,p} \tag{4.152}$$

and

$$Q = Q_n + Q_p = \frac{3}{4\pi}\sqrt{\frac{16\pi}{5}} AR^2 \beta, \tag{4.153}$$

where $R = 1.2A^{1/3}$. The deformation parameter β is related with to quadrupole moments:

$$\beta_p = \sqrt{5\pi}\frac{Q_p}{3ZR^2}, \tag{4.154}$$

$$\beta_n = \sqrt{5\pi}\frac{Q_n}{3NR^2}. \tag{4.155}$$

4.6 Optimal Parameter Sets for the Relativistic Mean-Field Model

The relativistic approach considered above is an effective theory for calculating nuclear observables in the framework of the mean-field and no-Fermi-sea approximation. The first calculations made within the RMF approach used a linear model (without self-interactions of mesonic fields). However, it was shown that in this case it is possible to obtain only a qualitative description of the nuclear bulk properties. The introduction of nonlinear self-interactions (σ^3 and σ^4 terms in the first stage) was an essential step in obtaining a quantitative understanding of the nuclear observables that was at least not worse than that obtained in the framework of highly sophisticated nonrelativistic models. An extended version of the theory (extended by including the nonlinear sigma self-coupling terms) is now used as a standard model for the relativistic description of finite nuclei; various parameterizations for this version have been developed, namely NL1 [103], NL2 [146], NLC [77], NL-SH [147], NL3 [148], and NLRa [149]. The relativistic mean-field theory has been quite successful in describing finite nuclei, including deformed ones, at the line of β-stability; with the same parameters, the separation energies of neutrons and protons, the binding energies, the mean square charge radii, the electric quadrupole moments, and the deformation parameters of the proton and neutron distributions have been evaluated for even–even nuclei with $16 \leq A \leq 256$. The relativistic mean-field theory has also been applied to unstable nuclei [96, 114, 150, 151]. The results compare extremely well with the existing data.

The set NL1 can be regarded as the single best set capable of reproducing the ground-state properties consistently better for both spherical and deformed nuclei over the entire periodic table. However, although the RMF theory using NL1 has been shown to be successful, there are several problems regarding its application to nuclei and dense matter. First of all, the parameter set NL1 was determined from the properties of the stable nuclei only and was not fully constrained by unstable nuclei. The resulting symmetry energy of nuclear matter obtained using NL1 turned out to be 44 MeV, which is very large as compared with the value in the mass formula ($\sim 33\,\mathrm{MeV}$). The negative sign of the self-coupling constant for the σ^4 term has been a defect that causes some trouble, as mentioned above. When one applies the equation of state obtained using NL1 to neutron star profiles, the maximum mass of the neutron stars is too large, because of too strong a repulsive vector-meson contribution at high densities, compared with maximum mass obtained from the EOS in microscopic many-body frameworks.

On the other hand, a recent publication by Brockmann and Machleidt [53] provides strong support for the relativistic description of nuclear matter. These authors have shown that the relativistic Brueckner–Hartree–Fock theory with a realistic nucleon–nucleon interaction derived from nucleon–nucleon scattering data gives the saturation property of nuclear matter for the first time. Furthermore, the resulting vector and scalar potentials V and S are close to those obtained with the RMF theory with the Bonn potential. The difference is due to the many-body effect, i.e. the Fock term and the Brueckner term. In [152] the coupling constants of the RMF theory have been made slightly density-dependent by adjusting them to fit the RBHF results. This density-dependent RMF theory (originally called the relativistic density-dependent Hartree (RDDH) theory) reproduces the nuclear properties of ^{16}O and ^{40}Ca extremely well.

How does the RMF theory using NL1 (and other parameterizations) compare with the RBHF theory? The results are shown in Fig. 4.1 for the vector and the scalar potentials V and S of nuclear matter as functions of density; the values of the parameters and the calculated properties of nuclear matter are given in Tables 4.3 and 4.4, respectively.

Some of the results for finite nuclei are presented in Tables 4.5 and 4.6 [116]. The vector potential of the RMF theory increases linearly with density and becomes stronger, while the vector potential of the RBHF theory bends

4.6 Optimal Parameter Sets for the Relativistic Mean-Field Model

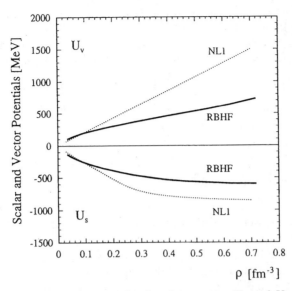

Fig. 4.1. Scalar and vector potentials of nuclear matter U_S and U_V, plotted as a function of the density ρ. The potentials obtained from the RBHF theory and from NL1 are depicted by *thick solid curves* and *dotted curves*, respectively [116]

Table 4.3. The parameters of the RMF Lagrangian determined by the least-squares fitting procedure, listed under TM1 for heavy nuclei and TM2 for light nuclei. For comparison, the parameters of NL1 and NL-SH are also listed under the respective columns [116]

	TM1	TM2	NL1	NL-SH
M (MeV)	938.0	938.0	938.0	939.0
m_σ (MeV)	511.198	526.443	492.250	526.059
m_ω (MeV)	783.0	783.0	795.359	783.0
m_ρ (MeV)	770.0	770.0	763.0	763.0
g_σ	10.0289	11.4694	10.1377	10.444
g_ω	12.6139	14.6377	13.2846	12.945
g_ρ	4.6322	4.6783	4.9757	4.383
g_2 (fm^{-1})	−7.2325	−4.4440	−12.1724	−6.9099
g_3	0.6183	4.6076	−36.2646	−15.8337
c_3	71.3075	84.5318	0.0	0.0

downward with density. The scalar potential of the RMF theory seems to overestimate the RBHF results at high density in order to compensate for the strong repulsion in the vector channel. This is the reason for using the wrong sign for the σ^4 self-coupling constant. The difference in the vector potential between the two frameworks indicates that one needs a nonlinear term in the

Table 4.4. The properties of nuclear matter calculated by the RMF theory with TM1, TM2, NL1, and NL-SH, and by the RBHF theory. The saturation density is denoted by ρ_0, the energy by E/A, the incompressibility by K, the effective mass by M^*, and the symmetry energy at the saturation density by a_{sym} [116]

	TM1	TM2	NL1	NL-SH	RBHF
ρ_0 (fm^{-3})	0.145	0.132	0.152	0.146	0.190
E/A (MeV)	−16.3	−16.2	−16.4	−16.3	−15.6
K (MeV)	281	344	211	355	290
M^*/M	0.634	0.571	0.573	0.597	0.601
a_{sym} (MeV)	36.9	35.8	43.5	36.1	34.7

vector channel as well as in the scalar channel. So we are forced to introduce [116] a simple nonlinear term in the ω channel in the form $(\omega_\mu \omega^\mu)^2$ as in the paper by Bodmer [115] who discussed the role of nonlinear vector self-coupling in the equation of state. So, in [116] a term $(\omega_\mu \omega^\mu)^2$ was introduced into the RMF theory and the model parameters were obtained by a least-squares fitting procedure (as for the experimental data, the available data on unstable nuclei were also used in [116]). An attempt was made to find a parameter set which was valid for all nuclei. To do this, 19 nuclei, from C up to Pb (including some unstable nuclei), were involved in the fitting procedure; the binding energies and charge radii used are shown in Table 4.5. In performing the fitting procedure, it was found difficult to obtain a good parameter set for all nuclei, unless a set similar to NL1 with a negative g_3 value was chosen. Hence, two regions were defined for lighter nuclei, with proton numbers smaller than and larger than $Z = 20$. This separation into two regions was done with the idea that a larger surface effect in lighter nuclei would affect the coupling constants of the effective theory under consideration. In Table 4.5 the nuclei chosen for the least-squares fitting procedures are listed together with the binding energies per particle and the charge radii used for the two regions.

The results of the procedure of fitting to the experimental values of the binding energies and charge radii are listed in Table 4.5. For comparison, the results obtained from the RMF calculation using NL1 and also from the recent parameter set NL-SH are shown. The results of [116] are found to be almost equivalent to that obtained from NL1 and NL-SH. Note that those nuclei without numbers in the column "NL1" cannot be solved with NL1. The parameter sets have been named TM1 for the heavy nuclei (A$\geq$ 40) and TM2 for the light nuclei ($A < 40$); they are shown in Table 4.3. It is interesting to note that g_3 is found to be positive in TM1, which is due to the inclusion of the ω^4 term.

Let us discuss briefly the properties of nuclear matter obtained with TM1 [116]. We compare the vector and scalar potentials with those obtained from the RBHF and RMF theories using NL1 and NL-SH in Figs. 4.1–4.3, and find

4.6 Optimal Parameter Sets for the Relativistic Mean-Field Model   71

Table 4.5. Experimental data for the binding energy per particle (E/A) and charge radius (R_c) for various nuclei, used in the least-squares fitting procedure, and calculated values. The results obtained using TM1 for heavy nuclei and TM2 for light nuclei are listed under "TM". Results obtained using NL1 and NL-SH are also shown for comparison [116]

Nucleus	E/A (MeV)	TM	NL1	NL-SH	R_c (fm)	TM	NL1	NL-SH
^{40}Ca	8.55	8.62	8.56	8.51	3.45	3.44	3.48	3.43
^{48}Ca	8.67	8.65	8.60	8.66	3.45	3.45	3.47	3.44
^{58}Ni	8.73	8.64	8.70	8.70	3.77	3.76	3.71	3.73
^{90}Zr	8.71	8.71	8.73	8.71	4.26	4.27	4.27	4.25
^{116}Sn	8.52	8.53	8.53	8.52	4.63	4.61	4.61	4.59
^{124}Sn	8.47	8.45	8.46	8.46	4.67	4.67	4.66	4.64
^{184}Pb	7.78	7.81	7.84	7.81	–	5.41	5.41	5.37
^{196}Pb	7.87	7.87	7.89	7.90	–	5.47	5.47	5.43
^{208}Pb	7.87	7.87	7.89	7.90	5.50	5.53	5.52	5.49
^{214}Pb	7.77	7.77	7.75	7.79	–	5.59	5.57	5.55
^{8}C	3.10	3.10	3.76	3.71	–	3.05	3.42	3.14
^{12}C	7.68	7.61	–	7.47	2.46	2.39	–	2.39
^{14}C	7.52	7.55	–	7.62	2.56	2.44	–	2.45
^{20}C	5.94	5.99	–	6.03	–	2.47	–	2.49
^{14}O	7.05	7.07	–	7.13	–	2.70	–	2.68
^{16}O	7.98	7.92	7.95	8.04	2.74	2.67	2.74	2.65
^{22}O	7.36	7.47	7.36	7.46	–	2.66	2.72	2.65
^{28}Si	8.45	8.47	8.25	8.28	3.09	3.07	3.03	3.04
^{34}Si	8.33	8.37	8.32	8.36	–	3.16	3.15	3.12
^{40}Ca	8.55	8.48	8.56	8.51	3.45	3.50	3.48	3.43
^{48}Ca	8.67	8.70	8.60	8.66	3.45	3.50	3.47	3.44

that TM1 provides potentials close to the RBHF values. The vector potential for TM1 does not grow so fast as that for NL1. As a consequence, the scalar potential for TM1 is also moderate and compares quite well with that of the RBHF theory. The repulsive effect is very strong in the case of NL1, which causes a large deviation from the result of the RBHF theory. NL-SH has almost the same properties as NL1. The curve obtained with TM1 gives a moderate rise of the energy, and E/A is found to be close to the value obtained with the RBHF theory.

Let us emphasize that the energy per particle of nuclear matter is softened considerably for TM1 as compared with the value for NL1. This finding agrees with the results of Bodmer who studied the role of the ω^4 term in the EOS extensively. This softening of the EOS is caused by the weakening of the

Table 4.6. The neutron skin ΔR (the difference between the proton and neutron root-mean-square radii) for several nuclei obtained with TM1, NL1, and NL-SH, in units of fm [116]. The experimental data have been extracted from the 800 MeV proton elastic-scattering data [153]

Nuclei	TM1	NL1	NL-SH	Expt
^{90}Zr	0.110	0.128	0.103	0.09(7)
^{116}Sn	0.161	0.200	0.156	0.15(5)
^{124}Sn	0.263	0.313	0.253	0.25(5)
^{208}Pb	0.271	0.321	0.266	0.14(4)

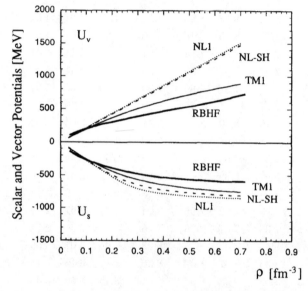

Fig. 4.2. Scalar and vector potentials of nuclear matter U_S and U_V, plotted as a function of the density ρ as in Fig. 4.1. The potentials obtained from the RBHF theory and from NL1 and NL-SH are depicted by *thick solid, dotted,* and *dashed curves*, respectively. The results obtained from the RMF theory with TM1 are shown by *thin solid curves* [116]

vector potential from a linear behavior at high densities due to the ω^4 term, and this is exactly what is demanded by the results of the RBHF theory. In fact, the density-dependent RMF approach of Brockmann and Toki, where the coupling constants decrease with density, provides a good description of ^{16}O and ^{40}Ca. It is then possible to make the attractive contribution of the σ meson weaker than in NL1; the attractive nature of this contribution appears in the sign of g_3 in TM1. This negative g_3 has been a problem in the use of NL1, although the RMF theory with NL1 is successful in describing

4.6 Optimal Parameter Sets for the Relativistic Mean-Field Model 73

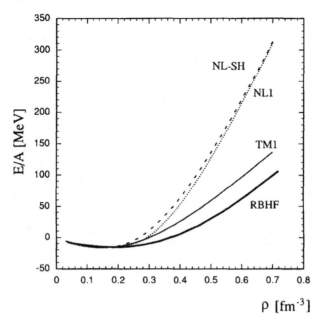

Fig. 4.3. Energy per particle E/A for nuclear matter as a function of the density ρ. The results obtained from the RBHF theory are denoted by the *thick solid* curve, while the results obtained from the RMF theory with TM1, NL1, and NL-SH are shown by *thin solid*, *dotted*, and *dashed* curves, respectively

nuclear properties when solutions can be obtained. The new parameter set TM1 provides a similar behavior to the RBHF theory. We believe the RBHF theory is realistic up to a certain density, typically $\rho \leq 2\rho_0$, and in particular the results obtained with the Bonn A potential are the ones to rely on, since this is the first case to reproduce the saturation property of nuclear matter. We should expect, however, that some many-body correlations and a change of hadron properties will set in as the matter density increases to higher values such that $\rho \geq 2\rho_0$, for the RBHF theory.

The saturation properties are shown in Table 4.4. We compare the values obtained with TM1, TM2, NL1, NL-SH, and the RBHF theory. The saturation densities for TM1 and TM2 are found to be smaller than the standard value. The energy per particle is close to $-16\,\text{MeV}$. The incompressibility for TM1 is found to be $K = 281\,\text{MeV}$, which is larger than that ($K = 211\,\text{MeV}$) for NL1. Again this larger value is similar to the value obtained from the RBHF theory ($K = 290\,\text{MeV}$). The symmetry energy is found to be $a_{\text{sym}} = 37\,\text{MeV}$, as compared with $a_{\text{sym}} = 44\,\text{MeV}$ for NL1. This smaller value may be obtained by providing smaller neutron skins for $N \neq Z$ nuclei. The symmetry energy for the RBHF theory is $a_{\text{sym}} = 35\,\text{MeV}$, while the result of Möller and Nix [154] obtained from their mass formula is $a_{\text{sym}} = 32.73\,\text{MeV}$.

The use of unstable-nuclear-beam facilities allows us to study the nuclear structure of nuclei far from the stability line. Such studies of exotic nuclei display a large diversity of phenomena which cannot be observed in stable nuclei. Examples are the neutron halos and the neutron skins for nuclei close to the neutron drip line [155, 156]. Many more are expected to be found.

As a result of the use of these facilities, the quality and quantity of experimental data on nuclear structure properties have increased over recent years. Systematic studies of isotopic shifts of nuclei across major shell closures are providing rigorous tests of the theoretical understanding of nuclear physics. Variations of shape are found such as changes from a spherical to an oblate shape or from a prolate to a triaxial shape. Several different shapes are being found in some nuclei, such as ^{185}Au, where four different coexisting shape bands have now been identified [157].

In [95, 96, 150], RMF theory has been applied to unstable nuclei. The description of the ground-state properties of stable and unstable nuclei with axial deformations in terms of RMF theory has been also studied, and the results are found to be very satisfactory [158]. In [114], one more degree of freedom has been included and the RMF theory has been extended to include triaxial deformations. In [116], the calculations were performed with TM1 for unstable nuclei also and compared with the existing data. The results are similar to those for NL1. The agreement with the data is remarkable. Concerning the neutron skins for nuclei far from the stability line, the neutron skins calculated with TM1 are somewhat smaller than those calculated with NL1 owing to the smaller symmetry energy for TM1. It is important also to test RMF models at finite temperature and different densities, this has been done in [159–161].

5 Electromagnetic Interactions of Nucleons in the Relativistic Framework

5.1 The Vector Dominance Model and the Nuclear Coulomb Potential

In the nucleus, there is, besides the vector and scalar meson fields, the Coulomb field. As is well known, the nucleons are not point particles, and therefore to construct the Coulomb field of a nucleus it is necessary to take into account the electromagnetic form factors of the nucleons. This can be done phenomenologically, by folding the proton and neutron densities obtained in the RMFA with the intrinsic form factors of the free proton or neutron, i.e. by replacing the electromagnetic current operator by [162]

$$J_\mu = \frac{1+\tau_3}{2}\gamma_\mu \to J_\mu = \sum_x \left\{ f_{1,x}(q)\gamma_\mu - \frac{f_{2,x}(q)}{4M}[\gamma_\mu, \gamma_\nu]q^\nu \right\}, \qquad (5.1)$$

where $f_{1,x}$ is the nucleon charge form factor, $f_{2,x}$ is the form factor of the anomalous magnetic moment, and q_ν is the four-vector of the exchanged momentum. This is an effective electromagnetic operator in the zeroth order. This operator should be applied both for the ground state and for transition moments.

The equation given above ignores the fact that the nucleon may change its internal structure in the nuclear medium. Such modifications may originate, for example, from the quark structure of the nucleon [163–165] and are indicated, for example, by the EMC effect [166, 167]. The behavior of composite nucleons in scalar and vector mean fields has been investigated in [168, 169].[1]. The results of these investigations suggest a justification for the use of the Dirac equation in the way in which it has been used in the RMF theories and in theories of relativistic nuclear scattering (see below).

In this section we consider an approach [170] where the electromagnetic structure of nucleons is taken into account on the basis of Sakurai's vector dominance model (VDM) [171]. In the framework of this model, the nucleons interact with the electromagnetic field through the fields of the ρ and ω mesons (see Fig. 5.1).

---
[1] The description of a spin-half particle with internal structure in an external electric field is considered in [172, 173].

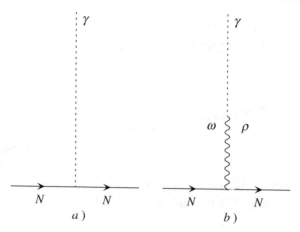

Fig. 5.1. Interaction of the electromagnetic field with (a) point nucleons and (b) in the vector dominance model (VDM)

The Lagrangian of the relativistic shell model with inclusion of vector meson dominance is

$$\mathcal{L} = \overline{\psi_\tau} \left(i\gamma^\mu \partial_\mu - M \right) \psi_\tau - g_\sigma \overline{\psi_\tau} \psi_\tau \varphi - g_\omega \overline{\psi_\tau} \gamma^\mu \psi_\tau \omega_\mu$$
$$- \frac{f_\omega}{2M} \overline{\psi_\tau} \sigma^{\mu\nu} \psi_\tau \partial_\mu \omega_\nu - 2\tau g_\rho \overline{\psi_\tau} \gamma^\mu \psi_\tau \rho_\mu - 2\tau \frac{f_\rho}{2M} \overline{\psi_\tau} \sigma^{\mu\nu} \psi_\tau \partial_\mu \rho_\nu$$
$$+ \frac{1}{2} \partial_\mu \varphi \partial^\mu \varphi - \frac{1}{2} m_\sigma^2 \varphi^2 - \frac{1}{2} \partial_\mu \omega_\nu \partial^\mu \omega^\nu$$
$$+ \frac{1}{2} m_\omega^2 \omega_\mu \omega^\mu - \frac{1}{2} \partial_\mu \rho_\nu \partial^\mu \rho^\nu + \frac{1}{2} m_\rho^2 \rho_\mu \rho^\mu - \frac{1}{2} \partial_\mu A_\nu \partial^\mu A^\nu$$
$$+ \frac{1}{2} \frac{e^2}{4} \left(\frac{m_\rho^2}{g_\rho^2} + \frac{m_\omega^2}{g_\omega^2} \right) A_\mu A^\mu - \frac{em_\omega^2}{2g_\omega} A^\mu \omega_\mu - \frac{em_\rho^2}{2g_\rho} A^\mu \rho_\mu \,, \quad (5.2)$$

where ψ_τ is the nucleon field, $\tau = \pm 1/2$ is the projection of the isospin of the nucleon; A^μ ($\mu = 0, 1, 2, 3$) is the electromagnetic field; φ, ω^μ, and ρ^μ are the fields of the scalar, ω, and ρ are mesons; and g_σ, g_ω, g_ρ, and e are the corresponding coupling constants. For the vector fields the tensor coupling with the nucleons is also taken into account; f_ω and f_ρ are constants of the gradient coupling, and $\sigma^{\mu\nu} = i/2[\gamma^\mu, \gamma^\nu]$. It will be shown below that the constants f_ω and f_ρ can be expressed in terms of the anomalous gyromagnetic ratios of the nucleons. It is readily verified that the Lagrangian (5.2) is invariant with respect to the gauge transformation

$$\psi' = e^{-i(1/2+\tau)\phi} \psi \,, \qquad \omega'_\mu = \omega_\mu + \frac{1}{g_\omega} \partial_\mu \phi \,,$$
$$\rho'_\mu = \rho_\mu + \frac{1}{g_\rho} \partial_\mu \phi \,, \qquad A'_\mu = A_\mu + \frac{1}{e} \partial_\mu \phi \,. \quad (5.3)$$

5.1 The Vector Dominance Model and the Nuclear Coulomb Potential

Within the framework of the traditional Lagrange–Euler scheme we obtain the system of coupled equations

$$\left[\gamma^\mu(i\partial_\mu - g_\omega\omega_\mu - 2\tau g_\rho\rho_\mu) - \left(\frac{f_\omega}{2M}\right)\sigma^{\mu\nu}\partial_\mu\omega_\nu\right.$$
$$\left. -2\tau\left(\frac{f_\rho}{2M}\right)\sigma^{\mu\nu}\partial_\mu\rho_\nu - M - g_\sigma\varphi\right]\psi_\tau = 0, \quad (5.4)$$

$$(-\partial_\mu \cdot \partial^\mu - m_\sigma^2)\varphi = g_\sigma\overline{\psi_\tau}\psi_\tau, \quad (5.5)$$

$$(-\partial_\mu \cdot \partial^\mu - m_\omega^2)\omega^\nu =$$
$$-g_\omega(\bar\psi_\tau\gamma^\nu\psi_\tau) + (f_\omega/2M)\partial_\mu(\bar\psi_\tau\sigma^{\mu\nu}\psi_\tau) - (em_\omega^2/2g_\omega)A^\nu, \quad (5.6)$$

$$(-\partial_\mu \cdot \partial^\mu - m_\rho^2)\rho^\nu =$$
$$-g_\text{p}(2\tau\bar\psi_\tau\gamma^\nu\psi_\tau) + (f_\rho/2M)\partial_\mu(2\tau\bar\psi_\tau\sigma^{\mu\nu}\psi_\tau) - (em_\rho^2/2g_\rho)A^\nu, \quad (5.7)$$

$$\left[-\partial_\mu \cdot \partial^\mu - \frac{e^2}{4}\left(\frac{m_\omega^2}{g_\omega^2} + \frac{m_\rho^2}{g_\rho^2}\right)\right]A^\nu = -\frac{em_\omega^2}{2g_\omega}\omega^\nu - \frac{em_\rho^2}{2g_\rho}\rho^\nu. \quad (5.8)$$

As is evident from these equations, the source of the electromagnetic field in the approach considered here is the fields of the vector mesons ω and ρ. This is the main peculiarity of the vector meson dominance model which distinguishes it from the traditional point-nucleon approximation, in which the vector proton density $\bar\psi\gamma^0\psi$ is the source of the Coulomb potential (see Fig. 5.1). The electromagnetic field, in turn, in the present approach, is one of the sources of the vector meson fields.

The relativistic shell model [59, 74, 101] is based on replacement of the operators of the nucleon currents on the right-hand sides of (5.5)–(5.8) by their expectation values for the ground state of the nucleus. Only the time components of the fields ($\nu = 0$) are nonzero, since there are no spatial components of the currents in the ground states of even–even nuclei. The fields are static, since in the absence of external fields the nuclear densities do not depend on the time. Therefore, in what follows we shall be interested in static solutions of (5.4)–(5.8).

In this case, (5.4)–(5.8) take the form

$$\left[-i\boldsymbol\alpha \cdot \boldsymbol\nabla + \beta(g_\sigma\varphi + M) + g_\omega\omega + 2\tau g_\rho\rho\right.$$
$$\left. -\frac{i}{2M}\beta\boldsymbol\alpha(f_\omega\boldsymbol\nabla\omega + 2\tau f_\rho \cdot \boldsymbol\nabla\rho)\right]\psi_\tau = E_\tau\psi_\tau, \quad (5.9)$$

$$(\Delta - m_\sigma^2)\varphi = g_\sigma(\rho_\text{s,p} + \rho_\text{s,n}), \quad (5.10)$$

$$(\Delta - m_\omega^2) \omega_0 = -g_\omega (\rho_{V,p} + \rho_{V,n}) - \frac{f_\omega}{2M} \text{div} (s_p + s_n) - \frac{em_\omega^2}{2g_\omega} C, \quad (5.11)$$

$$(\Delta - m_\rho^2) \rho_3^0 = -g_\rho (\rho_{V,p} - \rho_{V,n}) - \frac{f_\rho}{2M} \text{div} (s_p - s_n) - \frac{em_\rho^2}{2g_\rho} C, \quad (5.12)$$

$$\left(\Delta - \frac{e^2 m_\omega^2}{4g_\omega^2} - \frac{e^2 m_\rho^2}{4g_\rho^2} \right) C = -\frac{em_\omega^2}{2g_\omega} \omega_0 - \frac{em_\rho^2}{2g_\rho} \rho_3^0, \quad (5.13)$$

where ω_0, ρ_3^0, and $C = A_0$ are the time components of the fields ω^ν, ρ^ν, and A^ν; $\rho_V = (\rho_{V,n} + \rho_{V,p})$, where $\rho_{V,(p,n)}$ are the proton and neutron vector densities; $\rho_{s(p,n)}$ are the proton and neutron scalar densities; the spin density s is defined by the relation

$$s = i \langle \bar{\psi} \boldsymbol{\alpha} \psi \rangle. \quad (5.14)$$

The symbol $\langle \ldots \rangle$ denotes an average over the ground state of the nucleus. Within the framework of the scheme considered here, the pseudoscalar field does not have a source. Therefore, we do not include the contribution from the π mesons. However, one cannot exclude the possibility of excitation of a nonstationary π meson field; an investigation of this problem for relativistic nuclear matter will be performed below.

We shall now show that the fields ω_0, ρ_3^0, and C behave asymptotically like the Coulomb field. For this purpose we rewrite (5.11) and (5.12) in the form

$$\omega_0 = \frac{g_\omega}{m_\omega^2} (\rho_{V,p} + \rho_{V,n}) + \frac{e}{2g_\omega} C + \frac{f_\omega}{2M m_\omega^2} \text{div} (s_p + s_n) + \frac{1}{m_\omega^2} \Delta \omega^0, \quad (5.15)$$

$$\rho_3^0 = \frac{g_\rho}{m_\rho^2} (\rho_{V,p} - \rho_{V,n}) + \frac{e}{2g_\rho} C + \frac{f_\rho}{2M m_\rho^2} \text{div} (s_p - s_n) + \frac{1}{m_\rho^2} \Delta \rho_3^0. \quad (5.16)$$

Substituting (5.15) and (5.16) into (5.13), we find

$$\Delta \left(C + \frac{e}{2g_\rho} \rho_3^0 + \frac{e}{2g_\omega} \omega_0 \right)$$
$$= -e\rho_{V,p} - \frac{ef_\rho}{4Mg_\rho} \text{div} (s_p - s_n) - \frac{ef_\omega}{4Mg_\omega} \text{div} (s_p + s_n). \quad (5.17)$$

Taking into account that $\Delta r^{-1} = 0$ for $r \to \infty$ we find from (5.15)–(5.17)

$$C_\infty = \frac{a}{r}, \quad \rho_{3,\infty}^0 = \frac{ea}{2g_\rho r}, \quad \omega_{0,\infty} = \frac{ea}{2g_\omega r}. \quad (5.18)$$

To find the constant a, we calculate the integral of the two sides of (5.17) over a sphere with a large radius. Using (5.18), we find

$$-4\pi a \left(1 + \frac{e^2}{4g_\rho^2} + \frac{e^2}{4g_\omega^2} \right) = -eZ, \quad (5.19)$$

5.1 The Vector Dominance Model and the Nuclear Coulomb Potential

where Z is the charge of the nucleus. Bearing in mind that $g_\rho^2/(4\pi) = 0.644$, and $g_\omega^2/(4\pi) = 14.059$ (see below), we obtain

$$\frac{e^2}{4g_\rho^2} + \frac{e^2}{4g_\omega^2} = 0.003 , \quad (5.20)$$

i.e. the contribution of the $A\rho$ and $A\omega$ interactions to the charge is only 0.3%. Note that (in the units considered here) the experimental value of the electron charge e_{exp} is given by the relation

$$\frac{e_{\text{exp}}^2}{4\pi} = \frac{e^2}{4\pi}\left(1 + \frac{e^2}{4g_\rho^2} + \frac{e^2}{4g_\omega^2}\right)^{-1} = \frac{1}{137} . \quad (5.21)$$

We split the vector meson fields ω_0 and ρ_3^0 into two components: a nuclear component ω_{0N}, ρ_{3N}^0 and a Coulomb ω_{0C}, ρ_{3C}^0:

$$\omega_0 = \omega_{0C} + \omega_{0N} , \qquad \rho_3^0 = \rho_{3C}^0 + \rho_{3N}^0 . \quad (5.22)$$

Here the source of the fields ω_{0C} and ρ_{3C}^0 is the Coulomb potential C, and ω_{0N} and ρ_{3N}^0 are determined by means of the equations

$$(\Delta - m_\omega^2)\,\omega_{0N} = -g_\omega\,(\rho_{V,p} + \rho_{V,n}) - \frac{f_\omega}{2M}\,\text{div}\,(s_p + s_n) , \quad (5.23)$$

$$(\Delta - m_\rho^2)\,\rho_{3N}^0 = -g_\rho\,(\rho_{V,p} - \rho_{V,n}) - \frac{f_\rho}{2M}\,\text{div}\,(s_p - s_n) , \quad (5.24)$$

$$\Delta\left(C + \frac{e}{2g_\omega}\omega_{0c} + \frac{e}{2g_\rho}\rho_{3c}^0\right) = -\frac{em_\omega^2}{2g_\omega}\omega_{0N} - \frac{em_\rho^2}{2g_\rho}\rho_{3N}^0 = -e\rho_c(r) , \quad (5.25)$$

where $\rho_c(r)$ is the nuclear charge distribution given by the relation

$$\rho_c(r) = \frac{m_\omega^2}{2g_\omega}\omega_{0N}(r) + \frac{m_\rho^2}{2g_\rho}\rho_{3N}^0(r) , \quad (5.26)$$

and the mean square charge radius of the nucleus is

$$r_c^2 = \frac{1}{Z}\int \rho_c(r)r^2\,dV .$$

Below we shall use also the Hartree–Coulomb potential

$$eC_{\text{H}} = eC + \frac{e^2}{2g_\omega}\omega_0 + \frac{e^2}{2g_\rho}\rho_3^0 , \quad (5.27)$$

and the Coulomb potentials acting from the nucleus on the leptons, eC_{L}, and on the baryons, eC_{N}, where

$$eC_{\text{L}} = eC , \qquad eC_{\text{N}} = g_\omega\omega_{0C} + 2\tau g_\rho\rho_{3C}^0 . \quad (5.28)$$

Indeed, as is evident from (5.17) the quantity eC_H satisfies the ordinary Poisson equation to within the accuracy of small terms containing div s of the order v^2/c^2 (see the definition of s). Further, as is evident from (5.4) the nucleons interact with vector fields, and therefore eC_N is a Coulomb component of these fields. Finally, the quantity $\rho_c(r)$ which is determined by the relation (5.26) is a source of the lepton Coulomb potential to within the accuracy of a small contribution from the electromagnetic interaction of hadrons with the charge (see (5.20) and (5.25)).

In the VDM, the gradient coupling constants f_ω and f_ρ are determined by the anomalous magnetic moments of the nucleons. Let us consider one nucleon in an external long-wave electromagnetic field. In this case, from (5.6) and (5.7) we find $g_\omega \omega_\mu = 1/2\, eA_\mu$, and $g_\rho \rho_\mu = 1/2\, eA_\mu$, and from (5.4) we find

$$\left\{ \gamma^\mu \left[i\partial_\mu - e\left(\frac{1}{2} + \tau\right) A_\mu \right] - \frac{e}{2M}\left(\frac{f_\omega}{2g_\omega} + 2\tau \frac{f_\rho}{2g_\rho}\right) \sigma^{\mu\nu} \partial_\mu A_\nu \right\} \psi_\tau = 0 , \tag{5.29}$$

from which we obtain

$$\frac{f_\omega}{2g_\omega} = -0.06 \quad \text{and} \quad \frac{f_\rho}{2g_\rho} = 1.85 . \tag{5.30}$$

The values of these ratios show that the anomalous magnetic moments of the nucleons are determined mainly by the isovector component.

It is interesting to note that within the framework of the VDM, a neutron inside the nucleus also feels an action of the Coulomb potential, which is equal to the difference between the Coulomb components of the ω and ρ meson fields $g_\omega \omega_{0C} - g_\rho \rho^0_{3C}$. The value of the neutron Coulomb potential in the nucleus is small and depends very weakly on the nuclear charge Z [170].

5.1.1 Results of Calculations

The self-consistent system (5.9)–(5.13) determines the properties of nuclei through the constants g_ω, $G_\rho = 2g_\rho$, g_σ and m_σ.[2] The calculations of the ground-state properties of the ^{16}O, ^{40}Ca, ^{48}Ca nuclei presented in [170] give the following "medium" values for these constants:

$$\frac{g_\omega^2}{4\pi} = 14.059 , \qquad \frac{G_\rho^2}{4\pi} = 2.575,$$

$$\frac{g_\sigma^2}{4\pi} = 6.81 , \quad m_\sigma = 475 \text{ MeV} . \tag{5.31}$$

---

[2] Note that in the calculations of the properties of the nuclear ground state using the VDM in [170], the nonlinear self-interaction terms were not taken into account. The scheme was extended in [174] (in relativistic self-consistent investigations of the Okamoto–Nolen–Schiffer anomaly) to include scalar meson self-interactions.

5.1 The Vector Dominance Model and the Nuclear Coulomb Potential

Table 5.1. The r.m.s. radii of the distributions of the protons, neutrons, and charge (in fm). The experimental data given in brackets are taken from [177–179]

Nucleus	r_p	r_n	$r_n - r_p$	r_c
^{16}O	2.77	2.72	−0.05 (0.0)	2.83 (2.73)
^{40}Ca	3.43	3.37	−0.06 (0.05 ± 0.05)	3.48 (3.48)
^{48}Ca	3.42	3.62	0.2 (0.2 ± 0.05)	3.44 (3.47)

The "vacuum" values of these constants have been determined by analyzing NN scattering in the model of the one-boson exchange [34, 43]. These values are:

$$\frac{g_\omega^2}{4\pi} = 12.85 \pm 1.29 , \qquad \frac{g_\sigma^2}{4\pi} = 7.14 ,$$

$$\frac{G_\rho^2}{4\pi} = 2.470 \pm 0.255 [175] , \qquad m_\sigma = 550 \text{ MeV} . \qquad (5.32)$$

In Table 5.1 we give the results of the calculations of the mean square radii of the distributions of the protons, the neutrons, and the charge in the ^{16}O, ^{40}Ca, ^{48}Ca nuclei. The experimental values of these quantities, also given in Table 5.1, exhibit good agreement with the theory. The agreement between the theoretical charge distributions obtained in [170] and the experimental distributions [176] is entirely satisfactory.

For ^{48}Ca, the excitation energy of the isobar-analogue state was also calculated. This energy is given by

$$\Delta_c = \frac{2g_\rho}{N-Z} \int \rho_{3C}^0(r) \left(\rho_{V,n} - \rho_{V,p} \right) dV . \qquad (5.33)$$

The calculated value $\Delta_c = 7.041$ MeV agrees well with the experimental value $\Delta_c^{\text{exp}} = 7.180$ MeV [180].

The main point of this section is to show that the composite character of the nucleons (their electromagnetic structure) can be taken quite naturally into account in the framework of the RMF approach via the mechanism of the vector dominance model. To obtain a more detailed quantitative description of nuclear bulk properties, the nonlinear self-interaction terms also need to be taken into consideration in this case [174]. Note that vector meson dominance is considered at present as one of the two basic principles of low-energy QCD, the second one being chiral symmetry. Later, in Chap. 7, we consider implementations of the ideas of chiral symmetry in the nuclear-structure context.

5.2 Nuclear Magnetic Moments in the Relativistic Approach

The nonrelativistic isoscalar Schmidt values are particularly successful in reproducing the experimental magnetic moments of nuclei with a single particle or hole over a closed shell.

A relativistic expression for the single-particle magnetic moment was first derived in [181] (see also [182]). It is given in the following form, conventionally split into the contributions from the Dirac and Pauli interactions as

$$\mu = \mu_D + \mu_A, \quad (5.34)$$

where

$$\mu_D = \omega \left(\frac{2j+1}{j+1}\right) M \int_0^\infty r G(r) F(r) dr \quad (5.35)$$

has a nonzero value only for protons, while μ_A is determined by

$$\mu_A = -\omega \mu_a \left(\frac{j}{j+1}\right)^{1/2} \left[\left(\frac{j}{j+1}\right)^{\omega/2} \int_0^\infty G^2(r) dr \right.$$
$$\left. + \left(\frac{j}{j+1}\right)^{-\omega/2} \int_0^\infty F^2(r) dr \right], \quad (5.36)$$

and $\mu_a = 1.793$ for protons and -1.913 for neutrons. The wave functions G and F are solutions of the radial Dirac equation (4.60), (4.61), and $\omega(j+1/2) \equiv \kappa$ (see [58, 183]). In the nonrelativistic limit,

$$\int_0^\infty G^2(r) dr = 1, \quad \int_0^\infty F^2(r) dr = 0,$$

$$M \int_0^\infty r G(r) F(r) dr = \frac{\omega}{2}\left(j+\frac{1}{2}\right) - \frac{1}{4}, \quad (5.37)$$

and one obtains the Schmidt values for the single-particle magnetic moments:

$$\mu = \begin{bmatrix} j + 1/2 + \mu_a, & j = l + 1/2 \\ j - j/(j+1)\,(1/2 + \mu_a), & j = l - 1/2 \end{bmatrix}. \quad (5.38)$$

From (5.35) and (5.36) it can be clearly seen that the relativistic corrections to the Pauli term are small, since this term is dominated by the contribution of the squared value of the upper component $G(r)$ of the wave function. In contrast, the contribution of the direct part to the single-particle

5.2 Nuclear Magnetic Moments in the Relativistic Approach

magnetic moment is proportional to the lower component of the wave function. Owing to (4.88), the lower component is greatly strengthened because of the small value of the nucleon effective mass, this fact leading to an increase of the valence nucleon electromagnetic current. So the Dirac component of the magnetic moment is much more influenced by the relativistic corrections than the Pauli part is. This qualitative conclusion is supported in [58, 183, 184] by calculations carried out in the framework of a relativistic shell model with two strong fields $S(r)$ and $V(r)$ generating the observed value of the spin–orbit force in finite nuclei. These calculations were first carried out for the nuclei ^{15}N and ^{15}O. Later the calculations were repeated for nuclei with mass numbers $A = 17, 39$, and 41. All these calculations, performed in the single-valence-nucleon approximation, revealed a very strong contribution of the relativistic corrections ($\sim 110\%$), whereas all other effects (pion exchange currents, coupling to magnetic resonances, etc.) are of minor importance. So a problem appears when the isoscalar[3] nonrelativistic Schmidt values agree with the experimental magnetic moments for closed-LS-shell ± 1 nuclei reasonably well and, on the other hand, very strong relativistic corrections are generated destroying this agreement in the relativistic framework. Different solutions to this problem have been found by several authors [185–196]. For example, in [189] it is emphasized that while the velocity of a valence nucleon at the Fermi surface is given by

$$\boldsymbol{v} = \langle \bar{\psi} \boldsymbol{\alpha} \psi \rangle = \frac{\boldsymbol{p}_F}{E_F^*}, \qquad (5.39)$$

$$E^* = \sqrt{p^2 + M^{*2}} = E - V, \qquad (5.40)$$

$$M^* = M + S, \qquad (5.41)$$

(see (4.49) and (4.79) for notation), the nuclear convection current is not determined by (5.39). In an interacting many-body system, a careful distinction should be made in definitions of nuclear velocities and nuclear currents; for example, the nuclear current should satisfy gauge invariance (the continuity equation, in particular), this point being clarified in [197] by the meson exchange current method. Consider the Feynman pair graphs (Fig. 5.2).

In Fig. 5.2 it is supposed that particle 1 corresponds to a valence nucleon outside the Fermi sea, while particle 2 is related to a nucleon in the medium. In the lowest order, the mean-field approximation corresponds to the process shown in Fig. 5.2a. However, the diagram in Fig. 5.2a is not gauge invariant, but a combination of Figs. 5.2a and 5.2b will satisfy the continuity equation. So it is a requirement of gauge invariance that the external field interacts also

---

[3] The isovector moments are not in agreement with the Schmidt values. However, we shall not consider the problem of isovector moments here, they are strongly renormalized by one-pion exchange currents, Δ-isobars, and core polarization, which are theoretically uncertain and highly model-dependent.

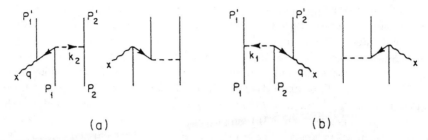

Fig. 5.2. The exchange current corrections investigated in [189, 197]. The *wavy line* corresponds to the external electromagnetic field, while the *dashed line* corresponds either to a σ or to an ω meson: (**a**) mean-field approximation, lowest order; (**b**) the lowest order particle-antiparticle buble in the relativistic RPA series

with nucleons in the medium, and the result for the exchange current for the convection current in the nuclear medium is determined by the formula

$$\frac{\boldsymbol{p}_F}{M}\left(1 - \frac{S+V}{M}\right), \qquad (5.42)$$

which differs from $\boldsymbol{p}_F/M$, i.e. from the free convection current, only by a binding-energy correction. In (5.42) the scalar contribution is produced by the process represented in Fig. 5.2a, while the vector contribution is produced by the process in Fig. 5.2b and is related to the space-like component $\boldsymbol{\omega}$ of the ω meson field.

The essential role of medium corrections in the relativistic framework was discussed in [185] in the context of Fermi liquid theory and in [75] in the context of the relativistic RPA. In both cases similar solutions are obtained; however, the RPA method of [75] has more common features with the meson exchange current method of [189]. Figure 5.2b corresponds to the lowest order $N\bar{N}$ bubble considered in the relativistic RPA expansions. This correction of the lowest order, determined by use of the transverse Hartree correlation function [44], is estimated to be $\Pi_T^H = -\rho_V/E_F^*$ in nuclear matter.

The relationship between self-energy corrections and medium corrections can be established, using the language of quantum field theory, via the Ward identities. These identities have been utilized in treating the magnetic-moment problem in [198, 199], and it has been shown that corrections obtained in [75, 185] should satisfy the Ward identities. These results have been utilized in [188] to calculate magnetic moments on the basis of the Hartree model for finite nuclei.

Coming back to the problem of the convection current, let us note that the consistent current is given by [186, 200]

$$\frac{\boldsymbol{p}_F}{E_F^*}\left(1 + \frac{(g_\omega^2/m_\omega^2)\Pi_T^H}{1-(g_\omega^2/m_\omega^2)\Pi_T^H}\right) = \frac{\boldsymbol{p}_F}{E_F} \qquad (5.43)$$

rather than by (5.39). To obtain the RPA medium correction to all orders, the ring approximation has been used. In (5.39), $E_{\rm F} = \rho_{\rm V} g_\omega^2/m_\omega^2 + E_{\rm F}^*$ is the eigenvalue of (4.79); it differs from the energy of the free nucleon only by the binding energy. As can be seen, the convection current is very close to its free value. In the Fermi liquid theory, the correction discussed above is often referred to as "back-flow" [201]. It comes from the space part of the ω meson field.

Polarization effects can be treated in the relativistic RPA, as was discussed above (see also [202]). However, the back-flow effect, caused by the polarization of the core due to an external particle, can be taken into consideration easily in the relativistic Hartree approximation [191]. One should bear in mind that the mean field generated by the current of the valence nucleon does not possess time-reversal and rotational invariance. In other words, the baryon currents do not vanish, as is usually assumed in the Walecka model, even in the mean-field approach; these currents appear to be the sources of the space components of the vector (ω meson) field; and they change the structure of the Dirac equation and produce polarization effects in the relativistic spinors.[4] For nuclei with A odd, the Dirac equation includes the space part $g_\omega \boldsymbol{\omega}$ of the ω meson field

$$[\boldsymbol{\alpha}\left(-i\boldsymbol{\nabla} - g_\omega \boldsymbol{\omega}\right) + g_\omega \omega_0 + \beta\left(M - g_\sigma \sigma\right)]\psi_\kappa = E_\kappa \psi_\kappa , \qquad (5.44)$$

which satisfies the Klein–Gordon equation

$$\left(-\Delta + m_\omega^2\right)\boldsymbol{\omega} = g_\omega \boldsymbol{j}_{\rm V} , \qquad (5.45)$$

while the equations for the time components ω^0 and ρ_3^0, for the scalar field, and for the electromagnetic field remain unaltered (see (4.51)–(4.54)). So the system of equations (5.44), (5.45), (4.51)–(4.54) actually needs to be solved self-consistently in this case. In most of the investigations up to [191], the current terms of the ω meson field were not included in the self-consistent calculations. There was a simple reason for this. The difficulty is that these terms violate spherical symmetry and could not be taken into consideration in a theoretical scheme developed for the spherical case. However, in [92] a method and a code were developed for calculating deformed axially symmetric systems. The current terms (5.45) and the space-components $g_\omega \boldsymbol{\omega}$ were incorporated into the framework of the scheme developed in [92]. Cylindrical coordinates $(z, r_\perp, \varphi)$ were used. It was established that, owing to the axial symmetry, the cylindrical components of the vector fields do not depend on the angle φ. The total current in this case can be decomposed into two parts: (1) azimuthal currents $j_\varphi(z, r_\perp)$ on circular lines around the axis of symmetry, with $j_\perp = j_z = 0$, and (2) currents with no azimuthal components, with

---

[4] This effect is referred to as nuclear magnetism [84], since the situation in this case is similar to that in electrodynamics where a magnetic field $\boldsymbol{A}$ is generated by an electromagnetic current.

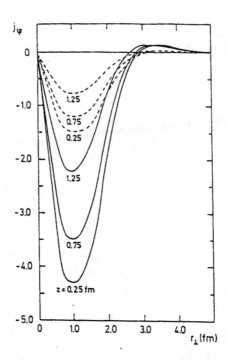

Fig. 5.3. The azimuthal current $j_\varphi(z, r_\perp)$ in the nucleus ^{15}N as a function of the radial distance $r_\perp$ for different values of the coordinate z along the symmetry axis. The *dashed lines* are calculated without polarization of the ^{16}O core; the *full lines* include in addition the polarization of the core. The units are 10^{-3}fm^{-3}. (From [84])

closed flow lines in planes containing the axis of symmetry. In both cases the vector potentials and currents are of the same type, the equations (5.45) being decoupled.

Finally, after convergence, a solution with a current of type (1) only was found [191].[5] In Fig. 5.3, this current is given for the ^{15}N nucleus. The azimuthal current $j_\varphi(z, r_\perp)$ calculated without polarization (one proton in the $1p_{1/2}$ shell is removed from the self-consistently calculated ^{16}O core) is represented by the dashed lines for several different values of the coordinate z. The continuous lines were calculated by self-consistent solution of the system of equations (5.44), (5.45), (4.51)–(4.54). As can be seen from this figure, the external hole generates a current in the core which causes the total current to be enhanced by a factor of three.

In Table 5.2, we present the isoscalar magnetic moments for four odd-A nuclei with mass numbers $A = 15$, 17, 39, and 41. The results were obtained as the arithmetic average of the two mirror nuclei with the same mass,

---

[5] The calculations were performed with the parameter set NL1 discussed above.

5.2 Nuclear Magnetic Moments in the Relativistic Approach

Table 5.2. The isoscalar magnetic moments for nuclei with particle numbers $A = 15, 17, 39$, and 41 [84]. The experimental values are compared with the Schmidt values for the nonrelativistic shell model, with relativistic calculations for a spherical potential (without polarization) (Sph. RMF), and with calculations in a deformed potential (with polarization) (Def. RMF) [191]

	Magnetic moment	15	17	39	41
μ	Experiment	0.22	1.41	0.71	1.92
	Schmidt	0.19	1.44	0.64	1.94
	Sph. RMF	0.32	1.57	0.94	2.21
	Landau[29]	0.19	1.41	0.64	1.91
	Def. RMF	0.18	1.48	0.64	1.97
μ_D	Schmidt	0.17	1.50	0.60	2.00
	Sph. RMF	0.30	1.63	0.91	2.27
	Def. RMF	0.15	1.54	0.60	2.03
μ_A	Schmidt	0.02	−0.06	0.04	−0.06
	Sph. RMF	0.02	−0.06	0.04	−0.06
	Def. RMF	0.03	−0.06	0.04	−0.06

$$\mu(T=0) = \frac{\mu(N,Z) + \mu(N+1, Z-1)}{2}. \tag{5.46}$$

In Table 5.2, the experimental values are given together with the nonrelativistic Schmidt values, the results obtained from the spherical RMF theory [191] (Sph. RMF), where only the core is calculated self-consistently, the values obtained by the Landau–Migdal approach, and the completely self-consistent results (Def. RMF), the effects of the nuclear magnetism included. The anomalous part of the isoscalar magnetic moments is very small. There is a cancellation effect of (negative) contributions of the polarization currents to the Dirac moment in the neutron nuclei and the (positive) contribution in the proton nuclei. So a very good agreement with the Schmidt values is finally obtained in the deformed RMF framework.

The conclusion of [191] is that it is not the mean-field approach which gives the bad values of isoscalar magnetic moments in relativistic models. The main reason is that the eigenfunctions of odd-mass nuclei are not time-reversal invariant, and for this reason the baryonic currents cannot be ignored in the RMF theory. If they are taken into consideration, they generate the space components of the ω meson field, which modify the nuclear eigenfunction and the resulting values of the magnetic moments. So, finally, for the isoscalar part of the magnetic moments in LS-closed-shell nuclei ±1 nucleon, one finds excellent agreement with the Schmidt values.

In the present section we have considered odd nuclei in connection with the problem of nuclear magnetic moments. As mentioned earlier, for the ground state of even–even nuclei, time-reversal invariance can be assumed. The single-particle states in this case are pairwise degenerate [203], and the contribution of each nucleon state to the baryonic current cancels with that of its time-reversed counterpart. In odd nuclei the situation is different. The odd nucleon spoils time-reversal invariance and generates the basic contribution to the currents. There is no degeneracy anymore, and there is also a contribution to the currents (see (5.45) and a similar equation for ρ meson) due to core polarization. For this reason, theoretical calculations for odd nuclei are more involved than those for even–even nuclei. However, such calculations have been carried out in [203] in the framework of the RMF theory. In that paper, another important aspect in connection with odd nuclei was investigated, namely, the relation between the spectrum of nucleon separation energies and that of single-nucleon energies in the neighboring even–even nucleus.

6 The Relativistic Approach to Nucleon–Nucleus Scattering

6.1 Energy Dependence of the Real Part of the Optical Potential

For nucleon energies of 500 MeV or more, the experimental data on elastic scattering are analyzed either in the framework of the Glauber theory [204] (the eikonal approximation in the multiple-scattering theory) or that of the Kerman, McManus, and Thaler theory [205] (the impulse approximation in the multiple-scattering theory). At lower energies, the optical model is the basic approach used for the theoretical analysis of nucleon–nucleus scattering.

In this section we shall consider the problem of the energy dependence of the real part of the optical potential [206–210]. The energy dependence of the nucleon–nucleus optical potential observed in experiments may be obtained in the framework of the conventional nonrelativistic descriptions by introducing a nonlocal interaction.

One of the advantages of the Dirac phenomenology in comparison with the Schrödinger picture is that in the former case the "intrinsic" energy dependence of the optical potential is revealed, which explains the observed change of sign of the real part of the optical potential from attractive to repulsive at intermediate proton energies.

In relativistic theories one has to solve a Dirac equation of the type (4.79). However, in the optical model, the calculation of proton–nucleus elastic scattering is carried out by solving the Schrödinger equation. Therefore, to compare the relativistic potential with the empirical potential, it is reasonable to write out the wave equation in the Schrödinger form. The transformation to such a form can be carried out by various methods and has been done by several authors [206, 211]. If one eliminates the lower component χ of the wave function in favor of the upper component φ and introduces a function ϕ in accordance with the equation

$$\phi = \sqrt{\frac{2M + \varepsilon}{D(r)}}\, \varphi \,, \qquad (6.1)$$

where

$$D(r) = 2M + S(r) - V(r) + \varepsilon \,, \qquad (6.2)$$

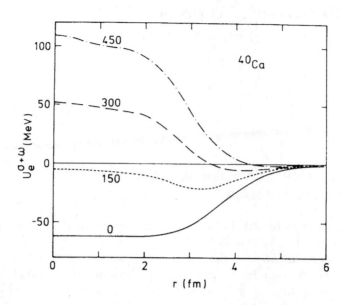

Fig. 6.1. Potential $U_e(r;\varepsilon)$ for ^{40}Ca for $\varepsilon = 0, 150, 300,$ and 450 MeV. (From [206])

The function ϕ will satisfy a local equation of Schrödinger type with a potential depending on energy [206, 212],

$$\left(\frac{p^2}{2M} + U_e(r;\varepsilon) + U_{s0}(r;\varepsilon)\frac{\boldsymbol{\sigma}\cdot\boldsymbol{l}}{r}\right)\phi = \frac{k^2}{2M}\phi\,, \qquad (6.3)$$

where k is the relativistic asymptotic momentum defined by

$$k^2 = 2M\varepsilon + \varepsilon^2\,. \qquad (6.4)$$

The potentials U_e and U_{s0} can be written as

$$U_e(r;\varepsilon) = S(r) + V(r) + (2M)^{-1}\left[S^2(r) - V^2(r)\right] + \frac{\varepsilon}{M}V(r)\,, \qquad (6.5)$$

$$U_{s0}(r;\varepsilon) = -[2MD(r)]^{-1}\frac{\mathrm{d}}{\mathrm{d}r}[D(r)]\,. \qquad (6.6)$$

Note that we have not included some small corrections to the potential $U_e(r;\varepsilon)$ in (6.3) since consideration of these corrections does not influence further discussion (these corrections have been taken into account in [206, 212, 213]).

Our further discussion in this section will be related to the case of the nucleus ^{40}Ca. In Fig. 6.1 we present the dependence of $U_e(r;\varepsilon)$ for $\varepsilon = 0$, 150, 300, and 450 MeV for this nucleus. From Fig. 6.1 it can be seen that $U_e(0;\varepsilon) \approx -60$ MeV for $\varepsilon = 0$; a corresponding calculation [206] shows that

6.1 Energy Dependence of the Real Part of the Optical Potential

at $\varepsilon \sim 160$ MeV the potential $U_e(0;\varepsilon)$ changes its sign and becomes positive at higher energies. It is known that the depth of the potential in the optical model is not determined very reliably; the differential cross sections are, to a greater extent, sensitive to the value of the volume integral (per nucleon)

$$J_{U_e}/A = 4\pi A^{-1} \int U_e(r;\varepsilon) r^2 \, dr \ . \tag{6.7}$$

As is shown in [206], in the Walecka model the value of J_{U_e}/A for the nucleus ^{40}Ca changes its sign at a nucleon energy equal to 270 MeV. This energy is much bigger than that at which the value of $U_e(0;\varepsilon)$ changes its sign. This means that not only the value but also the shape of the potential $U_e(r;\varepsilon)$ in the relativistic framework depends on the incident nucleon energy. So there exists a range of energies at which $U_e(0;\varepsilon)$ is positive, while the volume integral J_{U_e} of the potential $U_e(r;\varepsilon)$ remains negative: the Schrödinger equivalent potential has an attractive "pocket" at the nuclear surface.

Let us explain the mechanism that generates this effect in the relativistic theory. For this purpose we rewrite the potential (6.5) in the following form [206]:

$$U_e(r;\varepsilon) = U_e(r;0) + \frac{\varepsilon}{M} V(r) \ , \tag{6.8}$$

where

$$U_e(r;0) = [S(r) + V(r)] \frac{\mathcal{M}(r)}{M} \tag{6.9}$$

and

$$2\mathcal{M}(r) = 2M + S(r) - V(r) \ . \tag{6.10}$$

From (6.8) it can be seen that the potential in the equivalent Schrödinger equation depends linearly on the energy ε, the slope being determined by the coefficient $V(r)/M$. The potential $S(r)$ is attractive and $V(r)$ is repulsive, so the depth of the potential U_e decreases as the energy increases. This behavior corresponds to the experimental results. As seen from (6.8), an important role in the energy dependence of the optical potential is played by the repulsive potential $V(r)$ [206].

In [206] (see p. 2035), the forms of the two functions $U_e(r;0)$ and $V(r)$ are compared. From this comparison it can be seen that the radius of the repulsive energy-dependent component $\varepsilon/M \, V(r)$ in the equation for $U_e(r;\varepsilon)$ is much smaller than that of the component $U_e(r;0)$. In [206], it is shown for the case of the nucleus ^{40}Ca that the forms of the potentials $S(r)$ and $V(r)$ do not differ. This means that the difference between the forms of the potentials $U_e(r;0)$ and $V(r)$ is determined by the factor $\mathcal{M}(r)/M$. Since the radius of the second term in (6.8) is much less than that of the first term, the potential $U_e(r;\varepsilon)$ acquires its specific "wine-bottle-bottom" shape as the nucleon energy ε increases.

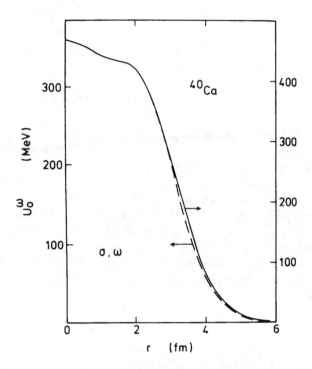

Fig. 6.2. Comparison of the forms of the potentials $V(r)$ (*dashed line, left scale*) and $S(r)$ (*continuous line, right scale*) for ^{40}Ca in the relativistic scattering theory [206]

6.2 Coulomb–Nuclear Interference Effects in Nucleon–Nucleus Scattering

In the case of protons, (6.3)–(6.5) contain a Coulomb potential $C(r)$ along with the repulsive vector field. Owing to the presence of a term of the form $[S^2 - V^2]$, the effective potential in (6.3) contains a term $-CV/M$, where M is the nucleon rest mass. Since the potential $Re\,V$ is about 300–400 MeV at the nuclear center, this Coulomb–nuclear interference (CNI) term brings about an appreciable attraction, which may lead to a sizable effect on properties such as the Coulomb energy of a proton bound in a potential well. This point was examined in [214], where it was found that the negative CNI potential is partly compensated by a strong energy dependence of the total effective potential, but that a lowering of the proton Coulomb energy still remains. However, a detailed comparison of Coulomb displacement energies in mirror nuclei calculated with relativistic and nonrelativistic self-consistent models shows that the differences between the nonrelativistic and relativistic results do not exceed 300–400 KeV [174]. To look for a more favorable case where the CNI effect would show up more clearly, it was suggested in [215]

6.2 Coulomb–Nuclear Interference Effects in Nucleon–Nucleus Scattering

that one should consider proton–nucleus and neutron–nucleus scattering at the same incident energy. Thus one would get rid of the energy dependence effect which is inherent in the Coulomb displacement energy problem. In a schematic case, it was found [215] that there was a large CNI effect on the low partial-wave phase shifts. This result was intriguing enough to motivate a more detailed study [216] and a reexamination of this question in a realistic situation, namely elastic nucleon–nucleus scattering at intermediate energies with realistic optical potentials.

If one considers an $N = Z$ target nucleus, the difference between the central potentials for protons and neutrons with the same incident energy ε is given in the relativistic theory by

$$U_{\mathrm{e}}^{(\mathrm{p})} - U_{\mathrm{e}}^{(\mathrm{n})} = C - \frac{CV}{M} + \Delta U_{\mathrm{cent}} \ . \tag{6.11}$$

In the nonrelativistic approach, the difference between the proton and neutron optical potentials at the same energy ε contains, besides the term C, a (complex) contribution ΔV_{C} called the Coulomb correction term [217]:

$$U_{\mathrm{opt}}^{(\mathrm{p})} - U_{\mathrm{opt}}^{(\mathrm{n})} = C + \Delta V_{\mathrm{C}} \ . \tag{6.12}$$

The term ΔV_{C} arises from the energy dependence of optical potentials and the fact that the Coulomb potential changes locally the kinetic energy of the incident proton. This Coulomb correction term would not exist if the nucleon optical potentials were energy-independent, while the CNI term that is discussed in [216] would still be present. Thus one must distinguish between the CNI potential $-CV/M$ of the relativistic approach and the Coulomb correction term ΔV_{C} of nonrelativistic optical potentials even though their volume integrals are found to be of comparable magnitude in some situations [218].

The question is whether effects due to the CNI term $-(CV)/M$ can be compensated by other effects linked to the term ΔU_{cent}. If this CNI effect really exists, it could be detected by performing proton and neutron elastic scattering at the same energy ε on an $N = Z$ nucleus and analyzing the measured angular distributions and spin observables by using the relativistic and nonrelativistic optical potentials. This should be preferably done in the intermediate energy range where both relativistic and nonrelativistic approaches are valid. Although such data exist for protons, the experimental situation for neutrons is far less well known and does not allow such a comparison (see e.g. [219–222]).

For this reason, in [216] a procedure was developed in which realistic pseudo-data were produced by the Dirac phenomenology and then analyzed with a nonrelativistic optical model. As a test case the target ^{40}Ca and the incident energy $\varepsilon = 160\,\mathrm{MeV}$ were chosen in [216]. Hama et al. [223] have given comprehensive parameterizations of Dirac optical potentials fitted to elastic scattering of protons at energies $65\,\mathrm{MeV} \leq \varepsilon \leq 1040\,\mathrm{MeV}$ off target nuclei in the range $40 \leq A \leq 208$. These parameterizations reproduce well the measured differential cross sections $d\sigma/d\Omega$, polarizations, and spin rotation parameters (see the next section for definitions). Using the parameter set 1 of [223] for the Dirac optical potentials S and V, the proton differential cross sections and spin observables were calculated by solving (6.3) (with the Coulomb potential and small missing corrections included). The calculated proton results were taken as pseudo-data points (see [223] for details). With the same potentials S and V, the corresponding neutron observables were also calculated and taken as pseudo-data points. Next, a nonrelativistic optical potential $U_{\rm opt}^{(p)}$ was found by a least-squares search added to a standard optical-model routine. The results for this $U_{\rm opt}^{(p)}$ reproduce perfectly well the proton pseudo-data. Then, in the spirit of the assumption that the Coulomb correction term $\Delta V_{\rm C}$ can be dropped, the neutron optical potential is uniquely determined to be $U_{\rm opt}^{(n)} = U_{\rm opt}^{(p)} - C$. This is consistent with the underlying assumption that both the relativistic optical potentials (S,V) and the nonrelativistic ones $(U_{\rm opt}^{(n)}, U_{\rm opt}^{(p)})$ are calculated at the same incident energy. This choice was made to single out the effects of $-(CV)/M$, if any. Further, the neutron observables were calculated with $U_{\rm opt}^{(n)}$. The agreement with the neutron pseudo-data is practically at the same level as that for protons, except for some deviations in the fit of the analyzing power A_y (which is equivalent to polarization) at large angles which are slightly more pronounced for neutrons. So these results demonstrate that the CNI term has no visible effect. There are two main reasons for this. Firstly, the potential $\Delta U_{\rm cent}$ defined by (6.11) contains several terms but the dominant one is $C\varepsilon/M$. For $\varepsilon \simeq 160$ MeV this term cancels about one-half of $\text{Re}(-CV/M)$ in the nuclear interior but, more importantly, this cancellation becomes more or less complete in the surface region, where $\text{Re}(V/M)$ approaches ε/M. Secondly, the absorptive potential has the effect that the inner region becomes less important since scattering occurs largely around the surface region where the above cancellation works best and where the value of CV/M is smaller anyway.

6.3 Relativistic Impulse Approximation

The relativistic impulse approximation was developed in [224–226] just after the first experimental data on elastic scattering of polarized protons (of energy 497 MeV) by nuclei were obtained [227, 228] (see also [229]). Analysis of these experimental data demonstrated that the impulse approximation in the theory of multiple scattering [205] describes fairly well the differential cross sections but leads to strong disagreement with experiment in reproducing the spin observables (the analyzing power A_y and spin rotation function Q). In [230, 231] it is shown that the situation changes drastically if one uses the relativistic version of the impulse approximation.

In this case the total cross section is calculated via the solution of the following Dirac equation:

$$[\boldsymbol{\alpha}\cdot\boldsymbol{p} + V(r) + \beta(M + S(r)) - 2\mathrm{i}\boldsymbol{\gamma}\boldsymbol{T}(r)]\psi = E\psi, \tag{6.13}$$

where the potentials are determined by folding the nuclear densities and the invariant NN scattering operator F,

$$F = \sum_{i=1}^{5} F_i(s,t,u)k_i. \tag{6.14}$$

Here, $k_1 = 1$ (scalar), $k_2 = \gamma_1^\mu \gamma_{2\mu}$ (vector), $k_3 = \sigma_1^{\mu\nu}\sigma_{2\mu\nu}$ (tensor), $k_4 = \gamma_1^5 \gamma_2^5$ (pseudoscalar), and $k_5 = \gamma_1^5 \cdot \gamma_2^5 \cdot \gamma_1^\mu \cdot \gamma_{2\mu}$ (axial vector), so one has

$$F = F_\mathrm{S} + \gamma_1^\mu \gamma_{2\mu} F_\mathrm{V} + \sigma_1^{\mu\nu}\sigma_{2\mu\nu} F_\mathrm{T} + \gamma_1^5 \gamma_2^5 F_\mathrm{PS} + \gamma_1^5 \gamma_1^\mu \gamma_2^5 \gamma_{2\mu} F_\mathrm{PV}. \tag{6.15}$$

The arguments s, t, and u are the Mandelstam invariants and they may be expressed in terms of the proton momentum $\boldsymbol{p}$ and the momentum transfer $\boldsymbol{q}$ for on-mass-shell kinematics. It is found empirically that the amplitudes F_S, F_V, and F_T are much larger than any amplitudes obtained in a nonrelativistic decomposition using Galilean-invariant operators [77]. If a polarized nucleon is scattered by a nonpolarized spin-zero target nucleus, the contribution to the potential is determined only by the scalar, vector, and tensor terms:

$$S(r) = -\frac{4\pi \mathrm{i} p}{M} F_\mathrm{S} \otimes \bar{\psi}\psi, \tag{6.16}$$

$$V(r) = -\frac{4\pi \mathrm{i} p}{M} F_\mathrm{V} \otimes \bar{\psi}\beta\psi, \tag{6.17}$$

$$\boldsymbol{T}(r) = -\frac{4\pi \mathrm{i} p}{M} F_\mathrm{T} \otimes \mathrm{i}\langle \bar{\psi}\boldsymbol{\alpha}\psi \rangle, \tag{6.18}$$

where the symbol $\otimes$ stands for the operation of folding, while $-(4\pi \mathrm{i} p)/M$ is a kinematic factor used to relate the conventional scattering amplitudes to Lorentz-invariant Feynman amplitudes in a frame where the nucleon has a momentum $\boldsymbol{p}$. The scattering amplitude for the nonpolarized nucleus has the following form:

$$F(\theta) = f(\theta) + \mathrm{i}\sigma_\mathrm{n} g(\theta), \tag{6.19}$$

where σ_n is the projection of the nucleon spin on the unit vector normal to the scattering plane. The functions f and g determine three values observed in the experiment [216], namely the total cross section

$$\sigma(\theta) = |f(\theta)|^2 + |g(\theta)|^2, \tag{6.20}$$

the analyzing power

$$A_y(\theta) = \frac{2}{\sigma(\theta)} \mathrm{Re}\left[f(\theta)(\mathrm{i}g(\theta))^*\right], \tag{6.21}$$

and the spin-rotation function

$$Q(\theta) = \frac{2}{\sigma(\theta)} \operatorname{Im} \left[f(\theta) \left(\mathrm{i} g(\theta) \right)^* \right] . \qquad (6.22)$$

In [230, 231], these values are calculated for the scattering of polarized protons of energy 497 MeV from the nuclei ^{40}Ca and ^{208}Pb. The potentials entering (6.13) are calculated in [232] using the results of the phase analysis carried out in [233]. We do not present figures comparing the relativistic calculations with the results obtained in the framework of the Kerman–McManus–Thaler theory: one can find such figures elsewhere [230]. We mention here only that the use of the relativistic impulse approximation leads to a "dramatic" improvement in the agreement between theory and experiment for spin observables. For the most recent results on this subject, see [234].

6.4 p̄–Nucleus Scattering

The relativistic impulse approximation can be applied also to elastic p̄–nucleus scattering; see, for example, [235]. In that paper, the authors consider the scattering of antiprotons at an energy of 46.8 MeV from ^{12}C utilizing, invariant N̄N amplitudes taken from [236] and nuclear densities obtained in a framework of a relativistic shell model [237]. Good agreement with experiment [238] is obtained.

Various aspects of the p̄–nucleus interaction have been considered in [61, 239]. In those papers the real part of the p̄–nucleus potential is worked out on the basis of an NN potential of the OBEP type using the fact that NN and NN̄ interactions are related by the G-parity rule. Let the NN potential be given by [34]

$$V_{\mathrm{NN}} = \sum_i V_i , \qquad (6.23)$$

where V_i is a t-channel meson exchange contribution. In this case the NN̄ potential can be obtained by the following formula [240, 241]:

$$V_{\mathrm{N\bar{N}}} = \sum_i G_i V_i , \qquad (6.24)$$

where G_i stands for the G-parity of the exchanged meson. If a system i consists of n_i pions, the G-parity is given by

$$G_i = (-)^{n_i} . \qquad (6.25)$$

This transformation can be used in a quite general form [7], attributing G parity to every meson (for a system of mesons). The conventional OBE scheme involves π, η, ω, ρ, σ, and a_0. In accordance with the G-parity rule, the pseudoscalar–isovector meson (π), the vector–isoscalar meson (ω), and the scalar–isovector meson (a_0) change the sign of this interaction in the NN̄ channel, while the signs of their interaction via the exchange of η, σ, and

ρ mesons remain unchanged. This rule leads to a single-particle central potential in the p̄–nucleus system that is much deeper [−(500–800) MeV] than in the nucleon–nucleus case [61]. This result was obtained in [61] using several different OBE models and the G-parity transformation. The large value of the central p̄–nucleus potential is connected with the fact that the two strong fields S and V add coherently in the case of antiprotons and produce a very strong central p̄–nucleus interaction, while the spin–orbit force becomes strongly reduced in the p̄–nucleus interaction in comparison with the nucleon–nucleus case (in [61] the contribution of Galilean-noninvariant components of the two-body spin–orbit interaction is also discussed and calculated for the NN̄ -system).

In [239–242], potential resonances in antiproton–nucleus scattering are studied, calculations being carried out for ^{12}C, ^{16}O, ^{40}Ca and ^{208}Pb. A resonance structure in the strength function for large-angle elastic scattering is obtained. The effect is highly sensitive to the parameters chosen for the potential. The resonance cross section falls off with increasing A. The basic reason for the appearance of resonances is the strong attraction in the real part of the optical potential, which results in the capture of the antiprotons and their orbiting around the nucleus, followed by a decay of the resonant system. From the calculations, it follows that the resonance effect is intensified if $R_V > R_W$ or $a_V > a_W$, where R and a stand for the radius and diffuseness of the optical potential, and the subscripts V and W refer to the real and imaginary parts, respectively. Under these conditions, an antiproton "pocket" of attraction appears [242].[1] At the same time, an increase of the depth of the imaginary part of the optical potential from 100 to 200 MeV reduces the cross section for backscattering by about an order of magnitude, although the resonance pattern is retained. As pointed out in [242], the elimination of the resonances would require $R_V < R_W$, $a_V < a_W$, and very strong absorption. Interesting aspects of the p̄–nucleus interaction are discussed also in [243, 244].

In [243], a relativistic Hartree approach is developed which describes the bound states of nucleons and antinucleons consistently (see also Ap-

---

[1] Because of the annihilation interaction of antiprotons with the nucleons of the nucleus, the imaginary part of the optical potential, $W(r)$, may also have a large depth. At present, $W(0)$ has not been determined reliably; a study of the widths of antiprotonic atoms yields values in the range 100–200 MeV. Since the radius of the annihilation interaction is short (of the order of $1/M$), it may be expected that the radius R_W and the diffuseness a_W of the imaginary part of the optical potential will be the same as for the nuclear density distribution.

Since the radius of the NN interaction in the meson theory of the nuclear forces is determined by the Compton wavelength of the exchange mesons, the shape parameters of the real part of the optical potential (the radius R and the diffuseness a) differ from the corresponding parameters of the nucleon distribution. In particular, we have $R_V = R + \Delta$, where the average value $\Delta = 0.17$ fm was used in [239] for all nuclei.

pendix A5). In this reference the authors make an attempt to go beyond the no-sea approximation (see also [88]) .

Closing this chapter we conclude that the main features of the approach described above are the relativistic invariance of the equations of motion and the presence of two strong fields: an attractive field $S(r)$ with the transformation properties of a world scalar, and a repulsive field $V(r)$ transforming like a four-vector. The vector field is related to the ω meson, $V(r)$ being the time component of the four-vector field ω^μ. However, in the experimental data, there is no scalar meson with a mass ~ 500 MeV, but such a meson is necessary for successful operation of the approach described above. At the present stage one may treat the attractive scalar field $S(r)$ as a convenient tool for simulating the 2π contribution. The presence of these two fields in the nucleus can now be considered as a firmly established fact, the isovector component of the nuclear potential being related to the ρ meson. Such a relativistic framework appears to be very fruitful even at the mean-field level (in relation to the saturation effect, spin–orbit interaction, energy dependence of the real part of the optical potential, "dramatic" improvement of the agreement between theory and experiment for spin observables, etc.)

7 Pion Dynamics and Chiral Symmetry

7.1 Pionic Excitations in Nuclear Matter

In traditional nonrelativistic nuclear physics the nucleus is considered to consist of point-like nucleons interacting via NN forces which can be determined from NN scattering. About 30 years ago or so, however, other degrees of freedom were introduced into nuclear physics [245]. The sensational one of these was pion condensation (a transition to a state with a nonzero amplitude of the pion field), in which the energy of a pion becomes zero in nuclear matter owing to the interaction of the pion with nuclear medium, which also includes the delta isobar [246–249].

At present there is no experimental evidence for pion condensation at normal nuclear densities. In the nonrelativistic approach [246–249], correlation effects such as the repulsive spin–isospin interaction are the cause of the stability against pion condensation. However, in the nonrelativistic framework spin–isospin correlations are introduced phenomenologically, i.e. the stability problem encounters another problem concerning the nature of this interaction. The relativistic solution to the problem of the stability of nuclear matter [39, 250–252] contains several specific features connected purely with the relativistic treatment of the problem.

It is well known that the RPA method corresponds to small-amplitude oscillations of the self-consistent field when the influence of these oscillations on the single-particle states can be taken into account by perturbation theory (linear response). Clearly the stability problem for the ground state is meaningful for small oscillations only, since large-amplitude vibrations must surely affect the structure of the ground state. Therefore we can confine ourselves to the RPA when considering the stability problem. Note that such a situation occurs in pionic atoms when the Bohr radius of a pion orbit substantially exceeds the nuclear size. This is why the nuclear interaction leads only to relatively small corrections to the Coulomb interaction in this case. Thus we see that the existence of a small extension parameter in the pion–nucleus interaction has no analogy in the interaction between free particles.

Let us write down the model Lagrangian including the pion, nucleon, and Δ-isobar fields [39, 253], as well as the mean nuclear fields of the ω and σ mesons (only isoscalar fields are present in the static ground state owing to isotopic symmetry):

7 Pion Dynamics and Chiral Symmetry

$$\mathcal{L} = \frac{1}{2}\partial_\mu \boldsymbol{\pi} \cdot \partial^\mu \boldsymbol{\pi} - \frac{1}{2}m_\pi^2 \boldsymbol{\pi}^2 + \frac{x}{1+x} \mathrm{i} g_\pi \bar{\psi}\gamma_5 \boldsymbol{\tau} \cdot \boldsymbol{\pi}\psi$$

$$+ g_\pi \frac{1}{2M(1+x)} \bar{\psi}\gamma_5\gamma^\mu \boldsymbol{\tau}\partial_\mu \boldsymbol{\pi}\psi + \bar{\psi}\gamma^\mu(\mathrm{i}\partial_\mu - \omega_\mu)\psi - \frac{\mu_0}{2M}\bar{\psi}\sigma^{\mu\nu}\omega_{\mu\nu}\psi$$

$$- (M+S)\bar{\psi}\psi + \frac{g_{\pi N\Delta}}{2M}(\bar{\Delta}_\mu T^\dagger \widehat{Q}^{\mu\nu}\psi + \bar{\psi}T\widehat{Q}^{\nu\mu}\Delta_\mu)\partial_\nu \boldsymbol{\pi}$$

$$+ \bar{\Delta}^\mu \left[\left(-\gamma_\beta P_\Delta^\beta + M_\Delta^*\right)g_{\mu\nu} + P_\mu^\Delta \gamma_\nu + P_\nu^\Delta \gamma_\mu \right.$$

$$\left. - \gamma_\mu \cdot \gamma_\beta \cdot P_\Delta^\beta \cdot \gamma_\nu - M_\Delta^* \gamma_\mu \gamma_\nu\right] \cdot \Delta^\nu , \tag{7.1}$$

$$\sigma^{\mu\nu} = \frac{\mathrm{i}}{2}[\gamma^\mu, \gamma^\nu] , \qquad P_\Delta^\mu = \mathrm{i}\partial^\mu - \omega^\mu , \tag{7.2}$$

$$\widehat{Q}^{\mu\nu} = g^{\mu\nu} - \left(Z + \frac{1}{2}\right)\gamma^\mu \gamma^\nu , \qquad M_\Delta^* = M_\Delta + S . \tag{7.3}$$

Here, the bold variables denote vectors in isospin space; ψ, Δ^μ, and $\boldsymbol{\pi}$ are the amplitudes of the nucleon, Δ_{33}-isobar, and pion fields, respectively; and $(1/2)\boldsymbol{\tau}$, $\boldsymbol{t}_\Delta$, and $\boldsymbol{T}$ are the isospin operators for the nucleon, the Δ-isobar and the transition $(3/2 \to 1/2)$ in the isospin space:

$$\left\langle \frac{1}{2}t \middle| T_{-k} \middle| \frac{3}{2}t_\Delta \right\rangle = \left\langle \frac{3}{2}t_\Delta \middle| T_k^\dagger \middle| \frac{1}{2}t \right\rangle = C\left(\frac{1}{2}1\frac{3}{2}; tkt_\Delta\right). \tag{7.4}$$

The part of the Lagrangian related to the free field of the Δ-isobar is written in accordance with the formalism developed by Johnson and Sudarshan [254], while the interaction of the Δ-isobar and the pion field ensures coupling only to the component with spin 3/2 [255–270]. The parameter Z is arbitrary, however, the results obtained below do not depend on this value. μ_0 is the isoscalar anomalous nucleon magnetic moment. In (7.1) it is supposed that the average nuclear field modifies the Δ-isobar in the same way as a nucleon.

The pion–nucleon interaction may be introduced in a pseudovector or a pseudoscalar form [83, 269, 270]. In the nonrelativistic framework these two types of interaction are identical. However, in the relativistic approach they are different. Our starting Lagrangian (7.1) allows mixed couplings for the pion [39].[1] The parameter x in (7.1) is introduced in such a way that the

---

[1] Such a possibility was considered also in [39, 271] in a totally different context of low-energy NN scattering and in a relativistic calculation of the triton binding energy. The chiral linear model, Fubini terms included, also corresponds to the case where mixing of ps and pv interactions is adopted (see below).

We should mention also some papers [272] where ps–pv mixing is used in calculations for pionic atoms. It is possible to modify the nucleon field according to the unitary Weinberg transformation [273]. By this transformation (see Appendix A8) the ps interaction is converted into a pv coupling. In [274], a generalized Weinberg transformation is introduced. This depends on a free parameter x (mixing parameter) which regulates what portion of the ps interaction is converted into a pv one.

vacuum Lagrangians corresponding to different values of this parameter are equivalent (the two types of πNN coupling being identical in this case). However, in the nuclear medium the value of x becomes important.

Two possibilities for the choice of the constants g_π and $g_{\pi N\Delta}$ may be considered:

$$\text{I}: \begin{cases} \dfrac{g_\pi^2}{4\pi} = 14 \\ \left(\dfrac{g_{\pi N\Delta}}{g_\pi}\right)^2 = 3 \end{cases} ; \quad \text{II}: \begin{cases} \dfrac{g_\pi^2}{4\pi} = 14.5 \\ \left(\dfrac{g_{\pi N\Delta}}{g_\pi}\right)^2 = 3.376 \end{cases} . \quad (7.5)$$

Both possibilities are often used in the literature. The value of the ratio $(g_{\pi N\Delta}/g_\pi)^2$ in the first case corresponds to the quark model [83, 275, 276]. In the second case this value is derived from the probability of the process $\Delta \to N\pi$ in the vacuum [277].

7.2 Equations of Motion for a Pion Field in a Nuclear Medium

Let us write out the Euler–Lagrange equations for the pion field using the Lagrangian (7.1) [39, 253]:

$$\partial^\mu \partial_\mu \boldsymbol{\pi} + m_\pi^2 \boldsymbol{\pi} = \frac{x}{1+x} \mathrm{i} g_\pi \langle \bar{\psi}\gamma_5 \boldsymbol{\tau}\psi \rangle_\mathrm{g} - \frac{g_\pi}{2M(1+x)} \partial_\mu \langle \bar{\psi}\gamma_5 \gamma^\mu \boldsymbol{\tau}\psi \rangle_\mathrm{g}$$

$$- \frac{g_{\pi N\Delta}}{2M} \partial_\mu \left(\langle \underline{\bar{\Delta}}^\mu \boldsymbol{T}^\dagger \psi \rangle_\mathrm{g} + \langle \bar{\psi} \boldsymbol{T} \underline{\Delta}^\mu \rangle_\mathrm{g} \right) , \quad (7.6)$$

where $\underline{\Delta}^\mu = \widehat{Q}^{\mu\nu}\Delta_\nu$, $\underline{\bar{\Delta}}^\mu = \bar{\Delta}_\nu Q^{\nu\mu}$, and $\langle \ldots \rangle_\mathrm{g}$ stands for the average value over the nuclear ground state, perturbed by the pion field entering the right-hand side of (7.6). Notice that if the influence of the pion field on the nuclear ground state is not taken into account, the right-hand side of (7.6) equals zero since the pseudoscalar density, the pseudovector current, and the Δ-isobar field are not present in the ground state.

Let us consider the time dependence of the pion field amplitude,

$$\boldsymbol{\pi}^{t_3} = \pi^{t_3}(\boldsymbol{r})\mathrm{e}^{-\mathrm{i}\omega t}\boldsymbol{e}_{t_3} + \pi^{t_3 *}(\boldsymbol{r})\mathrm{e}^{\mathrm{i}\omega t}\boldsymbol{e}_{-t_3} , \quad (7.7)$$

where t_3 is the isospin projection of the pion, $\boldsymbol{e}_\pm = (1/\sqrt{2})(\boldsymbol{e}_x \pm \mathrm{i}\boldsymbol{e}_y)$, $\boldsymbol{e}_0 = \boldsymbol{e}_z$, $\pi^\mp = (1/\sqrt{2})(\pi_x \pm \mathrm{i}\pi_y)$, and $\pi^0 = \pi_z$. Substituting the amplitude (7.4) into (7.6), we find

$$\left[\omega^2 + \nabla^2 - m_\pi^2\right] \pi^{t_3}(\boldsymbol{r}) = -\mathrm{i}\frac{x}{1+x} g_\pi \langle \bar{\psi}\gamma_5 \tau_{-t_3}\psi \rangle_{\mathrm{g}+}$$

$$+ g_\pi \frac{1}{2M(1+x)} \partial_\mu \langle \bar{\psi}\gamma_5 \gamma^\mu \tau_{-t_3}\psi \rangle_{\mathrm{g}+}$$

$$+ \frac{g_{\pi N\Delta}}{2M} \partial_\mu \left[\langle \underline{\bar{\Delta}}^\mu T_{-t_3}\psi \rangle_{\mathrm{g}+} + \langle \bar{\psi} T_{-t_3} \underline{\Delta}^\mu \rangle_{\mathrm{g}+} \right] , \quad (7.8)$$

where $\langle\ldots\rangle_{\mathrm{g}+}$ stands for the positive-frequency part of $\langle\ldots\rangle_{\mathrm{g}}$, $\tau_0 = \tau_z$, $\tau_{\pm 1} = (\tau_x \pm i\tau_y)/\sqrt{2}$, and similarly for T_{t_3}.

To calculate $\langle\ldots\rangle_{\mathrm{g}+}$, we use the fact that the amplitude of the pion field is small and that the linear-response approximation (LRA) can be utilized when considering the problem of pion condensation. In accordance with the LRA, we have

$$\langle \bar{\psi}\hat{O}\psi\rangle_{\mathrm{g}+} \simeq \langle\bar{\psi}_0 \cdot \hat{O} \cdot \delta\psi(\omega,t_3)\rangle_{\mathrm{g}} + \langle \delta\bar{\psi}(-\omega,-t_3)\,\hat{O}\psi_0\rangle_{\mathrm{g}}$$
$$= \langle\bar{\psi}_0 \hat{O}\,\delta\psi(\omega,t_3)\rangle_{\mathrm{g}} + \langle\bar{\psi}_0 \hat{O}\,\delta\psi(-\omega,-t_3)\rangle_{\mathrm{g}}^*, \quad (7.9)$$

$$\langle\bar{\psi}\hat{O}\underline{\Delta}^\mu\rangle_{\mathrm{g}+} \simeq \langle\bar{\psi}_0 \hat{O}\underline{\Delta}^\mu(\omega,t_3)\rangle_{\mathrm{g}}, \quad (7.10)$$

$$\langle\underline{\bar{\Delta}}^\mu \hat{O}\psi\rangle_{\mathrm{g}+} \simeq \langle\underline{\bar{\Delta}}^\mu(-\omega,-t_3)\hat{O}\psi_0\rangle_{\mathrm{g}} = \langle\bar{\psi}_0 \hat{O}\underline{\Delta}^\mu(-\omega,-t_3)\rangle_{\mathrm{g}}^*, \quad (7.11)$$

where $\hat{O}$ is an operator with the property $\hat{O}^\dagger = \beta\hat{O}\beta$; ψ_0 are the nonperturbed nucleon wave functions, $\delta\psi(\omega,t_3)$ and $\Delta^\mu(\omega,t_3)$ are, respectively, the variation of the nucleon amplitude and Δ-isobar amplitude induced by the pion field $\boldsymbol{\pi}(\omega,t_3) = \pi^{t_3}(\boldsymbol{r},t)\boldsymbol{e}_{t_3} = \pi^{t_3}(\boldsymbol{r})\mathrm{e}^{-i\omega t}\boldsymbol{e}_{t_3}$; and $\delta\psi(-\omega,-t_3)$ and $\Delta^\mu(-\omega,-t_3)$ are induced by the field $\boldsymbol{\pi}(\omega,t_3)^* = \pi^{t_3}(\boldsymbol{r})^*\mathrm{e}^{i\omega t}\boldsymbol{e}_{-t_3}$. In the LRA, these quantities are calculated in a linear pion field approximation. Taking into account (7.11), (7.7) takes the following form:

$$[\omega^2 + \nabla^2 - m_\pi^2]\pi^{t_3} = -\frac{x}{1+x}ig_\pi\langle\bar{\psi}_0\gamma_5\tau_{-t_3}\delta\psi(\omega,t_3)\rangle_{\mathrm{g}}$$
$$+ g_\pi \frac{1}{2M(1+x)}\partial_\mu\langle\bar{\psi}_0\gamma_5\gamma^\mu\tau_{-t_3}\,\delta\psi(\omega,t_3)\rangle_{\mathrm{g}}$$
$$+ \frac{g_{\pi N\Delta}}{2M}\partial_\mu\langle\bar{\psi}_0 T_{-t_3}\underline{\Delta}^\mu(\omega,t_3)\rangle + \mathrm{c.c.}\begin{pmatrix}\omega\to-\omega\\ t_3\to-t_3\end{pmatrix}, \quad (7.12)$$

where the symbol

$$\mathrm{c.c.}\begin{pmatrix}\omega\to-\omega\\ t_3\to-t_3\end{pmatrix}$$

means that the expression standing before this symbol is complex conjugated (the pion field amplitude excluded), the procedure being accompanied by the replacements $\omega \to -\omega$, $t_3 \to -t_3$.

Let us now calculate the changes of the nucleon fields induced by the pion field. Consider the equation for a nucleon in the presence of the pion field:

$$(\omega + E + M - h_{\mathrm{N}})\,\delta\psi(\omega,t_3)$$
$$= \left(-ig_\pi \frac{x}{1+x}\beta\gamma_5\boldsymbol{\tau}\cdot\boldsymbol{\pi}(\omega,t_3) - g_\pi\frac{\beta\gamma_5\gamma^\mu\boldsymbol{\tau}\partial_\mu\boldsymbol{\pi}(\omega,t_3)}{2M(1+x)}\right)\psi$$
$$- \frac{g_{\pi N\Delta}}{2M}\boldsymbol{T}\beta\underline{\Delta}^\mu\partial_\mu\boldsymbol{\pi}(\omega,t_3), \quad (7.13)$$

where E is the energy of the respective nucleon state calculated from the boundary of the continuous spectrum, and h_{N} is the nuclear Hamiltonian, including the nuclear mean fields:

7.2 Equations of Motion for a Pion Field in a Nuclear Medium

$$h_N = -i\boldsymbol{\alpha} \cdot \boldsymbol{\nabla} + V + \beta(M+S) \,. \tag{7.14}$$

In the LRA, for the induced change of the nucleon field we obtain

$$\delta\psi(\omega, t_3) = \widehat{G}(\omega + E)\left(-ig_\pi \frac{x}{1+x}\beta\gamma_5 \pi^{t_3} - g_\pi \frac{1}{2M(1+x)}\beta\gamma_5 \gamma^\mu \partial_\mu \pi^{t_3}\right)\tau_{t_3}\psi_0 \,, \tag{7.15}$$

where $\widehat{G}(\varepsilon)$ is the nucleon Green function

$$\widehat{G}(\varepsilon) = \frac{1}{\varepsilon + M - h_N} \,; \tag{7.16}$$

$$\widehat{G}(\varepsilon)\psi_\tau(\xi) = \int d\xi' \, G_\tau(\varepsilon; \xi, \xi')\psi_\tau(\xi') \,, \tag{7.17}$$

where ξ includes the space and spin variables, and τ is the nucleon isospin projection.

To calculate the amplitude of the induced field of the Δ-isobar, let us write the corresponding equation defined by the Lagrangian density (7.1),

$$D_\Delta^{\nu\mu} \Delta_\mu = \frac{g_{\pi N\Delta}}{2M} \boldsymbol{T}^\dagger \widehat{Q}^{\nu\mu} \psi \partial_\mu \boldsymbol{\pi}(\omega, t_3) \,, \tag{7.18}$$

where the operator $D_\Delta^{\nu\mu}$ has the form [259]

$$D_\Delta^{\nu\mu} = (\gamma \cdot P_\Delta - M_\Delta^*) g^{\nu\mu} - P_\Delta^\nu \gamma^\mu - P_\Delta^\mu \gamma^\nu + \gamma^\nu \gamma \cdot P_\Delta \gamma^\mu + M_\Delta^* \gamma^\nu \gamma^\mu \,. \tag{7.19}$$

In the LRA, we have

$$\Delta^\mu = \widehat{G}_\Delta^{\mu\nu}(\omega + E)\frac{g_{\pi N\Delta}}{2M} T_{t_3}^\dagger \widehat{Q}_{\nu\alpha} \psi_0 \partial^\alpha \pi^{t_3} \,, \tag{7.20}$$

where $\widehat{G}_\Delta^{\mu\nu}(\varepsilon)$ is the Δ-isobar Green function:

$$\widehat{G}_\Delta^{\mu\nu}(\varepsilon) = \frac{1}{\gamma \cdot P_\Delta - M_\Delta^*} \Lambda^{\mu\nu} \,, \qquad P_\Delta^0 = \varepsilon + M - V \,, \tag{7.21}$$

$$\widehat{G}_\Delta^{\mu\nu}(\varepsilon) \Phi_{\tau\nu}(\xi) = \int d\xi' \, G_\Delta^{\mu\nu}(\varepsilon, \xi, \xi') \Phi_{\nu\tau}(\xi') \,. \tag{7.22}$$

Here $\Lambda^{\mu\nu}$ is the operator given by [255]

$$\Lambda^{\mu\nu} = g^{\mu\nu} - \frac{1}{3}\gamma^\mu\gamma^\nu - \frac{2}{3}\frac{P_\Delta^\mu P_\Delta^\nu}{M_\Delta^{*2}} + \frac{P_\Delta^\mu \gamma^\nu - P_\Delta^\nu \gamma^\mu}{3M_\Delta^*} \tag{7.23}$$

(on the mass shell this operator is equivalent to the projection operator onto the space with spin 3/2).

Substituting (7.15) and (7.20) into (7.12), we obtain

$$[\omega^2 + \nabla^2 - m_\pi^2]\pi^{t_3}(\boldsymbol{r}, t) = \frac{x}{1+x} g_\pi^2 \sum_i \bar{\psi}_i(\xi)\hat{P}_{t_3}\gamma_5 \int d\xi' \, G_{\tau_i}(\omega + E_i; \xi, \xi')$$

$$\times \left\{ -\frac{x}{1+x}\beta\gamma_5\pi^{t_3}(\boldsymbol{r}', t) + \frac{i}{2M(1+x)}\beta\gamma_5\gamma^\nu \partial_\nu[\pi^{t_3}(\boldsymbol{r}', t)] \right\} \psi_i(\xi')$$

$$+ g_\pi^2 \frac{1}{2M(1+x)} i\partial_\mu \sum_i \bar{\psi}_i(\xi)\hat{P}_{t_3}\gamma_5\gamma^\mu \int d\xi' \, G_{\tau_i}(\omega + E_i; \xi, \xi')$$

$$\times \left\{ -\frac{x}{1+x}\beta\gamma_5\pi^{t_3}(\boldsymbol{r}', t) + \frac{i}{2M(1+x)}\beta\gamma_5\gamma^\nu \partial_\nu[\pi^{t_3}(\boldsymbol{r}', t)] \right\} \psi_i(\xi')$$

$$+ \frac{g_{\pi N\Delta}^2}{4M^2} \partial_\mu \sum_i \bar{\psi}_i(\xi)\hat{Q}_{t_3}\hat{Q}^{\mu\alpha} \int d\xi' \, G_{\alpha\beta}^\Delta(\omega + E_i; \xi, \xi')$$

$$\times \hat{Q}^{\beta\nu}\psi_i(\xi') \partial_\nu[\pi^{t_3}(\boldsymbol{r}', t)] + \text{c.c.} \begin{pmatrix} \omega \to -\omega \\ t_3 \to -t_3 \end{pmatrix}. \tag{7.24}$$

In (7.24) $\sum_i$ stands for the sum over occupied nucleon states, and the operators in isospin space $\hat{P}_{t_3}$ and $\hat{Q}_{t_3}$ have been introduced, where

$$\hat{P}_{t_3} = \tau_{-t_3}\tau_{t_3}, \qquad \hat{Q}_{t_3} = T_{-t_3}T_{t_3}^\dagger, \tag{7.25}$$

$$\hat{P}_\pm \begin{vmatrix} n \\ p \end{vmatrix} = 2 \begin{vmatrix} n \\ p \end{vmatrix}, \quad \hat{P}_\pm \begin{vmatrix} p \\ n \end{vmatrix} = 0, \quad \hat{P}_0 \begin{vmatrix} n \\ p \end{vmatrix} = \begin{vmatrix} n \\ p \end{vmatrix}, \tag{7.26}$$

$$\hat{Q}_\pm \begin{vmatrix} n \\ p \end{vmatrix} = C^2\left(\frac{1}{2}1\frac{3}{2}; \mp\frac{1}{2}, \pm1, \pm\frac{1}{2}\right) \begin{vmatrix} n \\ p \end{vmatrix} = \frac{1}{3}\begin{vmatrix} n \\ p \end{vmatrix}, \tag{7.27}$$

$$\hat{Q}_\pm \begin{vmatrix} p \\ n \end{vmatrix} = C^2\left(\frac{1}{2}1\frac{3}{2}; \pm\frac{1}{2}, \pm1, \pm\frac{3}{2}\right) \begin{vmatrix} p \\ n \end{vmatrix} = \begin{vmatrix} p \\ n \end{vmatrix}, \tag{7.28}$$

$$\hat{Q}_0 \begin{vmatrix} n \\ p \end{vmatrix} = C^2\left(\frac{1}{2}1\frac{3}{2}; \mp\frac{1}{2}, 0, \mp\frac{1}{2}\right) \begin{vmatrix} n \\ p \end{vmatrix} = \frac{2}{3}\begin{vmatrix} n \\ p \end{vmatrix}. \tag{7.29}$$

Introducing the polarization operator $P_{t_3}(\omega; \boldsymbol{r}, \boldsymbol{r}')$ for a pion in the nucleus, (7.25)–(7.29) can be written in the following form:

$$[\omega^2 + \nabla^2 - m_\pi^2]\pi^{t_3}(\boldsymbol{r}) = \int d^3r' \, P_{t_3}(\omega; \boldsymbol{r}, \boldsymbol{r}')\pi^{t_3}(\boldsymbol{r}'). \tag{7.30}$$

Equation (7.30) for a pion in the nuclear medium is an integrodifferential equation. To obtain its solution is a difficult technical problem. This problem can be solved, however, for the particular case of infinite isotopically symmetric nuclear matter, to investigate its stability against pion condensation.

7.3 Pionic Polarization in Infinite Nuclear Matter

Since we are considering a pion in an the isotopically symmetric nuclear medium, the pion polarization operator does not depend on its charge, and

7.3 Pionic Polarization in Infinite Nuclear Matter

for definiteness we examine the case of a neutral pion. Let us write (7.24) for this case:

$$[q_0^2 - \boldsymbol{q}^2 - m_\pi^2]\pi^0(\boldsymbol{r},t) = (\Pi_N(q) + \Pi_\Delta(q))\,\pi^0(\boldsymbol{r},t)\,, \tag{7.31}$$

where $q = (q_0, \boldsymbol{q})$ is the pion four-momentum, and $\Pi_N(q)$ and $\Pi_\Delta(q)$ are the contributions of nucleon–hole and Δ-isobar–hole excitations to the polarization operator. The values of $\Pi_N(q)$ and $\Pi_\Delta(q)$ are given by

$$\Pi_N(q) = \frac{g^2}{(1+x)^2} \sum_i \psi_i^\dagger(\xi)\, \beta\gamma_5 \left(x + \frac{1}{2M}\gamma q\right)$$
$$\times \int d\xi'\, G(q_0 + E_i; \xi, \xi')\beta\gamma_5 \left(-x + \frac{1}{2M}\gamma q\right)\psi_i(\xi')$$
$$+ \text{c.c.}(q_0 \to -q_0)\,, \tag{7.32}$$

$$\Pi_\Delta(q) = \frac{-g_{\pi N\Delta}^2}{6M^2} q^\mu \sum_i \psi_i^\dagger(\xi)\,\beta$$
$$\times \int d\xi'\, \widehat{Q}_{\mu\alpha} G_\Delta^{\alpha\beta}(q_0 + E_i; \xi, \xi')\,\widehat{Q}_{\beta\nu}\psi_i(\xi')\, q^\nu + \text{c.c.}(q_0 \to -q_0)\,. \tag{7.33}$$

First, we calculate the nucleon part $\Pi_N(q)$ of the polarization integral $\Pi_N(q)$. It is convenient to represent this value in the following form:

$$\Pi_N(q) = 2\frac{g_\pi^2}{(1+x)^2} \int_{|\boldsymbol{p}|\leq p_F} \frac{d^3p}{(2\pi)^3} Sp\left[\beta\gamma_5\left(x+\frac{1}{2M}\gamma\cdot q\right)\right.$$
$$\times \frac{1}{p_0 + q_0 - \boldsymbol{\alpha}\cdot(\boldsymbol{p}+\boldsymbol{q}) - \beta M^*}\beta\gamma_5\left(-x+\frac{1}{2M}\gamma\cdot q\right)$$
$$\left.\times \frac{p_0 + \boldsymbol{\alpha}\cdot\boldsymbol{p} + \beta M^*}{2p_0}\right] + (q_0 \to -q_0)\,, \tag{7.34}$$

where use has been made of the nucleon Green function in momentum space,

$$\widehat{G}(p) = \frac{1}{p_0 - \boldsymbol{\alpha}\cdot\boldsymbol{p} - \beta M^*}\,, \tag{7.35}$$

and the projection operator onto the states with positive energy,

$$\widehat{P} = \frac{p_0 + \boldsymbol{\alpha}\cdot\boldsymbol{p} + \beta M^*}{2p_0}\,. \tag{7.36}$$

In (7.34)–(7.36), $M^* = M + S$ is the effective nucleon mass for an infinite nuclear system, and $p_0 = e_{\boldsymbol{p}} - V = (\boldsymbol{p}^2 + M^{*2})^{1/2}$, where $e_{\boldsymbol{p}}$ is the eigenvalue of the nucleon wave function.

Equation (7.34) can be rewritten easily in a relativistically covariant form, using for this purpose the properties of the Dirac matrices and those of the trace, as well as the fact that the nucleons are on shell:

$$\Pi_N(q) = \frac{g_\pi^2}{(1+x)^2} \int_{|\boldsymbol{p}|\leq p_F} \frac{d^3p}{(2\pi)^3} \frac{1}{p_0 q(q+2p)}$$
$$\times Sp\left\{ \left(x - \frac{\gamma q}{2M}\right)[-(p+q)\gamma + M^*] \right.$$
$$\left. \left(-x + \frac{\gamma q}{2M}\right)(p\cdot\gamma + M^*) \right\} + (q_0 \to -q_0). \qquad (7.37)$$

For the trace in (7.37), we obtain

$$Sp\left\{ \left(x - \frac{\gamma q}{2M}\right)[-(p+q)\gamma + M^*]\left(-x + \frac{\gamma q}{2M}\right)(p\cdot\gamma + M^*) \right\}$$
$$= \left(2x^2 + \frac{p\cdot q}{M^2}\right) q\cdot(q+2M) - 2q^2\left(x + \frac{M^*}{M}\right)^2. \qquad (7.38)$$

Substituting this equation into (7.37), we obtain:

$$\Pi_N(q) = \frac{4g_\pi^2 x^2}{(1+x)^2} \int_{|\boldsymbol{p}|\leq p_F} \frac{d^3p}{(2\pi)^3 p_0} - \frac{2g_\pi^2}{(1+x)^2}\left(x + \frac{M^*}{M}\right)^2$$
$$\times q^2 \int_{|\boldsymbol{p}|\leq p_F} \frac{d^3p}{(2\pi)^3}\frac{1}{p_0}\left[\frac{1}{q(q+2p)} + \frac{1}{q(q-2p)}\right]. \qquad (7.39)$$

In the second term of the second integral, the replacement $\boldsymbol{p} \to -\boldsymbol{p}$ has been carried out.

The first integral can easily be expressed in terms of a scalar nucleon density

$$\rho_S = \sum_i \bar{\psi}_i \psi_i = 4M^* \int_{|\boldsymbol{p}|\leq p_F} \frac{d^3p}{(2\pi)^3 p_0}. \qquad (7.40)$$

Finally, we obtain the following equation for the polarization operator:

$$\Pi_N(q) = \frac{g_\pi^2 x^2}{(1+x)^2 M^*}\rho_S - \frac{2g_\pi^2}{(1+x)^2}\left(x + \frac{M^*}{M}\right)^2$$
$$\times q^2 \int_{|\boldsymbol{p}|\leq p_F} \frac{d^3p}{(2\pi)^3 p_0}\left[\frac{1}{q(q+2p)} + \frac{1}{q(q-2p)}\right], \qquad (7.41)$$

or

7.4 Contribution of the Δ_{33} Resonance to the Pion Polarization Operator

$$\Pi_N(q) = \frac{g_\pi^2 x^2}{(1+x)^2 M^*} \rho_S + \frac{g_\pi^2}{(1+x)^2}\left(x + \frac{M^*}{M}\right)^2$$

$$\times \frac{q_0^2 - q^2}{q} \frac{1}{(2\pi)^3} \int_0^{P_F} \frac{p\,dp}{p_0} \ln\left|\frac{(q_0^2 - q^2 - 2pq)^2 - 4p_0^2 q_0^2}{(q_0^2 - q^2 + 2pq)^2 - 4p_0^2 q_0^2}\right|, \quad (7.42)$$

where the three-dimensional notations $q^2 = \mathbf{q}^2$ and $pq = |\mathbf{p}||\mathbf{q}|$ have been introduced in (7.42).

7.4 Contribution of the Δ_{33} Resonance to the Pion Polarization Operator

Now we calculate the Δ-isobar part of the polarization operator determined by (7.33). To write out the Δ-isobar Green function, we consider the Lagrangian for a free Δ-isobar, following [254]:

$$\mathcal{L}_\Delta = \bar{\Delta}^\mu\Big[\big(-\gamma p_\Delta + M_\Delta\big)g_{\mu\nu} - A\big(p_\nu^\Delta \gamma_\mu + p_\mu^\Delta \gamma_\nu\big)$$
$$+ K\gamma_\mu \gamma p_\Delta \gamma_\nu - T M_\Delta \gamma_\mu \gamma_\nu\Big]\Delta^\nu, \quad (7.43)$$

where

$$K = -\frac{1}{2}\left(3A^2 + 2A + 1\right), \quad T = \frac{1}{4}\left[(1+3A)^2 + 3(1+A)^2\right], \quad (7.44)$$

and the parameter A is arbitrary. The form of the representation of $\mathcal{L}_\Delta$ is determined by field-theoretic considerations connected with possibility of obtaining a quantized field for the Δ-isobar. The equation of motion corresponding to the Lagrangian (7.43) is given by

$$(-\gamma \cdot p_\Delta + M_\Delta)\Delta_\mu + \Big[-A\big(p_\mu^\Delta \gamma_\nu + p_\nu^\Delta \gamma_\mu\big)$$
$$+ K\gamma_\mu \gamma p_\Delta \gamma_\nu - T M_\Delta \gamma_\mu \gamma_\nu\Big]\Delta^\nu = 0. \quad (7.45)$$

The Green function for (7.45) has the following form [254, 259]:

$$G_{\mu\nu}^\Delta = \frac{\gamma \cdot p_\Delta + M_\Delta}{p_\Delta^2 - M_\Delta^2}\Bigg\{g_{\mu\nu} - \frac{A}{3(2A+1)^2}\frac{p_\mu^\Delta \gamma_\nu}{M_\Delta} - \frac{3A+2}{3(2A+1)}\frac{\gamma_\mu p_\nu^\Delta}{M_\Delta}$$
$$- \frac{2A}{3(2A+1)}\frac{p_\mu^\Delta p_\nu^\Delta}{M_\Delta^2} + \frac{(3A+1)(A+1)}{6(2A+1)^2}\frac{\gamma_\mu \gamma p_\Delta \gamma_\nu}{M_\Delta}$$
$$+ \frac{A(A+1)}{3(2A+1)^2}\frac{p_\mu^\Delta \gamma p_\Delta \gamma_\nu}{M_\Delta^2} - \frac{A+1}{3(2A+1)}\frac{\gamma_\mu \gamma p_\Delta p_\nu^\Delta}{M_\Delta^2}$$
$$- \left[\frac{1+3A+3A^2}{3(2A+1)^2} - \frac{A+1}{6(2A+1)}\frac{p_\Delta^2}{M_\Delta^2}\right]\gamma_\mu \cdot \gamma_\nu\Bigg\}. \quad (7.46)$$

It is possible to show (see [254]) that for the point transformation

$$\Delta_\mu \to \Delta_\mu + a\gamma_\mu\gamma^\nu \Delta_\nu , \qquad A \to \frac{A-2a}{1+4a} \qquad (7.47)$$

the Lagrangian density (7.43) remains invariant. Since the transformation (7.47) does not affect the components with spin 3/2, this means that all values of the parameter A are physically equivalent: the results of calculations that can be compared with experiment do not depend on its value. For this reason, the value of A may be chosen on the basis of considerations of convenience. As can be seen from the equation for the Green function (7.46), $A = -1$ is just such a type of choice; the Green function in this case is

$$G^\Delta_{\mu\nu} = \frac{\gamma \cdot p + M_\Delta}{p_\Delta^2 - M_\Delta^2} \Lambda_{\mu\nu} , \qquad (7.48)$$

where the operator $\Lambda_{\mu\nu}$ is determined by (7.23).

We choose the $\pi N\Delta$ interaction Lagrangian in the form

$$\mathcal{L}_{\pi N\Delta} = \frac{g_{\pi N\Delta}}{2M}[\bar{\Delta}_\mu \boldsymbol{T}^\dagger(g^{\mu\nu} - y(A)\gamma^\mu\gamma^\nu)\psi \\ + \bar{\psi}\boldsymbol{T}(g^{\mu\nu} - y(A)\gamma^\nu\gamma^\mu)\Delta_\mu]\partial_\nu\boldsymbol{\pi} . \qquad (7.49)$$

We require the interaction Lagrangian $\mathcal{L}_{\pi N\Delta}$ to be invariant under the transformation (7.47) [254], since this transformation does not affect the components of the amplitude of the Δ field with spin 3/2, and for this reason the physical values should remain unaltered.

From the relation

$$\mathcal{L}_{\pi N\Delta}\left(\Delta'_\mu = \Delta_\mu + a\gamma_\mu\gamma^\nu\Delta_\nu; A' = \frac{A-2a}{1+4a}\right) = \mathcal{L}_{\pi N\Delta}(\Delta_\mu; A) , \qquad (7.50)$$

one can easily derive the following identity:

$$y\left(\frac{A-2a}{1+4a}\right) = \frac{a+y(A)}{1+4a} , \qquad (7.51)$$

valid for all values of the parameter a (except $a = -1/4$).

Making the replacement $b = (A-2a)/(1+4a)$, we obtain

$$y(b) = \frac{A + 2y(A) + b[4y(A) - 1]}{4A+2} . \qquad (7.52)$$

From the latter identity, it follows that $y(A)$ may be a linear function only:

$$y(A) = \alpha A - Z . \qquad (7.53)$$

Substituting (7.53) into (7.51), we find

7.4 Contribution of the Δ_{33} Resonance to the Pion Polarization Operator

$$\alpha = -\frac{1+4Z}{2} . \qquad (7.54)$$

From this it follows that the $\pi N\Delta$ interaction is given by

$$\mathcal{L}_{\pi N\Delta} = \frac{g_{\pi N\Delta}}{2M} \left(\overline{\Delta}_\mu \boldsymbol{T}^\dagger \widehat{O}^{\mu\nu} \psi + \overline{\psi} \boldsymbol{T} \widehat{Q}^{\nu\mu} \Delta_\mu \right) \partial_\nu \boldsymbol{\pi} , \qquad (7.55)$$

where

$$\widehat{Q}^{\mu\nu} = g^{\mu\nu} + \left(\frac{1+4Z}{2} A + Z \right) \gamma^\mu \gamma^\nu . \qquad (7.56)$$

The region of the allowed values of the parameter Z is not restricted by the considerations given above. However, from the point of view of quantum field theory, the most preferable values are $Z = \pm 1/2$ [259]. In [255] the case $Z = -1/4$ is considered, which is notable because of the fact that the $\pi N\Delta$ interaction does not contain any arbitrary parameter A. In [255], πN scattering at small energies was investigated and it was established that $Z = -0.45 \pm 0.20$ leads to agreement with the experiment.

Note that in contrast to the parameter A, the values of which do not influence the final results that can be compared with experiment, these results do depend on the parameter Z. In the nonrelativistic approach, the results do not depend on Z. The reason for this is the fact that in this case the Δ-isobar is considered as being on shell. But on the mass shell, the Rarita–Schwinger condition is valid:

$$\gamma_\mu \cdot \Delta^\mu = 0 , \qquad (7.57)$$

which follows from the identity

$$\gamma_\mu \cdot \Lambda^{\mu\nu} = 0 . \qquad (7.58)$$

For this reason, the second term in the matrix $\widehat{Q}^{\mu\nu}$ (7.56) makes no contribution to the $\pi N\Delta$ vertex function.

As was mentioned above, the value $A = -1$ is mostly suitable. In this case the interaction (7.55) takes a form corresponding to our initial Lagrangian (7.1). This value of A will be used below in calculations of the polarization operator. Starting from (7.33) and (7.48), we write down the Δ-isobar contribution to the pion polarization operator in the following form:

$$\Pi_\Delta(q) = \frac{g_{\pi N\Delta}^2}{3M^2} \int_{|\boldsymbol{p}| \leq p_F} \frac{d^3 p}{(2\pi)^3} q^\mu q^\nu Sp \left\{ (g_{\mu\alpha} - y\gamma_\mu \gamma_\alpha) \frac{\gamma p_\Delta + M_\Delta^*}{-p_\Delta^2 + M_\Delta^{*2}} \right.$$
$$\left. \times \Lambda^{\alpha\beta} (g_{\beta\nu} - y\gamma_\beta \gamma_\nu) \frac{\gamma p + M^*}{2p_0} \right\} + (q_0 \longrightarrow -q_0) , \quad (7.59)$$

where we have taken into account the interaction of the Δ-isobar with the scalar nuclear field via its effective mass $M_\Delta^* = M_\Delta + S$. The vector field does not enter at all into the equation for the polarization operator for infinite nuclear matter (it is eliminated simply by shifting the reference level of single-particle energies). In (7.59) $p_\Delta^\mu = p^\mu + q^\mu$ and $p_0 = (\boldsymbol{p}^2 + M^{*2})^{1/2}$; for convenience, the notation $y = Z + 1/2$ is introduced. After involved calculations, one can obtain the following for $\Pi_\Delta(q)$ [253]:

$$\Pi_\Delta = \frac{4}{9}\left(\frac{g_{\pi N\Delta}}{M}\right)^2 \int_{|\boldsymbol{p}|\leq p_F} \frac{d^3p}{(2\pi)^3} \frac{1}{(M_\Delta^{*2} - p_\Delta^2)p_0}\left\{(1-y^2)q^2(p\cdot p_\Delta)\right.$$

$$-\frac{1}{M_\Delta^{*2}}(p_\Delta\cdot q)^2(p_\Delta\cdot p) + y^2\frac{p_\Delta^2}{M_\Delta^{*2}}q^2(p_\Delta\cdot p)$$

$$+ 2y(1-y)(p_\Delta\cdot q)(p\cdot q)\left(\frac{p_\Delta^2}{M_\Delta^{*2}} - 1\right)$$

$$+ (1+y-2y^2)M^*M_\Delta^* q^2 - \frac{M^*}{M_\Delta^*}(p_\Delta\cdot q)^2 + y(2y-1)\frac{M^*}{M_\Delta^*}p_\Delta^2\cdot q^2\right\}$$

$$+ (q_0 \longrightarrow -q_0). \qquad (7.60)$$

Finally, the dispersion equation for the pion in the nuclear medium has the following form:

$$q_0^2 - q^2 - m_\pi^2 - \Pi(q_0, q, \rho, x, Z) = 0, \qquad (7.61)$$

where the three-dimensional notation has been used, and the polarization operator depends on the density, the mixing parameter x (of the ps and pv couplings), and the parameter Z of the $\pi N\Delta$ interaction. So we have

$$\Pi(q_0, q, \rho, x, Z) = \Pi_N(q_0, q, \rho, x) + \Pi_\Delta(q_0, q, \rho, Z), \qquad (7.62)$$

where Π_N and Π_Δ are defined by (7.42) and (7.60), respectively. Thus, the question of stability against pion condensation is reduced to the question of whether (7.61) has a solution (at given values of the parameters ρ, x, Z) at $q_0 = 0$. To answer this question, it is sufficient to use the fact that the parameter x enters (7.61) in a comparatively simple way (rather than to solve the transcendental equation (7.61)). Equation (7.61) may be written down at $q_0 = 0$ as follows:

$$\left[q^2 + m_\pi^2 + \Pi_\Delta(q, \rho, Z)\right](1+x)^2$$

$$+ \frac{g_\pi^2 \rho_S}{M^*}x^2 - g_\pi^2\left(+\frac{M^*}{M}\right)^2\frac{q}{2\pi^2}J(q,\rho) = 0, \qquad (7.63)$$

where

7.4 Contribution of the Δ_{33} Resonance to the Pion Polarization Operator

$$J(q,\rho) = \int_0^{p_F} \frac{p\,dp}{p_0} \ln\left|\frac{q+2p}{q-2p}\right|$$

$$= \left(e_F - \frac{1}{2}(q^2 + 4M^{*2})^{1/2}\right) \ln\left|\frac{q+2p_F}{q-2p_F}\right| + q \ln\frac{p_F + e_F}{M^*}$$

$$+ \frac{1}{2}(q^2 + 4M^{*2})^{1/2} \ln\left|\frac{2M^{*2} - p_F\cdot q + e_F(q^2 + 4M^{*2})^{1/2}}{2M^{*2} + p_F\cdot q + e_F(q^2 + 4M^{*2})^{1/2}}\right|,$$

$$e_F = (p_F^2 + M^{*2})^{1/2}. \tag{7.64}$$

We present below some of the results obtained from investigating the stability problem.

1. *Pure pseudovector coupling, the Δ-isobar not being taken into account.* The main point of the relativistic treatment in this case is the appearance of a factor $(M^*/M)^2$ in front of the polarization integral (this factor is absent in the nonrelativistic treatment). The details of the stability behavior in this case depend on the choice of the parameters C_σ^2 and C_ω^2; see Appendix A2. For the values given in that Appendix, nuclear matter becomes unstable at a rather small density, $\rho_1 = 0.066$ fm^{-3}. It becomes stable again at a density $\rho_2 = 0.2075$ fm^{-3}, which slightly exceeds the equilibrium value $\rho_0 = 0.19$ fm^{-3}. However, by a suitable choice of the parameters C_σ and C_ω (M^* becomes modified by this choice), the system may be made stable at all densities. Such a behavior is due to the fact that the polarization integral enters (7.61) with the square of the effective mass as a factor (as seen from Appendix A2, M^* decreases rapidly with increasing ρ_S).

2. *Pure pseudoscalar coupling, the Δ-isobar not being taken into account.* As can be seen from (7.61), the pure pseudoscalar πN interaction mechanism leads to a pion mass renormalization in nuclear matter. The effective pion mass in this case becomes

$$\left(m^*_{\pi\text{pol}}\right)^2 = m_\pi^2 + \frac{g_\pi^2}{M^*}\rho_S. \tag{7.65}$$

The origin of this effect is related to a well-known difficulty in the pseudoscalar πN interaction theory. In that theory, an anomalously large S-wave πN scattering term of the form g_π^2/M arises in the second-order πN scattering amplitude. In the case of the πN scattering, this unphysical result is considered as an indication of the fact that the perturbation theory with respect to the πN coupling constant does not apply for free particles, and it is usually hoped that this term is quenched in higher orders. From the formal point of view, our result is also a second-order one. But really this is a consequence of the first-order perturbation theory with respect to the pion field amplitude, i.e. the smallness parameter, which has no analogy in the interaction between free particles. For this reason our result is of physical significance.

Using the equilibrium values of ρ_S and M^* (see Appendix A2) and $g_\pi^2/(4\pi) = 14$, we obtain $g_\pi^2 \rho_S/M^* = 12.385$ fm^{-2}, whereas $m_\pi^2 = 0.552$ fm^{-2}. So the effective pion mass becomes comparable with the free nucleon mass ($M = 4.75$ fm^{-1}) in the ground state of nuclear matter. It is worthwhile mentioning in this connection that the nonrelativistic limits for both the pseudoscalar and the pseudovector πN interaction Lagrangians are the same. This is the reason why neither the stability of nuclear matter at all densities for the pseudovector interaction nor the pion mass renormalization for the pseudoscalar case can be established within a nonrelativistic treatment of nuclei.

In this case the system remains stable at small densities because of the pion mass renormalization. It becomes unstable at $\rho = 0.194$ MeV, which is very close to $\rho_0 = 0.19$ fm^{-3}.

3. *Interference between pseudoscalar and pseudovector interactions, no Δ-isobar contribution.* Maps of the stability of nuclear matter can be found in [39]. The stability can be ensured by interference between the pseudovector and pseudoscalar mechanisms:

$$\mathcal{L}_{\pi N} = \frac{\mathcal{L}_{pv} + x\mathcal{L}_{ps}}{1+x}, \qquad (7.66)$$

x being the mixing parameter. As can be seen from the maps of stability [39], the proposed mechanism can ensure the stability of nuclear matter for any density value. It was mentioned above that the pseudovector and pseudoscalar interaction Lagrangians are the same in the nonrelativistic limit,

$$\mathcal{L}_{\pi N}^{nonrel} = \frac{g_\pi}{2M} \sum_\tau 2\tau \psi_\tau^\dagger \boldsymbol{\sigma} \psi_\tau \boldsymbol{\nabla} \pi, \qquad (7.67)$$

where τ is the nucleon isospin projection, and $\boldsymbol{\sigma}$ is the Pauli matrix. So the proposed stability mechanism is of purely relativistic origin.

4. *Inclusion of the $\Delta(1236)$ isobar.* The maps of stability obtained in [39] show that the stability behavior is changed significantly when the $\Delta(1236)$ isobar is included. Firstly, the threshold densities for the pure interactions become significantly smaller. These densities are 0.041 fm^{-3} and 0.109 fm^{-3} for the pseudovector and pseudoscalar mechanisms, respectively, both being less than ρ_0. Thus both interaction mechanisms lead to instability of the ground state of nuclear matter.
Secondly, the system becomes unstable at $\rho > 1.37$ fm$^{-3} = \rho_{max}$ irrespective of the value of the mixing parameter. However, $\rho_{max} = 7.2\rho_0$, and hence the proposed mechanism (of interference between ps and pv interactions) can ensure stability in a rather wide density region. The value of the mixing parameter x can be determined from a fit to pionic-atom data and πN experimental data (see [272] and references therein).
The spin-3/2–isospin-3/2 Δ-isobar plays an essential role in fitting pion–nucleon and nucleon–nucleon scattering data. Further discussion of the main features of the πNΔ interaction can be found in [278–283].

The collective pionic modes generated by the coupling of the pion to the delta-hole configuration are considered in [284]. These collective modes (pisobars) may play a very important role in strongly excited hadronuclear matter obtained in relativistic heavy-ion collisions.

7.5 Basic Equations of the Linear σ and σ–ω Models

Models respecting chiral symmetry [171, 270, 285–293] are of outstanding importance in strong-interaction physics. After the pioneering work of Lee, Wick, and Margulies [294, 295], many attempts have been made to describe nuclear structure using relativistic chiral models. In the present section, we consider some consequences arising from the assumption that the Lagrangian for interacting meson and baryon fields is invariant under chiral symmetry in addition to the conventional isotopic symmetry. We shall restrict ourselves also to the linear realization of chiral symmetry [296–300], since in this case the notion of the scalar field is evident; this fact enables us to make a link between the chiral approach and the nuclear Dirac phenomenology.

The linear σ model is a very fruitful implementation of the idea of chiral symmetry [296–298]. In accordance with [171], we write down the Lagrangian of the linear σ model in the following form:

$$\mathcal{L}_\sigma = \mathrm{i}\bar{\psi}\gamma^\mu \partial_\mu \psi + \frac{1}{2}\partial_\mu \boldsymbol{\pi} \partial^\mu \boldsymbol{\pi} + \frac{1}{2}\partial_\mu \sigma \cdot \partial^\mu \sigma - \frac{\mu}{2}(\sigma^2 + \boldsymbol{\pi}\cdot\boldsymbol{\pi})$$
$$- \frac{\lambda}{2}(\sigma^2 + \boldsymbol{\pi}\cdot\boldsymbol{\pi})^2 - g_\sigma \bar{\psi}(\mathrm{i}\gamma_5 \boldsymbol{\tau}\cdot\boldsymbol{\pi} + \sigma)\psi$$
$$+ \frac{C}{2}\left[-\bar{\psi}\gamma_5\gamma_\mu \boldsymbol{\tau}\psi(\sigma\partial^\mu \boldsymbol{\pi} - \boldsymbol{\pi}\cdot\partial^\mu \sigma) + \bar{\psi}\gamma_\mu \boldsymbol{\tau}\psi\cdot(\boldsymbol{\pi}\times\partial^\mu \boldsymbol{\pi})\right] . \quad (7.68)$$

$\mathcal{L}_\sigma$ describes interacting nucleon, π, and σ fields. The Lagrangian (7.68) is invariant both under the rotation in isotopic space

$$\delta\psi = -\frac{\mathrm{i}}{2}\boldsymbol{\beta}\boldsymbol{\tau}\psi , \quad (7.69)$$

$$\delta\boldsymbol{\pi} = -\boldsymbol{\pi}\times\boldsymbol{\beta} , \quad (7.70)$$

$$\delta\sigma = 0 , \quad (7.71)$$

and under the chiral rotation

$$\delta\psi = \frac{\mathrm{i}}{2}\boldsymbol{\varepsilon}\cdot\boldsymbol{\tau}\gamma_5\psi , \quad (7.72)$$

$$\delta\boldsymbol{\pi} = -\boldsymbol{\varepsilon}\sigma , \quad (7.73)$$

$$\delta\sigma = \boldsymbol{\varepsilon}\cdot\boldsymbol{\pi} . \quad (7.74)$$

This symmetry of the Lagrangian ensures the existence of two conserved currents: the vector current, owing to the isotopic symmetry and the axial current, owing to the chiral symmetry. In accordance with Noether's theorem, these currents are given by

$$V_\mu = \sum_i \frac{\partial \mathcal{L}}{\partial(\partial^\mu \chi_i)} \cdot \frac{d\chi_i}{d\beta} = \frac{1}{2}\bar{\psi}\gamma_\mu \boldsymbol{\tau}\psi + \frac{1}{2}(\boldsymbol{\pi} \times \partial_\mu \boldsymbol{\pi} - \partial_\mu \boldsymbol{\pi} \times \boldsymbol{\pi})$$
$$+ \frac{C\sigma}{2}\bar{\psi}\gamma_5\gamma_\mu \boldsymbol{\tau}\psi \times \boldsymbol{\pi} - \frac{C}{2}\bar{\psi}\gamma_\mu \boldsymbol{\tau}\psi \times \boldsymbol{\pi}, \quad (7.75)$$

$$A_\mu = \sum_i \frac{\partial \mathcal{L}}{\partial(\partial^\mu \chi_i)} \frac{d\chi_i}{d\varepsilon} = \frac{1}{2}\bar{\psi}\gamma_5\gamma_\mu \boldsymbol{\tau}\psi - (\sigma \cdot \partial_\mu \boldsymbol{\pi} - \partial_\mu \sigma \cdot \boldsymbol{\pi})$$
$$+ \frac{C}{2}\bar{\psi}\gamma_5\gamma_\mu \boldsymbol{\tau}\psi(\boldsymbol{\pi} \cdot \boldsymbol{\pi} + \sigma^2) - \frac{C}{2}\boldsymbol{\pi} \times (\bar{\psi}\gamma_5\gamma_\mu \boldsymbol{\tau}\psi \times \boldsymbol{\pi})$$
$$- \frac{C\sigma}{2}\bar{\psi}\gamma_\mu \boldsymbol{\tau}\psi \times \boldsymbol{\pi}, \quad (7.76)$$

where χ_i are the amplitudes of the fields entering the Lagrangian. The normalization of the currents is chosen to satisfy the commutation relations of the current algebras. From (7.75) and (7.76), we obtain the following for the axial nucleon form factor in the absence of the static pion field (we assume the vector form factor $g_V = 1$):

$$g_A = 1 + C\sigma^2. \quad (7.77)$$

Let us write down the equation of motion for the σ field corresponding to the Lagrangian (7.68), with the pion field being absent:

$$\Box\sigma + \mu\sigma + 2\lambda\sigma^3 = -g_\sigma \bar{\psi}\psi. \quad (7.78)$$

Equation (7.78) for the vacuum has the form

$$\sigma(\mu + 2\lambda\sigma^2) = 0. \quad (7.79)$$

This equation has three solutions:

$$\sigma = 0, \quad (7.80)$$

$$\sigma = \pm\left(-\frac{\mu}{2\lambda}\right)^{1/2}. \quad (7.81)$$

To establish which of these solutions is stable with respect to small fluctuations of the σ field, we write down the equation for the energy density in the case of the vacuum corresponding to (7.68):

$$E(\sigma) = \frac{\mu\sigma^2}{2} + \frac{\lambda\sigma^4}{2}. \quad (7.82)$$

It is easy to verify that for $\mu > 0$ the solution (7.80) is stable, while for $\mu < 0$ the solutions (7.81) are stable. The first case, where $\mu > 0$ (Lee–Wick mode), does not correspond to physical reality. If σ were equal to zero in the vacuum, the Lagrangian (7.68) would correspond to a massless nucleon field and σ and π fields with equal masses. Moreover, the axial nucleon form factor would be equal to unity ($g_A = 1$), i.e. to a value which contradicts experiment. The physically realized case is the solution where $\mu < 0$, with spontaneous breaking of the chiral symmetry (the Goldstone mode). Let us calculate the amplitude of the σ field starting from the vacuum value $\sigma_0 \neq 0$ and write down the Lagrangian (7.68) in terms of the field φ, where $\varphi = \sigma - \sigma_0$ [253]:

$$\mathcal{L}_\sigma = i\bar{\psi}\gamma^\mu\partial_\mu\psi - M\bar{\psi}\psi + \frac{1}{2}\partial_\mu\boldsymbol{\pi}\cdot\partial^\mu\boldsymbol{\pi} + \frac{1}{2}\partial_\mu\varphi\cdot\partial^\mu\varphi$$
$$- \frac{1}{2}m_\sigma^2\varphi^2 - g_\sigma\bar{\psi}(i\gamma_5\boldsymbol{\tau}\cdot\boldsymbol{\pi} + \varphi)\psi - \frac{m_\sigma^2}{2M}g_\sigma\cdot\varphi(\varphi^2 + \boldsymbol{\pi}\cdot\boldsymbol{\pi})$$
$$- \frac{m_\sigma^2}{8M^2}g_\sigma^2(\varphi^2 + \boldsymbol{\pi}\cdot\boldsymbol{\pi})^2 - g_\sigma\frac{C\sigma_0^2}{2M}\bar{\psi}\gamma_5\gamma_\mu\boldsymbol{\tau}\psi\partial^\mu\boldsymbol{\pi}$$
$$- \frac{C}{2}\bar{\psi}\gamma_5\gamma_\mu\boldsymbol{\tau}\psi(\varphi\cdot\partial^\mu\boldsymbol{\pi} - \boldsymbol{\pi}\cdot\partial^\mu\varphi)$$
$$+ \frac{C}{2}\bar{\psi}\gamma_\mu\cdot\boldsymbol{\tau}\psi(\boldsymbol{\pi}\times\partial^\mu\boldsymbol{\pi}) + \frac{m_\sigma^2}{8g_\sigma^2}M^2 , \quad (7.83)$$

where the following notation has been introduced:

$$\sigma_0^2 = -\frac{\mu}{2\lambda}, \quad M = g_\sigma\sigma_0, \quad m_\sigma^2 = \mu + 6\lambda\sigma_0^2 = -2\mu . \quad (7.84)$$

From (7.84) it follows that, after spontaneous breaking of the chiral symmetry, the nucleon field acquires a mass $M = g_\sigma\sigma_0$ (in what follows we consider $g_\sigma > 0$ and $\sigma_0 = \sqrt{-\mu/(2\lambda)} > 0$ for definiteness), the φ field also appears to be a massive field, and the π meson plays the role of the Goldstone boson. Taking into account the equivalence of pseudoscalar and pseudovector πNN couplings in vacuum, we conclude that the effective constant for the πNN interaction is given by

$$g_\pi = g_\sigma(1 + C\sigma_0^2) . \quad (7.85)$$

Let us write down the axial current taking into account the spontaneous symmetry breaking. From (7.76) we obtain

$$\boldsymbol{A}_\mu = \frac{1}{2}\bar{\psi}\gamma_5\gamma_\mu\boldsymbol{\tau}\psi(1 + C\sigma_0^2) - \sigma_0\partial_\mu\boldsymbol{\pi}$$
$$- \frac{C\sigma_0}{2}\bar{\psi}\gamma_\mu\boldsymbol{\tau}\psi\times\boldsymbol{\pi} - (\varphi\cdot\partial_\mu\boldsymbol{\pi} - \partial_\mu\varphi\cdot\boldsymbol{\pi})$$
$$+ \frac{C}{2}\bar{\psi}\gamma_5\gamma_\mu\boldsymbol{\tau}\psi(\boldsymbol{\pi}\cdot\boldsymbol{\pi} + 2\sigma_0\cdot\varphi + \varphi^2)$$
$$- \frac{C}{2}\boldsymbol{\pi}\times(\bar{\psi}\gamma_5\gamma_\mu\boldsymbol{\tau}\psi\times\boldsymbol{\pi}) - \frac{C\cdot\varphi}{2}\bar{\psi}\gamma_\mu\boldsymbol{\tau}\psi\times\boldsymbol{\pi} . \quad (7.86)$$

The first term on the right-hand side of (7.86) corresponds to the renormalized axial nucleon current, and the second term describes the pion decay in vacuum $\pi^\pm \to \mu^\pm + \nu_\mu$, the pion decay constant f_π being determined by the relation

$$\langle 0 | A_\mu^\alpha | \pi^\beta \rangle = iq_\mu f_\pi \delta_{\alpha\beta}, \qquad (7.87)$$

where α and β are the components in the isotopic spin space. Comparing (7.86) and (7.87), we obtain

$$f_\pi = \sigma_0. \qquad (7.88)$$

From (7.77), (7.86), (7.88), and from the second formula (7.84), it is possible to derive the equation

$$\frac{g_\pi}{g_A} = \frac{M}{f_\pi}, \qquad (7.89)$$

known as the Goldberger–Treiman relation [171].

The presence of a massless particle, identified with the pion, is a necessary consequence of the spontaneous chiral-symmetry breaking. To introduce pions with a physical mass, one needs dynamical breaking of the symmetry. Usually this is done via the term linear in the σ field, so that the Lagrangian has the following form:

$$\mathcal{L}_\sigma + \varepsilon\sigma, \qquad (7.90)$$

where the parameter ε determines the extent of the chiral-symmetry breaking. Since experiment is in agreement with the idea of an approximate chiral symmetry (one of the manifestations of this idea is the small value of the pion mass on the hadronic scale), the additional term $\varepsilon\sigma$ in (7.90) should be considered as a small one. The presence of this term slightly shifts the equilibrium value of the σ field in the vacuum σ_0; because of this point alone, the pion field acquires its nonzero mass. The Lagrangian of a system of nucleon, σ, and π fields after spontaneous chiral-symmetry breaking has the following form (similar to (7.83)):

$$\begin{aligned}\mathcal{L}'_\sigma &= \mathcal{L}_\sigma + \varepsilon\sigma = i\bar\psi\gamma^\mu\partial_\mu\psi - M\bar\psi\psi + \frac{1}{2}\partial_\mu\boldsymbol{\pi}\partial^\mu\boldsymbol{\pi} - \frac{1}{2}m_\pi^2\boldsymbol{\pi}\cdot\boldsymbol{\pi} \\ &\quad + \frac{1}{2}\partial_\mu\varphi\cdot\partial^\mu\varphi - \frac{1}{2}m_\sigma^2\varphi^2 - g_\sigma\bar\psi(i\gamma_5\boldsymbol{\tau}\cdot\boldsymbol{\pi} + \varphi)\psi \\ &\quad - \frac{m_\sigma^2 - m_\pi^2}{2M}g_\sigma\varphi(\varphi^2 + \boldsymbol{\pi}\cdot\boldsymbol{\pi}) - \frac{m_\sigma^2 - m_\pi^2}{8M^2}g_\sigma^2(\varphi^2 + \boldsymbol{\pi}\cdot\boldsymbol{\pi})^2 \\ &\quad - g_\sigma\frac{C\sigma_0^2}{2M}\bar\psi\gamma_5\gamma_\mu\boldsymbol{\tau}\psi\partial^\mu\boldsymbol{\pi} - \frac{C}{2}\bar\psi\gamma_5\gamma_\mu\boldsymbol{\tau}\psi(\varphi\cdot\partial^\mu\boldsymbol{\pi} - \boldsymbol{\pi}\cdot\partial^\mu\varphi) \\ &\quad + \frac{C}{2}\bar\psi\gamma_\mu\boldsymbol{\tau}\psi\cdot\boldsymbol{\pi}\times\partial^\mu\boldsymbol{\pi} + \frac{m_\sigma^2 - 5m_\pi^2}{8g_\sigma^2}M^2,\end{aligned} \qquad (7.91)$$

where

$$M = g_\sigma\sigma_0, \quad m_\sigma^2 = \mu + 6\lambda\sigma_0^2 = -2\mu + 3\frac{\varepsilon}{\sigma_0}, \quad m_\pi^2 = \mu + 2\lambda\sigma_0^2 = \frac{\varepsilon}{\sigma_0}. \qquad (7.92)$$

The equations for the vector and axial currents do not change when the term $\varepsilon\sigma$ is added to the Lagrangian (see (7.75) and (7.76)). However, since the

7.5 Basic Equations of the Linear σ and σ–ω Models

Lagrangian (7.90) is not chirally invariant, the axial current is not conserved exactly. For the fourdivergence of

$$\boldsymbol{A}_\mu = \sum_i \frac{\partial \mathcal{L}}{\partial(\partial^\mu \chi_i)} \frac{\mathrm{d}\chi_i}{\mathrm{d}\varepsilon} , \qquad (7.93)$$

we have

$$\partial^\mu \boldsymbol{A}_\mu = \sum_i \left[\partial^\mu \left(\frac{\partial \mathcal{L}}{\partial \chi_i^\mu} \right) \frac{\mathrm{d}\chi_i}{\mathrm{d}\varepsilon} + \frac{\partial \mathcal{L}}{\partial \chi_i^\mu} \cdot \frac{\mathrm{d}\chi_i^\mu}{\mathrm{d}\varepsilon} \right]$$

$$= \sum_i \left(\frac{\partial \mathcal{L}}{\partial \chi_i} \cdot \frac{\mathrm{d}\chi_i}{\mathrm{d}\varepsilon} + \frac{\partial \mathcal{L}}{\partial \chi_i^\mu} \cdot \frac{\mathrm{d}\chi_i^\mu}{\mathrm{d}\varepsilon} \right) = \frac{\mathrm{d}\mathcal{L}}{\mathrm{d}\varepsilon}, \quad (7.94)$$

the Euler–Lagrange equation

$$\frac{\partial}{\partial x_\mu} \left(\frac{\partial \mathcal{L}}{\partial \chi_i^\mu} \right) - \frac{\partial \mathcal{L}}{\partial \chi_i} = 0 \qquad (7.95)$$

being utilized.

For the Lagrangian (7.90), (7.94) has the following form (after (7.69)–(7.71) are taken into account):

$$\partial^\mu \boldsymbol{A}_\mu = \varepsilon \boldsymbol{\pi} . \qquad (7.96)$$

In accordance with (7.88) and (7.92), this can be rewritten in the following form:

$$\partial^\mu \boldsymbol{A}_\mu = m_\pi^2 f_\pi \boldsymbol{\pi} . \qquad (7.97)$$

This is the basic equation of the hypothesis of partially conserved axial current (the PCAC hypothesis).

Let us now discuss the problem of the ground state of nuclear matter. For this purpose we include the vector–isoscalar ω field also in the chirally symmetric Lagrangian (at present we restrict ourselves to the case of an isotopically symmetric nuclear system; for this reason, there is no necessity to introduce the ρ meson field). Note that the inclusion in the Lagrangian of the extra terms connected with the ω field does not lead to any modifications of the expressions for the vector and axial currents, since the ω field is invariant under both isotopic and chiral rotations. The Goldberger–Treiman relation and the PCAC hypothesis remain valid also.

First we consider a "hard" method to introduce the ω field, where the ω meson mass is included in the theory by hand and is not connected with spontaneous breaking of chiral symmetry. In this case the Lagrangian is given by

$$\mathcal{L} = \mathcal{L}'_\sigma - \frac{1}{4} \omega^{\mu\nu} \cdot \omega_{\mu\nu} + \frac{m_\omega^2}{2} \omega^\mu \cdot \omega_\mu - g_\omega \bar{\psi} \gamma_\mu \psi \omega^\mu , \qquad (7.98)$$

where $\omega_{\mu\nu}$ is given by (4.3), and $\mathcal{L}'_\sigma = \mathcal{L}_\sigma + \varepsilon \sigma$ is a Lagrangian with dynamical breaking of the chiral symmetry.

Let us write down equations for the σ and ω fields corresponding to the Lagrangian (7.98), using for $\mathcal{L}_\sigma$ the starting representation (7.68) and utilizing also (7.92):

$$\Box\sigma + \frac{3m_\pi^2 - m_\sigma^2}{2}\sigma + \frac{m_\sigma^2 - m_\pi^2}{2M^2}g_\sigma^2(\sigma^2 + \boldsymbol{\pi}\cdot\boldsymbol{\pi})\sigma$$
$$+ \frac{C}{2}\bar{\psi}\gamma_5\gamma_\mu\boldsymbol{\tau}\psi\partial^\mu\boldsymbol{\pi} + \frac{C}{2}\partial^\mu(\bar{\psi}\gamma_5\gamma_\mu\boldsymbol{\tau}\psi\cdot\boldsymbol{\pi}) - m_\pi^2\frac{M}{g_\sigma} = -g_\sigma\bar{\psi}\psi\,,$$
$$\Box\omega^\nu - \partial^\nu\partial_\mu\omega^\mu + m_\omega^2\omega^\nu = g_\omega\bar{\psi}\gamma^\nu\psi\,. \tag{7.99}$$

For infinite symmetric nuclear matter in the static case, we obtain

$$m_\sigma^2 S + (m_\sigma^2 - m_\pi^2)\frac{S}{2}\left(\frac{3S}{M} + \frac{S^2}{M^2}\right) = -g_\sigma^2\rho_S\,, \quad m_\omega^2 V = g_\omega^2\rho_V\,, \tag{7.100}$$

where we have introduced the following nuclear fields: a scalar field $S = g_\sigma(\sigma - \sigma_0)$ and a vector one $V = g_\omega\omega^0$. Equations (7.100) are not coupled, and can be solved independently. The equation for the scalar field (being an algebraic equation of the third degree) has, in the general case, three solutions. Let us write down these solutions. (a) For $\rho_S \leq \rho_{\max}$, where

$$\rho_{\max} = \frac{Mm_\sigma^2}{g_\sigma^2 3\sqrt{3}}\left[\frac{(1 - 3m_\pi^2/m_\sigma^2)^{3/2}}{(1 - m_\pi^2/m_\sigma^2)^{1/2}} + \frac{3\sqrt{3}m_\pi^2}{m_\sigma^2}\right], \tag{7.101}$$

the solutions are

$$S_1 = 2\sqrt{-\frac{p}{3}}\cos\frac{\alpha}{3} - M\,, \tag{7.102}$$

and

$$S_{2,3} = -2\sqrt{-\frac{p}{3}}\cos\left(\frac{\alpha}{3} \pm \frac{\pi}{3}\right) - M\,. \tag{7.103}$$

(b) For $\rho_S > \rho_{\max}$, there exists only one solution:

$$S_3 = \left(-\frac{q}{2} + Q^{1/2}\right)^{1/3} + \left(-\frac{q}{2} - Q^{1/2}\right)^{1/3} - M\,. \tag{7.104}$$

In the above,

$$p = -M^2\left(3 - 2\frac{m_\sigma^2}{m_\sigma^2 - m_\pi^2}\right),\quad \cos\alpha = -\frac{q}{2\sqrt{(-p/3)^3}}\,,$$
$$q = \frac{2M^2}{m_\sigma^2 - m_\pi^2}(g_\sigma^2\rho_S - m_\pi^2 M)\,,\quad Q = \left(\frac{p}{3}\right)^3 + \left(\frac{q}{2}\right)^2\,. \tag{7.105}$$

To understand the physical meaning of these solutions, we analyze qualitatively their dependence on ρ_S for the case $m_\pi = 0$. In this case, from (7.98), it is easy to obtain for $\rho_S = 0$

7.5 Basic Equations of the Linear σ and σ–ω Models

$$S_1 = 0, \quad S_2 = -M, \quad S_3 = -2M. \tag{7.106}$$

From this result it is clear that the solutions S_1 and S_3 correspond to nonzero values of the σ field in vacuum: M/g_σ and $-M/g_\sigma$ depending on the choice of the sign in (7.81), and that the solution S_2 corresponds to the case (7.80). In other words, S_1 and S_3 realize a situation with spontaneous breaking of the chiral symmetry, and the solution S_2 corresponds to the chirally-invariant vacuum.

With an increase of ρ_S from 0 to the value $\rho_{max} = M m_\sigma^2/(g_\sigma^2 3\sqrt{3})$, S_1 decreases monotonically, while S_2 increases monotonically and reaches the same value $S_{max} = -M\left(1 - 1/\sqrt{3}\right)$. At higher densities the solutions S_1 and S_2 do not exist. The solution S_3 exists at all densities $\rho_S > 0$, decreasing monotonically with an increase of ρ_S. It is clear that only the solution S_1 (with a zero vacuum value) may correspond to the normal state of nuclear matter (in [300] abnormal solutions, obtained on the basis of the solution S_3, are investigated). Notice that at $\rho_S = \rho_{max}$ the solutions S_1 and S_2 have the peculiar feature that $\partial S_{1,2}/\partial \rho_S = \infty$; however, $\partial S_{1,2}/\partial \rho_V$ is finite for all allowed values of ρ_V.

If the physical value of the pion mass is introduced, qualitatively the situation remains the same. In this case the values ρ_{max} and $S_{max} = S_{1,2}(\rho_{max})$ are changed to a certain extent: ρ_{max} is given by (7.101) and S_{max} is given by

$$S_{max} = M\left[\frac{1}{\sqrt{3}}\left(\frac{m_\sigma^2 - 3m_\pi^2}{m_\sigma^2 - m_\pi^2}\right)^{1/2} - 1\right]. \tag{7.107}$$

Notice that the maximal value of the scalar field for the normal physical branch (solution S_1) does not depend strongly on the value of m_σ, and is equal to $S_{max} = -412\,\text{MeV}$ for $m_\sigma = 850\,\text{MeV}$ and $S_{max} = -445\,\text{MeV}$ for $m_\sigma = 500\,\text{MeV}$. The maximal values of the density are equal to 0.18 fm^{-3} and 0.06 fm^{-3}, respectively. From these results it follows that in the linear σ model, the values $m_\sigma < 800\,\text{MeV}$ are not acceptable, since for such values of m_σ the physical solution does not exist at normal nuclear densities.

Let us write down the expression for the binding energy per nucleon in nuclear matter corresponding to (7.98):

$$\begin{aligned}
\frac{E}{A} &= \frac{T^{00}}{\rho_V} - M = \frac{1}{\rho_V}\left(\sum_i \partial^0 \chi_i \frac{\partial \mathcal{L}}{\partial(\partial_0 \chi_i)} - \mathcal{L}\right) - M \\
&= \frac{1}{\rho_V}\frac{4}{(2\pi)^3}\int_{|\mathbf{p}|\leq p_F}(p^2 + M^{*2})^{1/2}\mathrm{d}^3 p - M + \frac{1}{2}g_\omega^2 \frac{1}{m_\omega^2}\rho_V \\
&\quad + \frac{1}{\rho_V}\left(\frac{m_\sigma^2 S^2}{2g_\sigma^2} + \frac{m_\sigma^2 - m_\pi^2}{2Mg_\sigma^2}S^3 + \frac{m_\sigma^2 - m_\pi^2}{8M^2 g_\sigma^2}S^4\right).
\end{aligned} \tag{7.108}$$

In [299, 300], an investigation of the saturation properties has been performed starting from (7.108), and it was established that the version of the

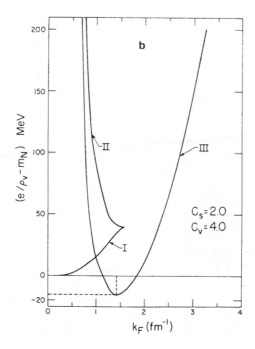

Fig. 7.1. The binding energy per particle as a function of the Fermi momentum. A cusp is clearly visible at $p_F \approx 1.5$ fm^{-1} [299, 301]

σ model considered here is not able to reproduce the saturation of nuclear matter, since the physical solution S_1 to be used in (7.108) ceases to exist (it bifurcates with the solution S_2) earlier than the point at which saturation is achieved. For the solutions S_2 and S_3, E becomes infinitely large when $\rho_V \to 0$; this leads to the well-known cusp catastrophe [301]. A typical energy-per-particle curve is shown in Fig. 7.1 [299, 301].

A possible solution to this unfavorable situation was suggested in [299, 300]. In these papers the ω field is introduced into the Lagrangian of the σ model as a massless field but interacts with the scalar σ field. The ω meson mass appears in this case because of spontaneous breaking of the global chiral symmetry, in the same way as it happens with the nucleons. In what follows we shall refer to this version of the σ model as to the σ–ω model. In accordance with [299], let us write down the Lagrangian of the σ–ω model in the following form (Boguta model):

$$\mathcal{L} = \mathcal{L}'_\sigma - \frac{1}{4}\omega^{\mu\nu} \cdot \omega_{\mu\nu} + \frac{1}{2}g_\omega^2(\sigma^2 + \boldsymbol{\pi} \cdot \boldsymbol{\pi})\omega_\mu \cdot \omega^\mu - g_\omega \bar{\psi}\gamma_\mu \psi \omega^\mu \,. \quad (7.109)$$

The ω field mass is determined by the value of the σ field and is equal to

$$m_\omega^* = g_\omega \cdot \sigma = g_\omega \left(\sigma_0 + \frac{S}{g_\sigma}\right) = m_\omega + \frac{g_\omega}{g_\sigma}S \,, \quad (7.110)$$

7.5 Basic Equations of the Linear σ and σ–ω Models

where m_ω is the ω meson mass in the Goldstone vacuum, and S is the scalar nuclear field. The equations for the nuclear scalar and vector fields for the Lagrange density (7.109) in the case of nuclear matter have the following form (for $\pi = 0$):

$$m_\sigma^2 S + \left(m_\sigma^2 - m_\pi^2\right) \frac{S}{2} \left(\frac{3S}{M} + \frac{S^2}{M^2}\right) = -g_\sigma^2 \rho_s + M^* V^2,$$

$$M^{*2} V = g_\sigma^2 \rho_V, \qquad (7.111)$$

where $M^* = M + S$.

In contrast to the version of the σ model considered earlier that uses the "hard" method to introduce the ω meson mass, in the σ–ω model the equations for the scalar and vector fields appear to be coupled. As shown in [299], because of this fact the system of equations (7.111) has the Goldstone solution ($S = 0$ for $\rho_V = 0$) at all densities or, in other words, the bifurcation of the Goldstone mode and the Lee–Wick mode ($S = -M$ for $\rho_V = 0$) does not occur anymore. In [299] it was argued that saturation can be obtained for nuclear matter in this model; the binding energy per nucleon being given by

$$\frac{E}{A} = \frac{1}{\rho_V} \frac{4}{(2\pi)^3} \int_{|p| \leq p_F} \left(p^2 + M^{*2}\right)^{1/2} d^3 p - M + \frac{1}{2} \frac{g_\sigma^2}{M^{*2}} \rho_V$$

$$+ \frac{1}{\rho_V} \left(\frac{m_\sigma^2 S^2}{2 g_\sigma^2} + \frac{m_\sigma^2 - m_\pi^2}{2 M g_\sigma^2} S^3 + \frac{m_\sigma^2 - m_\pi^2}{8 M^2 g_\sigma^2} S^4\right). \qquad (7.112)$$

The functional (7.112) has a minimum at a certain density, the saturation of nuclear forces being reproduced, in principle, in the σ–ω model. First of all, let us notice that the coupling constant g_ω enters neither (7.111) for the S and V fields nor the equation for E/A, this fact being related directly to the spatial homogeneity of nuclear matter. Thus, the problem of the binding energy of an infinite system in the σ–ω model contains only one free parameter: the mass of the scalar meson m_σ (the coupling constant g_σ is fixed by the observed values of the nucleon mass and the pion weak-decay constant f_π; see (7.84), (7.88). However, in [299] the value of g_σ was treated also as a free parameter of the model along with the value of m_σ. These two parameters were chosen to reproduce the equilibrium values of the density and binding energy per nucleon, equal to 0.145 fm^{-3} and -16 MeV, respectively. The following values of the parameters were obtained: $m_\sigma = 650$ MeV and $g_\sigma = 8.35$. It is clear that this value of g_σ agrees poorly with those obtained in the σ model using the experimental values for the nucleon mass M and pion weak-decay coupling constant f_π ($g_\sigma = M/f_\pi = 10.1$) and using the effective πNN coupling constant g_π and axial form factor g_A ($g_\sigma = g_\pi/g_A = \sqrt{14 \times 4\pi}/1.25 = 10.6$; see (7.85) and (7.77)). Further generalization of the model has been done in [302]. In this case the ω meson couplings with nucleons and with the σ field are introduced with independent constants g_ω and $\eta g_\omega \equiv G_\omega$. The Lagrangian has the form

$$\mathcal{L} = \mathcal{L}'_\sigma - \frac{1}{4}\omega^{\mu\nu}\omega_{\mu\nu} + \frac{1}{2}(\eta g_\omega)^2(\sigma^2 + \boldsymbol{\pi}\cdot\boldsymbol{\pi})\omega_\mu\omega^\mu - g_\omega\bar{\psi}\gamma_\mu\psi\omega^\mu , \qquad (7.113)$$

while the equations for the nuclear fields (for nuclear matter) are given by (for $\boldsymbol{\pi}=0$)

$$m_\sigma^2 S + (m_\sigma^2 - m_\pi^2)\frac{S}{2}\left(\frac{3S}{M} + \frac{S^2}{M^2}\right) = -g_\sigma^2\rho_{\rm s} + M^*\eta^2 V^2 , \qquad (7.114)$$

$$\eta^2 M^{*2} V = g_\omega^2 \rho_{\rm V} . \qquad (7.115)$$

The values of m_σ and η are treated as free parameters in [302]. The parameter g_σ is fixed by the equation $g_\sigma = g_\pi/g_{\rm A} = 10.6$. The empirical values of the equilibrium density and binding energy per nucleon ($\rho_0 = 0.17$ fm^{-3}, $E/A = -16$ MeV) are obtained for $m_\sigma = 883.6$ MeV and $\eta = 1.385$. Notice that the values of the scalar and vector fields obtained for these parameters are much smaller than those obtained in the Dirac phenomenology.

Further investigations of nuclear matter within the chiral σ–ω model are described in [303]. Let us now discuss briefly some results obtained for finite nuclei in the framework of a relativistic self-consistent calculation on the basis of the chiral σ–ω model.

7.6 Chiral σ–ω Model for Finite Nuclei

The results presented in this section were presented in [304–307]. They are based on the Lagrangian density given by (7.113), while the Euler–Lagrange equations (for $\boldsymbol{\pi}=0$) have the following form:

$$\left[\boldsymbol{\alpha}\cdot\boldsymbol{p} + \beta\cdot M^*(r) + V(r) + \frac{1}{2}e(1+\tau_3)A_0(r)\right]\psi_\lambda = E_\lambda\psi_\lambda ,$$

$$\nabla^2 M^*(r) + \frac{M^2 - M^{*2}(r)}{2M^2}\left[m_\sigma^2 M^*(r) + m_\pi^2(M - M^*(r))\times\frac{M^*(r)+2M}{M^*(r)+M}\right]$$
$$= g_\sigma^2\rho_{\rm S}(r) - M^*(r)\frac{G_\omega^2}{g_\omega^2}V^2(r) ,$$

$$\nabla^2 V(r) - \frac{G_\omega^2}{g_\sigma^2}M^{*2}(r)V(r) = -g_\omega^2\rho_{\rm V}(r) ,$$

$$\nabla^2 A_0(r) = -e\rho_{\rm p}(r) , \qquad (7.116)$$

where $M^*(r) = g_\sigma\sigma(r) = M + g_\sigma\varphi(r) = g_\sigma\sigma_0 + g_\sigma\varphi(r)$.

It should be noticed that the pion field equals zero in the Hartree approximation. However, even in this case the role of chiral symmetry is essential. This is revealed by the fact that the strength of the σN coupling, as well as the nonlinear φ terms, are uniquely determined by the πN coupling, and is also manifested by the structure of the equation for the scalar field, the chiral partner of the pion.

7.6 Chiral σ–ω Model for Finite Nuclei   123

In [304], (7.116) were solved self-consistently and nuclear bulk properties were calculated, the binding energy per nucleon being given by

$$\frac{E}{A} = \frac{1}{A}\left\{\sum_{\lambda<F}\varepsilon_\lambda + 4\pi\int_0^\infty r^2\,dr\left[\frac{1}{2g_\sigma^2}\left(\frac{dM^*}{dr}\right)^2 + \frac{m_\sigma^2}{2g_\sigma^2}(M^*-M)^2\right.\right.$$

$$-\frac{1}{2g_\omega^2}\left(\frac{dV}{dr}\right)^2 - \frac{m_\omega^2}{2g_\omega^2}\left(\frac{M^*}{M}\right)^2 V^2(r) + \frac{m_\sigma^2-m_\pi^2}{2M}\frac{1}{g_\sigma^2}(M^*-M)^3$$

$$\left.\left.+\frac{m_\sigma^2-m_\pi^2}{8M^2}\frac{1}{g_\sigma^2}(M^*-M)^4 - \frac{1}{2e^2}\left(\frac{dA_0}{dr}\right)^2\right]\right\}, \qquad (7.117)$$

where $\varepsilon_\lambda = E_\lambda - M$.

Two models were investigated in [304] for finite nuclei: Model I corresponds to the case $\eta = 1$ ($G_\omega = g_\omega$), while in Model II η was chosen equal to 1.385. We notice an interesting feature in both models: there exist several types of self-consistent solutions for one and the same set of fitting parameters, the system of (7.116) being strongly nonlinear. Consider ^{4}He nucleus. In this case three types of solutions were obtained: two types of abnormal solutions and one solution of normal type.

As for the abnormal solutions, there exists an (abnormal) solution with a negative value of the total binding energy (TBE) and an (abnormal) solution with a positive value of the TBE. We shall not discuss here the latter case in detail. The first type of abnormal solution demonstrates a shell-like structure, most of its mass being concentrated on the nuclear surface. The existence of this solution is connected with the possibility of the scalar field to have a kinked configuration.

Normal-type solutions, with a conventional density distribution, were also obtained for ^{4}He. The results for ^{16}O, ^{40}Ca, and ^{48}Ca are of the same character. The following conclusions were stated in [304]:

1. The normal-type solution can be generated in the framework of a relativistic chiral σ–ω model for finite nuclei.
2. The normal-type solution was used to calculate the following nuclear bulk properties: charge density distributions, r.m.s. radii, single-particle energies, and total binding energies for light and mid-weight nuclei. The values obtained are not far from the observed values.
3. In the case of ^{4}He, the normal-type solution for model I corresponds to the ground state of this nucleus.
4. The theory simultaneously also contains configurations of abnormal type, which for large values of m_σ may in some cases be much more strongly bound than those of normal type.
5. However, for more conventional values of m_σ (~ 500 MeV) only one possibility remains – that of a normal configuration.

In [305] the exchange terms were taken into consideration, the σ–ω model being treated in the Hartree–Fock approach. The σ–ω model includes the pion–nucleon interaction in a pseudoscalar form. So in [305] the self-consistent ground state was obtained in the relativistic Hartree–Fock theory for pseudoscalar coupling. The problem is that the structure of the σ–ω model demands $g_\pi = g_\sigma$, so the pion–nucleon interaction is reduced in this model. As seen from calculations [305], a reduction (by about 30%) of the pseudoscalar coupling appears to be sufficient to obtain a self-consistent Hartree–Fock solution. The details of the relativistic Hartree–Fock (RHF) theory will be discussed in the next chapter. Here we formulate only the main results obtained in [305] in the framework of RHF approach for the σ–ω model and discuss briefly the role of the exchange effects in this case:

1. The exchange terms play an essential role in bringing the symmetry energy close to the experimental value (for Model II, in particular).
2. The exchange effects decrease the binding energy of the abnormal solutions in finite nuclei, and pions (even with a reduced ps coupling) make these configurations unstable.
3. The small value of M^*, brought about by the π–nucleon interaction through the ps coupling, yields spin–orbit splittings closer to the experimental values (in normal solutions).
4. The spectrum and geometrical properties of normal solutions are remarkably improved if π and ρ mesons are considered.

Finally, we mention some important problems that remain unsolved in the Boguta chiral σ–ω model. In this model the ω meson is not introduced as a gauge field; for this reason the σ–ω interaction, the most important part of the model, is treated phenomenologically. The values of the σN and πN interactions appear to be too small and the Goldberger–Treiman relation is not satisfied. The scalar and vector fields are too weak to be compatible with the Dirac phenomenology. The predicted values of the incompressibility modulus are too large in the Boguta model. Further discussion of this model is given in [308–310]. The main conclusion is that models that include a light scalar meson (with the dual role of the chiral partner of the pion and the mediator of the intermediate-range NN interaction) and spontaneous symmetry breaking from a "Mexican-hat" potential do not reproduce quantitatively the basic ground-state properties of finite nuclei at the Hartree level. The situation for this model is improved to a certain extent in the RHF approximation by including the contributions of the pion (with pseudoscalar coupling) and of the ρ meson alongside those of the σ and ω fields. In this case the shell effects in nuclear density distributions (which are observed for that model in the Hartree approximation) are strongly attenuated and the charge density becomes remarkably close to the experimental data. However, Hartree–Fock calculations have in this case been carried out only for light and mid-weight nuclei (from ^{4}He up to ^{48}Ca) at present [305]. More complete Hartree–Fock calculations for heavy nuclei are in progress.

Meanwhile, several attempts have been done to develop workable chiral models that produce reasonable results for finite nuclei in the Hartree approximation. For example, in [308–310] a nonlinear realization (see also [311, 312]) of the chiral symmetry is considered, with the effective Lagrangian chosen by requiring the model to satisfy the low-energy theorems of broken scale invariance in QCD (at the tree level in the effective scalar field). This approach has been shown to describe finite nuclei successfully at the Hartree level.

Another approach has been developed in [313–315] (see also [316, 317]). It is based on an effective chiral Lagrangian (a generalization of the linear σ model) which incorporates the broken global chiral and scale symmetries of QCD. In this model the Lagrangian contains two scalar fields: the σ field, treated as a chiral partner of the pion, and the glueball field Φ_G (this latter field does not interact with nucleons directly).[2] The ω meson mass is formed phenomenologically and is strongly favored in this case to be of the form $\omega_\mu \omega^\mu \Phi_G^2$ by the bulk properties of nuclei; the mass of the ρ meson is produced in [313] by the same mechanism. The effective Lagrangian in [313] contains also a phenomenological term of the type $(\omega_\mu \omega^\mu)^2$, this version of the model being determined by six parameters (the ρ meson excluded). This model can provide a good description of finite nuclei, which is quite comparable to that obtained with the standard Walecka model extended to include nonlinear φ^3 and φ^4 terms.

7.7 Effective Gauge-Invariant Nuclear Lagrangian

In view of the great success achieved in describing electroweak interactions on the basis of elevating the global symmetry to a local one and the fundamental role of local symmetry in quantum chromodynamics, it appears to be important to formulate the nuclear chiral model in such a way that the local gauge symmetry related to the group $U(1) \times SU(2) \times SU(2)$ (which consists of phase transformations, rotations in isospin space, and chiral rotations) is taken into account. In this case the vector fields are considered as the gauge fields. However, most of the present approaches exhibit a deficiency in the context of nuclear structure in that they neglect the local chiral and isotopic symmetry when the mass term for the vector mesons is introduced. This deficiency was eliminated for the vector–isovector fields (i.e. for the ρ and a meson fields, the axial meson field a being the chiral partner of the ρ) in [321, 322], where the masses of vector–isovector mesons were produced via the Higgs mechanism.

In [323] a gauge model is developed where all vector fields appearing in the nuclear-structure context (ρ, a, ω, and A, where A is the electromagnetic field) are treated as Yang–Mills fields. Eight gauge fields are obtained in this case, the pion being the only Goldstone particle of the model.

---

[2] In [318] an approach was developed in which a scalar glueball was added to a modified Walecka-type Lagrangian. Another attempt was made in [313, 318, 319]. In this case the glueball potential field was introduced in the linear σ model; corresponding results on the quark level have been obtained in [320].

The underlying symmetry group is $U_{em}(1) \times U_\omega(1) \times SU(2) \times SU(2)$, with two $SU(2)$ groups relating to chiral and isotopic transformations and two $U(1)$ groups being phase transformations (these symmetries are considered to be local). In [323] the $U_\omega(1)$ symmetry group is introduced to incorporate the massive ω field into the model. To obtain a massive $\tilde{U}(1)$ gauge field, it appears to be necessary to introduce, besides the scalar field σ (the chiral partner of the pion), a new (pseudo)scalar field Σ related to the $U(1)$ symmetry and playing the role of a Higgs field for the ω meson. The nucleon mass in this model is formed by the interaction of nucleons with both the σ and the Σ fields. Note that in this approach the chiral σ meson is definitely not related to the 2π exchange. In this connection, the situation is similar to that in [324], where two scalar fields are introduced: one is the chiral partner of the pion and has a large mass, and the other is due to 2π exchange. The difference between the present model and that of [324] is that the second (pseudo)scalar Σ field is not related to 2π exchange either, since it is a Higgs field with a self-interaction ensuring a solution with a nonzero vacuum value of the field. Because 2π exchange undoubtedly plays an essential role in generating the attractive NN interaction, it should be also taken into account as a constituent of the model. In [323] a simpler way to do this is chosen by introducing one more scalar field (a third one) S_π on the basis of the conclusion (see [324], for example) that the isoscalar 2π exchange can be well reproduced by the exchange of a scalar meson with a mass between 500 and 600 MeV. In [325, 326] the model described above [322, 323] is applied to obtain the ground-state properties of nuclear matter and finite nuclei.

As a starting point, we consider a Lagrangian which includes the scalar field σ, the pseudoscalar field $\tilde{\boldsymbol{\pi}}$, the vector–isoscalar fields B and ω, the vector–isovector field $\tilde{\boldsymbol{\rho}}$, and the axial boson field $\tilde{\boldsymbol{a}}$, as well as the nucleonic fermion field ψ. The Lagrangian is locally invariant with respect to the symmetry group $U_{em}(1) \times U_\omega(1) \times SU(2) \times SU(2)$ (referred to as SG in what follows), consisting of two types of phase transformations $U(1)$, rotations in isotopic space, and chiral rotations. To ensure a spontaneous breaking of $U_{em}(1) \times U_\omega(1) \times SU(2) \times SU(2)$ symmetry, three further fields are introduced into the Lagrangian: the scalar–isoscalar real field σ (the chiral partner of the pion), and two Higgs fields, namely the (pseudo)scalar–isoscalar complex field Σ related to the $U_\omega(1)$ group, and the scalar–isospinor field Ψ which is transformed by the $U_{em}(1) \times SU(2) \times SU(2)$ group. At the initial level, our theory includes the nucleonic field ψ, the pionic field $\tilde{\boldsymbol{\pi}}$, and the σ fields, which are, so to speak, the basic degrees of freedom of the present chiral model. In addition, the four vector fields B_μ, ω_μ, $\tilde{\boldsymbol{\rho}}_\mu$, and $\tilde{\boldsymbol{a}}_\mu$ are gauge fields which elevate the global chiral symmetry to a local one; and, finally, the two Higgs fields Ψ and Σ provide the mass of the gauge vector fields ω, $\tilde{\boldsymbol{\rho}}$, and $\tilde{\boldsymbol{a}}$. Note that the $U_\omega(1)$ charge ζ of the Σ field is a free (integer) parameter of the theory.

7.7 Effective Gauge-Invariant Nuclear Lagrangian

The Lagrangian is given by the following equation:

$$\mathcal{L} = \bar{\psi}\left[i\gamma_\mu D^\mu - (\sigma + i\gamma_5 \boldsymbol{\tau} \cdot \tilde{\boldsymbol{\pi}})(g_\sigma + g_\Sigma |\Sigma|^2)\right]\psi$$
$$+ \frac{1}{2}\left[\mathcal{D}_\mu\tilde{\boldsymbol{\pi}} \cdot \mathcal{D}^\mu\tilde{\boldsymbol{\pi}} + \mathcal{D}_\mu\sigma \cdot \mathcal{D}^\mu\sigma + (\mathcal{D}_\mu\Sigma)^\dagger \cdot (\mathcal{D}^\mu\Sigma) + (\partial_\mu\Psi)^\dagger \cdot \mathcal{D}^\mu\Psi\right]$$
$$- \frac{1}{4}[B_{\mu\nu}B^{\mu\nu} + \omega_{\mu\nu}\omega^{\mu\nu} + \tilde{R}_{\mu\nu}\tilde{R}^{\mu\nu} + \tilde{A}_{\mu\nu}\tilde{A}^{\mu\nu}]$$
$$+ \frac{\mu_\sigma^2}{2}(\sigma^2 + \tilde{\boldsymbol{\pi}}^2) - \frac{\lambda_\sigma^2}{2}(\sigma^2 + \tilde{\boldsymbol{\pi}}^2)^2 + \frac{\mu_\Sigma^2}{2}|\Sigma|^2 - \frac{\lambda_\Sigma^2}{2}|\Sigma|^4$$
$$+ \frac{\mu_\Psi^2}{2}(\Psi^\dagger\Psi) - \frac{\lambda_\Psi^2}{2}(\Psi^\dagger\Psi)^2 - \frac{\nu^2}{2}(\Psi^\dagger \Gamma_5 \Psi)^2$$
$$+ C\bar{\psi}[\gamma_5(\tilde{\boldsymbol{\pi}} \cdot \mathcal{D}_\mu\sigma - \sigma \cdot \mathcal{D}_\mu\tilde{\boldsymbol{\pi}}) + (\tilde{\boldsymbol{\pi}} \times \mathcal{D}_\mu\tilde{\boldsymbol{\pi}})]\gamma^\mu \boldsymbol{\tau} \psi$$
$$+ D\Big[(\sigma \cdot \mathcal{D}_\mu\tilde{\boldsymbol{\pi}} - \tilde{\boldsymbol{\pi}}\mathcal{D}_\mu\sigma)(\sigma \cdot \mathcal{D}^\mu\tilde{\boldsymbol{\pi}} - \tilde{\boldsymbol{\pi}} \cdot \mathcal{D}^\mu\sigma)$$
$$+ (\tilde{\boldsymbol{\pi}} \times \mathcal{D}_\mu\tilde{\boldsymbol{\pi}})(\tilde{\boldsymbol{\pi}} \times \mathcal{D}^\mu\tilde{\boldsymbol{\pi}})\Big], \qquad (7.118)$$

where we have introduced the tensors

$$B_{\mu\nu} = \partial_\mu B_\nu - \partial_\nu B_\mu, \qquad (7.119)$$
$$\omega_{\mu\nu} = \partial_\mu \omega_\nu - \partial_\nu \omega_\mu, \qquad (7.120)$$
$$\tilde{R}_{\mu\nu} = \partial_\mu\tilde{\boldsymbol{\rho}}_\nu - \partial_\nu\tilde{\boldsymbol{\rho}}_\mu - 2g_\rho(\tilde{\boldsymbol{\rho}}_\mu \times \tilde{\boldsymbol{\rho}}_\nu + \tilde{\boldsymbol{a}}_\mu \times \tilde{\boldsymbol{a}}_\nu), \qquad (7.121)$$
$$\tilde{A}_{\mu\nu} = \partial_\mu\tilde{\boldsymbol{a}}_\nu - \partial_\nu\tilde{\boldsymbol{a}}_\mu - 2g_\rho(\tilde{\boldsymbol{\rho}}_\mu \times \tilde{\boldsymbol{a}}_\nu - \tilde{\boldsymbol{\rho}}_\nu \times \tilde{\boldsymbol{a}}_\mu), \qquad (7.122)$$

and the covariant derivatives

$$\mathcal{D}_\mu\psi = [\partial_\mu + ig_B \cdot B_\mu + ig_\omega \omega_\mu + ig_\rho(\tilde{\boldsymbol{\rho}}_\mu - \gamma_5\tilde{\boldsymbol{a}}_\mu)\boldsymbol{\tau}]\psi, \qquad (7.123)$$
$$\mathcal{D}_\mu\tilde{\boldsymbol{\pi}} = \partial_\mu\tilde{\boldsymbol{\pi}} + 2g_\rho(\tilde{\boldsymbol{\pi}} \times \tilde{\boldsymbol{\rho}}_\mu + \sigma\tilde{\boldsymbol{a}}_\mu), \qquad (7.124)$$
$$\mathcal{D}_\mu\sigma = \partial_\mu\sigma - 2g_\rho\tilde{\boldsymbol{\pi}} \cdot \tilde{\boldsymbol{a}}_\mu, \qquad (7.125)$$
$$\mathcal{D}_\mu\Sigma = \partial_\mu\Sigma + ig_\omega\zeta\Sigma\omega_\mu, \qquad (7.126)$$
$$\mathcal{D}_\mu\Psi = [\partial_\mu + ig_B B_\mu + ig_\rho(\tilde{\boldsymbol{\rho}}_\mu - \Gamma_5\tilde{\boldsymbol{a}}_\mu)\boldsymbol{T}]\Psi. \qquad (7.127)$$

The matrices Γ_5 and $\boldsymbol{T}$ act on Ψ in the same way as the matrices γ_5 and $\boldsymbol{\sigma}$ do with respect to the Dirac field ψ. Note that the Higgs field Σ may be either scalar or pseudoscalar because it enters the Lagrangian quadratically (this point is discussed in more detail in [323]).

The Higgs field Ψ has the form

$$\Psi = \begin{pmatrix} f_1 \\ f_2 \\ f_3 \\ f_4 \end{pmatrix}, \qquad (7.128)$$

where the components f_1 and f_2 have a parity opposite to that of the components f_3 and f_4. The fields f_1 and f_3 have the electric charge of the proton, whereas the fields f_2 and f_4 are neutral.

7 Pion Dynamics and Chiral Symmetry

In the Higgs sector, the symmetry group SG is represented by the field vector

$$\Omega = \begin{pmatrix} \Psi \\ \sigma \\ \widetilde{\pi} \\ \Sigma \end{pmatrix}, \qquad (7.129)$$

which has 14 real components, three of them being associated with the pionic Goldstone mode.

The following unitary gauge for the Higgs field vector is chosen in [323]:

$$\Omega = \begin{bmatrix} 0 \\ f_H + h_1 \\ 0 \\ h_2 \\ \sigma_0 + \varphi \\ \widetilde{\pi} \\ \Sigma_0 + \Phi \end{bmatrix}, \qquad (7.130)$$

where f_H, σ_0 and Σ_0 are the vacuum state values, and the fields h_1, h_2, φ, and Φ are the real, massive Higgs fields. The two scalar fields σ and Σ are presented here as the sums

$$\sigma = \sigma_0 + \varphi, \qquad \Sigma = \Sigma_0 + \Phi, \qquad (7.131)$$

φ and Φ being the fluctuations of these fields around their respective vacuum state expectation values.

To proceed to physical mesonic fields, one must diagonalize the part of the Lagrangian containing the B, $\widetilde{\rho}$, $\widetilde{a}$, and $\widetilde{\pi}$ fields. This can be done by introducing the electromagnetic field A_μ and renormalized isovector, axial, and pionic fields ρ_μ, a_μ, and π according to the relations given in [322, 323].

In [323] it is argued that the ω, ρ, and a fields have acquired masses because of spontaneous breaking of the chiral, isotopic, and phase symmetry. These masses vanish as the vacuum expectation values of the Higgs fields f_H, σ_0, and Σ_0 go to zero:

$$m_\rho = g_\rho f_H, \qquad m_\omega = g_\omega \zeta \Sigma_0, \qquad (7.132)$$

where g_ρ and g_ω are the ρ and ω meson coupling constants. After the transformations mentioned above have been done, our Lagrangian for the unitary gauge takes the form:

7.7 Effective Gauge-Invariant Nuclear Lagrangian

$$\mathcal{L} = \bar{\psi}\left\{\not{p} - \frac{e}{2}(1+\tau_3)\not{A} - g_\omega \cdot \not{\omega} - g_\rho \cdot \not{\rho}\cdot\boldsymbol{\tau} - g_\rho \cdot g_A \frac{m_a^2}{m_\rho^2}\gamma_5 \not{a}\cdot\boldsymbol{\tau} - \widetilde{M}\right.$$

$$\left. - \frac{g_A}{2f_\pi}\gamma_5\gamma^\mu\cdot\boldsymbol{\tau}\cdot\partial_\mu\boldsymbol{\pi}\right\}\psi - \frac{1}{4}\left(F_{\mu\nu}F^{\mu\nu} + \omega_{\mu\nu}\cdot\omega^{\mu\nu} + \boldsymbol{\rho}_{\mu\nu}\cdot\boldsymbol{\rho}^{\mu\nu}\right.$$

$$\left. + \boldsymbol{a}_{\mu\nu}\cdot\boldsymbol{a}^{\mu\nu}\right) + \frac{m_\omega^{*2}}{2}\omega_\mu\cdot\omega^\mu + \frac{m_\rho^2}{2}\boldsymbol{\rho}_\mu\cdot\boldsymbol{\rho}^\mu + \frac{m_a^2}{2}\boldsymbol{a}_\mu\cdot\boldsymbol{a}^\mu + \frac{1}{2}\partial_\mu\boldsymbol{\pi}\cdot\partial^\mu\boldsymbol{\pi}$$

$$+ \frac{1}{2}\partial_\mu\sigma\cdot\partial^\mu\sigma + \frac{1}{2}\partial_\mu\Sigma\cdot\partial^\mu\Sigma + \frac{1}{2}\partial_\mu h_1\cdot\partial^\mu h_1 + \frac{1}{2}\partial_\mu h_2\cdot\partial^\mu h_2$$

$$+ \frac{m_\sigma^2}{4}\sigma^2\left(1 - \frac{\sigma^2}{2\sigma_0^2}\right) + \frac{m_\Sigma^2}{4}\Sigma^2\left(1 - \frac{\Sigma^2}{2\Sigma_0^2}\right)$$

$$+ \frac{m_{h_1}^2}{4}\left[(f_H + h_1)^2 + h_2^2\right]\left[1 - \frac{(f_H + h_1)^2 + h_2^2}{2f_H^2}\right]$$

$$- \frac{m_{h_2}^2}{2f_H^2}(f_H + h_1)^2 h_2^2 + \mathcal{L}_{\rho,\mathrm{em}}, \tag{7.133}$$

where g_A is the axial nucleon form factor and the following notations are used:

$$\not{p} \equiv i\gamma^\mu\partial_\mu, \quad \not{\omega} \equiv \gamma^\mu\cdot\omega_\mu, \quad \not{\rho} \equiv \gamma^\mu\cdot\boldsymbol{\rho}_\mu, \quad \not{a} \equiv \gamma^\mu\cdot\boldsymbol{a}_\mu. \tag{7.134}$$

The Lagrangian (7.133) can also be represented in terms of the fields φ and Φ. Note that in contrast to the conventional linear σ model [296–298], in this case the value of σ_0 is not fixed by the pion weak-decay rate constant f_π ($f_\pi \approx 93\,\mathrm{MeV}$), because of the scaling of the pion field in our case. In the present approach we have

$$\sigma_0 \frac{m_\rho}{m_a}\sqrt{1 + 2D\sigma_0^2} = f_\pi. \tag{7.135}$$

Since the parameter $\mathcal{D}$ is free (see (7.118)), σ_0 is free as well. The present approach reproduces also the well-known relation [327] between the masses of the ρ and axial-vector mesons,

$$m_a^2 = m_\rho^2\left(1 - \frac{4g_\rho^2 f_\pi^2}{m_\rho^2}\right)^{-1}. \tag{7.136}$$

If the $\boldsymbol{a}$ field is identified with the observed a_1 meson (with the experimental mass $m_{a_1} \approx 1270\,\mathrm{MeV}$), one finds a slight discrepancy between the calculated and experimental values. Thus, by adopting $g_\rho = 2.84$, which leads to the proper isobaric-spin potential in the nucleus (see [77] and references therein), one obtains $m_{a_1} \approx 1060\,\mathrm{MeV}$. In accordance with (7.135), this gives the value $\sigma_0 = 128\,\mathrm{MeV}$ (for $\mathcal{D} = 0$), which will be utilized in what follows.

The nucleon mass
$$M = \left(g_\sigma + g_\Sigma \Sigma_0^2\right)\sigma_0 \tag{7.137}$$

is also created spontaneously, owing to the interaction of the nucleons with the scalar σ and Σ fields. For our further purposes it is useful to introduce the quantity ξ, which is the ratio of the portions of the nucleon mass formed by interaction with the σ and Σ fields separately:

$$1 - \xi = \frac{g_\sigma \sigma_0}{M}, \qquad \xi = g_\Sigma \frac{\Sigma_0^2}{M} \sigma_0 . \qquad (7.138)$$

The effective masses for the nucleon and ω meson are introduced via the following equations:

$$\widetilde{M} = \left(g_\sigma + g_\Sigma \Sigma^2\right) \sigma , \qquad (7.139)$$

$$m_\omega^* = g_\omega \zeta \Sigma . \qquad (7.140)$$

From (7.139) it can be seen that the effective mass of the ω meson is formed by $\Sigma^2 \omega^2$ interaction term in (7.133).

The antisymmetric tensors $\boldsymbol{\rho}_{\mu\nu}$ and $\boldsymbol{a}_{\mu\nu}$ are given by

$$\boldsymbol{\rho}_{\mu\nu} = \partial_\mu \boldsymbol{\rho}_\nu - \partial_\nu \boldsymbol{\rho}_\mu , \qquad \boldsymbol{a}_{\mu\nu} = \partial_\mu \boldsymbol{a}_\nu - \partial_\nu \boldsymbol{a}_\mu . \qquad (7.141)$$

Finally, the term $\mathcal{L}_{\rho,\text{em}}$ contains small electromagnetic effects due to renormalization of the neutral ρ meson component in the chiral theory considered here. Its role was discussed in [322, 323] and shown to be negligible in the nuclear-structure context. For this reason we shall neglect this term in our present calculations and consider the ρ meson contribution in a standard manner [77]. In what follows we suppose the Higgs masses m_{h_1} and m_{h_2} to be so large that there is no need to consider the excitations of the Ψ field.

7.8 Mean-Field Results for Nuclear Matter and Finite Nuclei

To investigate the main properties of the model suggested above, we use the mean-field approximation and consider the ground-state properties of symmetric infinite nuclear matter and spherical doubly magic nuclei (i.e. we consider the static limit). In this case only the ω, ρ, and nucleon fields and the Higgs fields σ, Σ survive. The Euler–Lagrange equations for the surviving fields, obtained from (7.133), read

$$\left\{-i\boldsymbol{\alpha} \cdot \boldsymbol{\nabla} + \beta \widetilde{M} + g_\omega \cdot \omega^0 + g_\rho \boldsymbol{\tau} \cdot \boldsymbol{\rho}_0 + e\frac{1+\tau_3}{2} A^0\right\} \psi_i = E_i \cdot \psi_i , \qquad (7.142)$$

$$\left\{-\Delta + m_\omega^{*2}\right\} \omega^0 = g_\omega \rho_V , \qquad (7.143)$$

$$\left\{-\Delta + m_\rho^2\right\} \rho_3^0 = g_\rho \rho_3 , \qquad (7.144)$$

$$-\Delta A(r) = e \rho_p , \qquad (7.145)$$

7.8 Mean-Field Results for Nuclear Matter and Finite Nuclei

$$\left\{-\Delta - \frac{m_\sigma^2}{2}\left(1 - \frac{\sigma^2}{\sigma_0^2}\right)\right\}\sigma = -\left(g_\sigma + g_\Sigma \Sigma^2\right)\rho_S, \tag{7.146}$$

$$\left\{-\Delta - \frac{m_\Sigma^2}{2}\left(1 - \frac{\Sigma^2}{\Sigma_0^2}\right)\right\}\Sigma = -2g_\Sigma \Sigma \sigma \rho_S + g_\omega^2 \zeta^2 \Sigma \omega_0^2. \tag{7.147}$$

The total energy density of the system is given by

$$\mathcal{H} = \psi^\dagger \left[\boldsymbol{\alpha}\left(-i\boldsymbol{\nabla}\right) + \beta\left(g_\sigma + g_\Sigma \Sigma^2\right)\sigma + g_\omega \omega_0\right]\psi$$
$$-\frac{m_\omega^{*2}}{2}\omega_0^2 + \frac{(\nabla\sigma)^2}{2} + \frac{(\nabla\Sigma)^2}{2} - \frac{(\nabla\omega_0)^2}{2}$$
$$-\frac{m_\sigma^2}{4}\sigma^2\left(1 - \frac{\sigma^2}{2\sigma_0^2}\right) - \frac{m_\Sigma^2}{4}\Sigma^2\left(1 - \frac{\Sigma^2}{2\Sigma_0^2}\right). \tag{7.148}$$

Equations (7.142)–(7.147) form a system of strongly nonlinear equations. The nonlinearities are produced by interactions and self-interactions of the two mesonic fields σ and Σ. Nonlinear terms in mesonic fields are also involved in the Dirac equation via the effective nucleon mass.

Let us remark here also that the model contains two scalar mesons σ and Σ directly in the Lagrangian (7.133), where σ is the chiral partner of the pion and Σ is the Higgs meson for the ω field, both of them taking part in the formation of the nucleon mass. The role of these two mesons is to maintain the chiral symmetry and to produce the proper density dependence of the solutions, i.e. to incorporate the nonlinearity properly. However, indirectly there is one more scalar meson in our approach. It is known [34] that there exists a strong interaction between two pions; for this reason one should take the corresponding correlated 2π exchange process into account in the NN interaction. The correlated $\pi\pi$S wave contribution can be well approximated by the exchange of a scalar–isoscalar boson and, in principle, this contribution may be calculated correctly via the interaction Lagrangian (see (7.133)).

We introduce here an additional scalar field S_π just to take into account the 2π isoscalar exchange, however, we do not calculate S_π on the basis of our Lagrangian but suppose that this field obeys a standard Klein–Gordon equation

$$(\Box + m_S^2)S_\pi = -g_S\overline{\psi}\psi, \tag{7.149}$$

where g_S and m_S are considered as fitting parameters.

It should be emphasized that the utilization of this field corresponds completely to our Lagrangian and needs no additional interaction terms to be introduced into (7.133). This point of view is similar to the philosophy adopted in [324].

Notice also that the use of S_π leads to a further modification of the nucleon effective mass,

$$\widetilde{M}^* = \widetilde{M} + S_\pi, \tag{7.150}$$

in our calculations.

In the general case we consider the masses of the scalar fields m_σ and m_Σ, the scalar ratio ξ (see (7.138)), and the ω meson coupling constant g_ω as free parameters of the model. The coupling constants g_σ and g_Σ and the vacuum value Σ_0 are obtained from (7.138) and (7.132) ensuring the right value of the free ω meson mass. The value of σ_0 is fixed at 128 MeV in accordance with (7.135).

The parameter ζ may take only integer values ($\zeta = 1, 2, 3, \ldots$) owing to its physical meaning, ζ being the $U_\omega(1)$ charge of the Σ field. From (7.142)–(7.147), written for nuclear matter, one can see that a change in the value of ζ can be compensated by a corresponding modification of other parameters considered as free. For definiteness, $\zeta = 1$ was used in [325] in the calculations for nuclear matter. It should be noticed that ζ does influence (7.142)–(7.147) for finite systems; however, as was established by calculations for finite nuclei, for $m_\Sigma > 600$ MeV the influence of ζ is negligible. For this reason we use $\zeta = 1$ for both nuclear matter and finite nuclei here.

The system of (7.142)–(7.147), (7.149) was solved numerically in [325] and self-consistent solutions were obtained for the ground state of nuclear matter and finite nuclei. The general case of finite masses (m_σ and m_Σ) is, for nuclear matter, connected with the solution of the following equations for the nucleon spinors and the ω^0, φ, Φ, and S_π fields:

$$\left(\slashed{p} - g_\omega \slashed{\omega} - \widetilde{M}\right)\psi = 0, \tag{7.151}$$

$$m_\omega^{*2}\omega_0 = g_\omega \rho_V, \tag{7.152}$$

$$\frac{m_\sigma^2}{2\sigma_0^2}\varphi^3 + \frac{3}{2}\frac{m_\sigma^2}{\sigma_0}\varphi^2 + m_\sigma^2\varphi + \left[g_\sigma + g_\Sigma(\Sigma_0 + \Phi)^2\right]\rho_S = 0, \tag{7.153}$$

$$\frac{m_\Sigma^2}{2\Sigma_0^2}\Phi^2 + \frac{m_\Sigma^2}{\Sigma_0}\Phi + 2g_\Sigma(\sigma_0 + \varphi)\rho_S - g_\omega^2\zeta^2\omega_0^2 = 0, \tag{7.154}$$

$$S_\pi = -\frac{g_S^2}{m_S^2}\rho_S. \tag{7.155}$$

Four models were considered in [325]. Two of these models (Models I and II) take into account the scalar field S_π and, in accordance with the comments made above, may be considered as realistic. Model I utilizes finite values for m_σ and m_Σ and contains six free parameters (g_ω, m_σ, m_Σ, ξ, g_S, m_S); in the case of Model II, where $m_\sigma = \infty$, five parameters are involved, and are determined from nuclear data. We should mention also that for infinite nuclear matter the ratio g_S/m_S enters the equations, rather than two parameters g_S and m_S independently (this ratio was taken as either equal or very close to the value used in the original Walecka model [74]). Models III and IV should be considered as illustrative; the contribution of S_π is not taken into account in this case. Model III considers finite values of m_σ and involves four parameters; Model IV uses $m_\sigma = \infty$, i.e. three parameters are taken as adjustable in this case. The sets of parameters for the various models are given in Table 7.1 along with calculated values of the "nuclear-matter observables" E/A,

7.8 Mean-Field Results for Nuclear Matter and Finite Nuclei

Table 7.1. Values of parameters, and quantities calculated for nuclear matter ($\zeta = 1$, $\sigma_0 = 128\,\text{MeV}$) (M = model)

	g_ω	m_σ (MeV)	m_Σ (MeV)	ξ
M I	12.585	950	1095.5	0.25
M II	11.974	∞	1220	0.25
M II*	11.974	∞	∞	0.25
M III	7.28	800	818.4	0.5
M IV	5.06	∞	413	0.4

	g_S/m_S (fm)	g_S	E/A (MeV)	ρ_0 (fm^{-3})	K (MeV)
M I	3.3	5.5	-16	0.16	180
M II	3.5	5.5	-16	0.155	276
M II*	3.5	5.5	-14.544	0.1833	511
M III	–	–	-16	0.16	287
M IV	–	–	-16	0.153	424

ρ_0, and K. These three values are not sufficient to determine uniquely all adjustable parameters. The values of the parameters given in Table 7.1 were determined not only from the nuclear-matter data but also from a procedure of fitting the bulk properties of finite nuclei.

Chiral models produce an incompressibility modulus K which is, as a rule, far from the phenomenologically desirable values (see, however, [309, 313]). We should emphasize that our approach gives the equilibrium values for nuclear matter (ρ_0, E/A, K, $\widetilde{M}$) corresponding to just these phenomenological values (see Table 7.1, Models I and II). Good results for the modulus K are obtained in [325] owing to the proper density dependence incorporated in this model, formed just by the σ and Σ fields. One can see this directly, by comparing the results in Table 7.1 for Models II and II*. In Model II the σ field is frozen ($\sigma = \sigma_0$, $m_\sigma = \infty$) but the Σ field is not ($m_\Sigma \neq \infty$). In this case one obtains reasonable values for all nuclear-matter observables, the modulus K in particular. In Model II*, both scalar fields σ and Σ are frozen ($m_\sigma = \infty$, $m_\Sigma = \infty$). Thus Model II* ($m_\omega^* = m_\omega$) corresponds completely to the Walecka model. So in Model II* saturation of nuclear matter is realized (as can be seen from Table 7.1), but one obtains too high a value of the incompressibility modulus (511 MeV).

Finite-nucleus properties have been obtained by solution of (7.142)–(7.147) and (7.149) for the S_π field, self-consistent Hartree calculations being carried out for five doubly magic spherical nuclei from ^{16}O up to ^{208}Pb for the four models described above. Calculations were performed for the four

Table 7.2. Comparison of the present Hartree calculations with experimental values. The total binding energy per particle and the proton spin–orbit splitting Δ_{LS} for the 1p shell (^{16}O) or 1d shell (^{40}Ca and ^{48}Ca) are given in MeV; the r.m.s. charge radii r_{c} are in fm

	^{16}O			^{40}Ca			^{48}Ca			^{90}Zr		^{208}Pb	
	$-E/A$	r_{c}	Δ_{LS}	$-E/A$	r_{c}	Δ_{LS}	$-E/A$	r_{c}	Δ_{LS}	$-E/A$	r_{c}	$-E/A$	r_{c}
M I	7.40	2.73	6.32	8.30	3.45	6.90	8.59	3.44	6.75	8.60	4.23	7.99	5.48
M II	6.88	2.72	5.5	7.90	3.45	5.98	8.17	3.47	5.61	8.27	4.24	7.73	5.48
M III	11.1	2.49	2.84	10.8	3.27	2.63	10.8	3.36	2.10	10.3	4.15	9.21	5.48
M IV	5.97	2.69	1.23	7.24	3.42	1.34	7.12	3.48	1.18	7.43	4.25	7.13	5.49
Experiment	7.98	2.73	6.3	8.55	3.48	7.2	8.67	3.47	4.3	8.71	4.27	7.87	5.5

sets of parameters given in Table 7.1; for nuclei with $N \neq Z$ the contribution of the ρ meson has been taken into account, with the ρN coupling constant g_ρ equal to 2.84. The parameters given in Table 7.1[3] were chosen, as was mentioned above, by a fitting procedure of the nuclear-matter observables and of the bulk properties of finite nuclei, i.e. the r.m.s. charge radii r_{c}, spin–orbit splittings Δ_{LS}, and the total binding energy per particle $-E/A$. The results of these calculations are given in Table 7.2.

On the whole, from all of the data presented in [325], one can see that Models III and IV do not survive and cannot be considered as working models. They produce results for the binding energy per particle and the r.m.s. charge radii of only qualitative character, and values of the spin–orbit splittings that are too small. Moreover, Model III shows strong shell effects in the density distributions (density oscillations), which are not observed experimentally. This is a consequence of the lack of an attractive field with a small mass.

---

[3] We should mention here also the values of Σ_0. The values for Models I–IV were calculated rather than treated as fitting parameters and are given, respectively, by $\Sigma_0 = 62.2$, 65.4, 107.6, and 155.8 MeV. Let us remark also that the value of $\xi = 0.25$ obtained for the realistic cases of Models I and II means that three-quarters of the nucleon mass M is formed by coupling to the σ field and one-quarter via coupling to the Σ field.

7.8 Mean-Field Results for Nuclear Matter and Finite Nuclei

In contrast, Models I and II have all of the features of reasonable working models. The results obtained with these models in the Hartree approximation give a good description of both nuclear matter and the bulk properties of finite nuclei. The density distributions obtained in this case reproduce the experimental distributions reasonably well. The aim of [325] was to investigate whether the models suggested there could be treated as working models, rather than to obtain optimal parameterizations for the models. So it is possible to say, in summary, that this aim was achieved, though an even better description of the bulk nuclear properties could be obtained.

In [325], a chiral approach developed earlier [322, 323] was investigated in the context of nuclear structure and an attempt was made to obtain the ground-state properties of nuclear matter and finite nuclei in the mean-field approximation. This approach treats all the vector fields involved as Yang–Mills fields, the pion being the only Goldstone particle of the theory. The two-pion exchange produced by the pion–nucleon interaction in the model Lagrangian is simulated by a scalar–isoscalar field S_π. Two further scalar fields are essential ingredients of this approach: the σ field, a chiral partner of the pion, and the Σ field, a Higgs field for the ω meson; these fields are responsible for maintaining the chiral symmetry and creating the non-linear structure of the equations of motion. They also provide a mechanism to suppress the value of the incompressibility modulus K to a desirable phenomenological value. The good values of K are a remarkable feature of the chiral approach developed in [322, 323, 325].

The two models mentioned above (Models I and II) were constructed within the framework of this chiral approach. Both of these models reproduce reasonably well the nuclear-matter observables and bulk properties of finite nuclei, the results being comparable to descriptions of the same type obtained within standard relativistic calculations. Model I contains six parameters, while Model II contains only five, the scalar field being frozen in the latter case (corresponding to an infinitely heavy σ meson). However, Model II gives results which are similar to those obtained with Model I (the density distributions in Model II are even better than in Model I, this fact being related to the larger values of K obtained in Model II in comparison with Model I). Both Model I and Model II may be utilized as working models.

Effective chiral Lagrangians for spin-1 mesons have been considered also in [52, 328]. A nonlinear realization of the chiral symmetry was adopted in those papers.

Another challenging approach to the problem considered in this section could be to proceed to a wider (though not so exact) symmetry group (such as $SU(3)$) [329].

The modern point of view is to combine the two basic principles of low-energy QCD (vector meson dominance (Chap. 5) and chiral symmetry) in a reasonably chosen effective Lagrangian [311, 330].

8 The Relativistic Hartree–Fock Approach

8.1 The Relativistic Hartree–Fock Lagrangian

In Chap. 4 we summarized the current status of the relativistic theory of nuclear structure with self-interactions of mesonic fields in the relativistic mean-field approximation. In the context of the discussions in this chapter, the relativistic mean-field theory corresponds to the relativistic Hartree approximation. In spite of the success of this theory, some shortcomings should be pointed out:

1. Too small a value of the symmetry energy is obtained if the coupling constant g_ρ takes its experimental value. In fact, important contributions due to exchange (Fock) terms and ρ–N tensor coupling are missing in this approach (see, however, Sect. 5.1).
2. Though the spin–orbit force is globally rather well described, this theory is unable to reproduce reasonably well the reduction of d-state splitting when going from ^{40}Ca to ^{48}Ca.
3. The mean-field approach cannot take into account the effective spin–spin interaction that arises from the identity of the nucleons through the antisymmetrization of the wave function. This effect is more important in spin-unsaturated systems [331].
4. The Hartree approximation, in principle, corresponds only to instantaneous interaction (without retardation).[1]

It is important to investigate the role of the Fock terms, since they are essentially associated with the isovector mesons, namely the pion and the ρ, first of all. Indeed, as was noted above, the isovector mesons do not contribute at all to the Hartree terms for $N = Z$ systems. The π meson, whose role might seem to be important, does not contribute in the Hartree approximation. The main interest in the Hartree–Fock (HF) model, which is more complicated in its practical realization, is associated with the possibility of taking the isovector π and ρ mesons into account (in the case of $N = Z$ nuclei, too) (see Appendix A9).

---

[1] Let us emphasize here that, strictly speaking, the Hartree approximation should be based on the free NN forces.

8 The Relativistic Hartree–Fock Approach

Relativistic Hartree–Fock calculations for finite nuclei were first performed in [57, 60]. In [60] the scheme of an approximate relativistic theory (ART) of the nucleus was developed on the basis of the meson NN potentials, specific attention was paid to the problem of the spin–orbit splittings in finite nuclei. In the first case [57], calculations were carried out for a vector–scalar nucleon–nucleon force model. In [57] a study was also made of the exchange potentials resulting from the pseudoscalar part (one-pion exchange) of the nucleon–nucleon interaction. It was shown that these terms appear to be important. However, the baryon self-energies become extremely large at normal nuclear density, if a ps coupling is utilized.

The relativistic Hartree–Fock method has been investigated further in [63, 64, 332]. Jaminon, Mahaux, and Rochus [333] studied infinite nuclear matter and concentrated on the single-nucleon optical potential. The next step was taken in [101]; the authors of that paper begin with a renormalizable relativistic quantum field Lagrangian and focus on the properties of nuclear and neutron matter (with and without pion exchange) (see also [334]). The most complete RHF calculations for finite nuclei have been performed in [335, 336]. The role of retardation effects was investigated in [337] for nuclear matter and in [338] for finite nuclei. A least-squares fit to nuclear ground-state properties has been carried out at the Ohio Supercomputer Center in Columbus [339]. Contributions to this field have also been made in [340–342]. Essential progress in developing the RHF method was achieved in [117, 305, 343], the nonlinear interactions and self-interactions of mesonic fields being incorporated into the framework of the RHF approach. In [344] an RHF approximation was developed for a globally chirally invariant model including σ, ω, π, ρ mesons and the axial meson a_1 (see Appendix A10). In [345] a method to derive density-dependent effective interactions was considered.

The effective Lagrangian density $\mathcal{L}$ of the model considered here is given by the sum of a free Lagrangian $\mathcal{L}_0(\psi, \varphi, \omega, \rho, \pi, A)$, an interaction Lagrangian $\mathcal{L}_{\text{int}}$ (nucleon–meson) responsible for interactions of nucleons with the various meson fields, and a nonlinear potential-energy functional U_{NL} (meson–meson) taking into account meson self-interactions and meson–meson interactions of isoscalar meson fields:

$$\mathcal{L} = \mathcal{L}_0(\psi, \varphi, \omega, \rho, \pi, A) + \mathcal{L}_{\text{int}} \text{ (nucleon–meson)} - U_{\text{NL}} \text{ (meson–meson)}. \quad (8.1)$$

The free Lagrangian density is given by

$$\begin{aligned}\mathcal{L}_0(\psi, \varphi, \omega, \rho, \pi, A) &= \overline{\psi}(i\gamma_\mu \partial^\mu - M)\psi - \frac{1}{2}m_\sigma^2 \varphi^2 + \frac{1}{2}(\partial_\mu \varphi \cdot \partial^\mu \varphi) \\&+ \frac{1}{2}m_\omega^2 \omega_\mu \omega^\mu - \frac{1}{4}\omega_{\mu\nu}\omega^{\mu\nu} + \frac{1}{2}m_\rho^2 \boldsymbol{\rho}_\mu \cdot \boldsymbol{\rho}^\mu - \frac{1}{4}\boldsymbol{\rho}_{\mu\nu} \cdot \boldsymbol{\rho}^{\mu\nu} \\&+ \frac{1}{2}(\partial_\mu \boldsymbol{\pi} \cdot \partial^\mu \boldsymbol{\pi} - m_\pi^2 \boldsymbol{\pi}^2) - \frac{1}{4}F_{\mu\nu} \cdot F^{\mu\nu}, \end{aligned} \quad (8.2)$$

the tensors $\omega_{\mu\nu}$, $\boldsymbol{\rho}_{\mu\nu}$, $F_{\mu\nu}$ and all other notation being defined earlier.

8.1 The Relativistic Hartree–Fock Lagrangian

The meson–nucleon interaction Lagrangian is written as

$$\mathcal{L}_{\text{int}}(\text{nucleon–meson}) = -g_\sigma \bar{\psi}\psi\varphi - g_\omega \bar{\psi}\gamma^\mu \omega_\mu \psi - \frac{f_\omega}{2M}\bar{\psi}\sigma^{\mu\nu}\partial_\mu \omega_\nu \psi$$
$$- g_\rho \bar{\psi}\gamma^\mu \boldsymbol{\rho}_\mu \cdot \boldsymbol{\tau}\psi - \frac{f_\rho}{2M}\bar{\psi}\sigma^{\mu\nu}\partial_\mu \boldsymbol{\rho}_\nu \cdot \boldsymbol{\tau}\psi - e\bar{\psi}\gamma^\mu \frac{1}{2}(1+\tau_3)A_\mu \psi$$
$$- \frac{f_\pi}{m_\pi}\bar{\psi}\gamma_5 \gamma^\mu \partial_\mu \boldsymbol{\pi} \cdot \boldsymbol{\tau}\psi . \tag{8.3}$$

As was discussed above, the πNN interaction Lagrangian $\mathcal{L}_{\pi NN}$ can be written into two possible forms. We can use either the pseudoscalar (ps) coupling,

$$\mathcal{L}^{\text{ps}}_{\pi NN} = -ig_\pi \bar{\psi}\gamma_5 \boldsymbol{\pi} \cdot \boldsymbol{\tau}\psi , \tag{8.4}$$

or the pseudovector (pv) coupling,

$$\mathcal{L}^{\text{pv}} = -\frac{f_\pi}{m_\pi}\bar{\psi}\gamma_5 \gamma_\mu \partial^\mu \boldsymbol{\pi} \cdot \boldsymbol{\tau}\psi . \tag{8.5}$$

Both couplings lead to the same one-pion exchange potential[2] in the nonrelativistic limit for on-shell nucleons if their corresponding coupling constants satisfy the equivalence relation

$$\frac{g_\pi}{2M} = \frac{f_\pi}{m_\pi} . \tag{8.6}$$

As for nuclear structure, it is known that the baryon self-energies become extremely large (about 40 times larger than their pseudovector counterpart) at the normal nuclear density if a pseudoscalar coupling is used, which has a drastic effect on the single-particle spectrum [57, 101] (see, however, Chap. 7). For this reason the pseudovector coupling is chosen in the RHF framework. Equation (8.3) includes also tensor couplings for the ω and ρ mesons (for the ω meson, the tensor coupling is small and is not considered in the numerical calculations described here). The presence of tensor couplings implies that the model Lagrangian is no longer renormalizable and that all physical observables should be calculated at the tree level.

Finally, the potential-energy functional U_{NL} (meson–meson) in (8.1) is taken in the form

$$U_{\text{NL}}(\text{meson–meson}) = U_{\text{NL}}(\varphi,\omega) = \frac{1}{3}\bar{b}M(g_\sigma\varphi)^3 + \frac{1}{4}\bar{c}(g_\sigma\varphi)^4$$
$$+ \bar{d}M(g_\sigma\varphi)(g_\omega^2 \omega_\mu \omega^\mu) - \frac{1}{4}\bar{e}(g_\sigma\varphi)^2(g_\omega^2 \omega_\mu \omega^\mu) - \frac{1}{4}\bar{f}(g_\omega^2 \omega_\mu \omega^\mu)^2 , \tag{8.7}$$

---

[2] We should mention that in the RHF method the pion field is taken into account explicitly in the one-pion approximation. However, the two-pion exchange is considered implicitly via the scalar–isoscalar field.

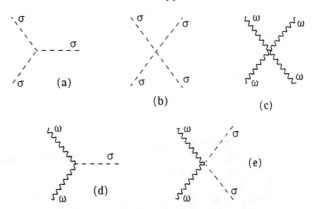

Fig. 8.1. Diagrammatic representation of interactions between isoscalar meson fields (the *dashed lines* correspond to scalar mesons, while the *wavy lines* correspond to the ω field). Cases (a), (b), and (c) represent σ^3, σ^4, and ω^4 self-interaction terms, respectively, and (d) and (e) represent $\sigma\omega^2$ and $\sigma^2\omega^2$ interaction terms, respectively

which is specified by five dimensionless parameters $\bar{b}$, $\bar{c}$, $\bar{d}$, $\bar{e}$, and $\bar{f}$. The form of $U_{\rm NL}(\varphi,\omega)$ given by (8.7) takes into account self-interactions of the isoscalar–scalar meson field (cubic and quartic terms in φ), quartic self-interactions of the isoscalar–vector meson field (the cubic self-interaction $\sim \omega^3$ is not introduced because of its parity), and terms which are responsible for interactions between isoscalar meson fields. The potential-energy functional $U_{\rm NL}$ reproduces the contribution of the many-body forces in the model considered here. Equation (8.7) is the simplest general form (compatible with relativistic-invariance requirements) containing σ and ω fields (see also [310]). The interactions and self-interactions between isoscalar meson fields are represented diagrammatically in Fig. 8.1.

The motivation for introducing the σ^3, σ^4, and ω^4 terms into the theory was described in Chap. 4, and the corresponding motivation for the $\sigma\omega^2$ and $\sigma^2\omega^2$ terms is described in Chap. 7.

The Euler–Lagrange equations for the various fields can be obtained from the Lagrangian density given by (8.1)–(8.3), (8.7). For the nucleon field, we have a Dirac equation

$$\left\{ i\gamma^\mu \partial_\mu - M - g_\sigma \varphi - g_\omega \gamma^\mu \omega_\mu - \frac{f_\omega}{2M}\sigma^{\mu\nu}\partial_\mu \omega_\nu - g_\rho \gamma^\mu \boldsymbol{\rho}_\mu \cdot \boldsymbol{\tau} \right.$$
$$\left. - \frac{f_\rho}{2M}\sigma^{\mu\nu}\partial_\mu \boldsymbol{\rho}_\nu \cdot \boldsymbol{\tau} - \frac{e}{2}\gamma^\mu(1+\tau_3)A_\mu - \frac{f_\pi}{m_\pi}\gamma_5 \gamma^\mu \partial_\mu \boldsymbol{\pi} \cdot \boldsymbol{\tau} \right\} \psi(x) = 0 \,, \quad (8.8)$$

$$\left(\partial^\mu \partial_\mu + m_\sigma^{*2}\right) \varphi = -g_\sigma \bar{\psi}\psi \,, \quad (8.9)$$

$$\left(\partial^\mu \partial_\mu + m_\omega^{*2}\right) \omega_\nu = g_\omega \bar{\psi}\gamma_\nu \psi - \frac{f_\omega}{2M}\partial^\mu(\bar{\psi}\sigma_{\mu\nu}\psi) \,, \quad (8.10)$$

$$\left(\partial^{\mu}\partial_{\mu}+m_{\pi}^{2}\right)\boldsymbol{\pi}=\frac{f_{\pi}}{m_{\pi}}\partial_{\mu}(\bar{\psi}\gamma_{5}\gamma^{\mu}\boldsymbol{\tau}\psi)\,,\tag{8.11}$$

$$\left(\partial^{\mu}\partial_{\mu}+m_{\rho}^{2}\right)\boldsymbol{\rho}_{\nu}=g_{\rho}\bar{\psi}\gamma_{\nu}\boldsymbol{\tau}\psi-\frac{f_{\rho}}{2M}\partial^{\mu}\left(\bar{\psi}\sigma_{\mu\nu}\boldsymbol{\tau}\psi\right)\,,\tag{8.12}$$

$$\partial^{\mu}\partial_{\mu}A_{\nu}-\frac{e}{2}\bar{\psi}(1+\tau_{3})\gamma_{\nu}\psi=0\,.\tag{8.13}$$

In 8.9 and 8.10, m_{σ}^{*} and m_{ω}^{*} are the effective masses of the scalar and ω mesons, respectively. They can be written in terms of the σ and ω_{μ} fields as

$$m_{\sigma}^{*2}=m_{\sigma}^{2}+\bar{b}g_{\sigma}^{2}M(g_{\sigma}\varphi)+\bar{c}g_{\sigma}^{2}(g_{\sigma}\varphi)^{2}+\bar{d}g_{\sigma}^{2}M\frac{(g_{\omega}\omega_{0})^{2}}{g_{\sigma}\varphi}-\frac{1}{2}\bar{e}g_{\sigma}^{2}(g_{\omega}\omega_{0})^{2}\,,$$

$$m_{\omega}^{*2}=m_{\omega}^{2}-2\bar{d}Mg_{\omega}^{2}(g_{\sigma}\cdot\varphi)+\frac{1}{2}\bar{e}g_{\omega}^{2}(g_{\sigma}\varphi)^{2}+\bar{f}g_{\omega}^{2}(g_{\omega}\omega_{0})^{2}\,,\tag{8.14}$$

ω_0 being the time component of the vector field ω_{μ}. As can be seen from (8.14), all meson–meson interactions manifest themselves via a corresponding contribution either to m_{σ}^{*} or to m_{ω}^{*} (or to both of them).

The effective Lagrangian (8.1)–(8.3), (8.7) includes interactions and self-interactions of isoscalar fields. The interactions of isovector fields can also be incorporated formally into the framework of the RHF approach. These interactions will lead to a density dependence of the masses of isovector fields, in particular. For example, one may take into account self-interactions of the ρ meson field of the type $\mathcal{L}_{\rho\rho}^{\rm SI}\sim(g_{\rho}^{2}\boldsymbol{\rho}_{\mu}\cdot\boldsymbol{\rho}^{\mu})^{2}$. However, since the ρ meson contribution itself is small in comparison with that of the ω meson in the nuclear-structure context, we may hope that including $\mathcal{L}_{\rho\rho}^{\rm SI}$ will not greatly influence the nuclear bulk properties. For this reason, this type of self-interactions is ignored in nuclear-structure problems at present. A discussion of the role of the ρ–ρ self-interactions at higher densities is given in [346]. We should mention that nonlinear terms proportional to $\pi^{2}\sigma^{2}$ and $\pi^{2}\omega_{\mu}\cdot\omega^{\mu}$ (the strength of these interactions may be determined from chiral models, see Chap. 7) can be taken into account in the HF approximation as a density-dependent contribution to the pion mass and by modifying the pion propagator in the same way as was discussed above for the σ and ω_{μ} fields [305]. However, it is known that polarization effects induced by π mesons in the nuclear medium also produce a density-dependent contribution to the pion mass (see Chap. 7). It can be shown that these two density-dependent contributions almost cancel each other. For this reason, the pion mass in the nuclear medium is taken to be equal to its bare mass in the RHF framework.

8.2 The Relativistic Hartree–Fock Approach for Symmetric Nuclear Matter

We start our considerations from the case where the potential-energy functional $U_{\rm NL}$ (meson–meson) in (8.1) is assumed to be zero:

$$U_{\text{NL}}(\text{meson–meson}) = 0 \ . \tag{8.15}$$

Here, we follow very closely the procedure described in [335]. For $U_{\text{NL}} = 0$, the effective meson masses are identical to the corresponding bare values.

We expand the nucleon field operators in the set of creation and annihilation operators:

$$\psi(\boldsymbol{r},t) = \sum_\alpha \left\{ \psi_\alpha(\boldsymbol{r}) \exp(-\mathrm{i}E_\alpha t) b_\alpha + \psi_{a\alpha}(\boldsymbol{r}) \exp(\mathrm{i}E'_\alpha t) d_\alpha^\dagger \right\} \ , \tag{8.16}$$

$$\psi^\dagger(\boldsymbol{r},t) = \sum_\alpha \left\{ \psi_\alpha^\dagger(\boldsymbol{r}) \exp(\mathrm{i}E_\alpha t) b_\alpha^\dagger + \psi_{a\alpha}^\dagger(\boldsymbol{r}) \exp(-\mathrm{i}E'_\alpha t) d_\alpha \right\} \ . \tag{8.17}$$

Here $\psi_\alpha(\boldsymbol{r})$ and $\psi_{a\alpha}(\boldsymbol{r})$ are complete sets of Dirac spinors, b_α and $b_\alpha^\dagger$ represent annihilation and creation operators for nucleons in a state α, and d_α and $d_\alpha^\dagger$ are the corresponding operators for antinucleons. Since in this section we are considering exchange corrections to the mean-field approximation, the d and $d^\dagger$ terms are omitted in (8.16) and (8.17). The neglected terms correspond to the self-consistent negative energy states.

The two-body interactions generated by meson exchanges are not instantaneous. However, in most of the papers on the RHF description of the nuclear ground state a simplifying assumption is made, namely that the time dependence of the meson fields is neglected (the time component of the four-momentum carried by the meson is neglected). By definition, the Hartree approximation is instantaneous. So this assumption has no consequence for the direct (Hartree) terms, while for the exchange (Fock) terms it amounts to neglecting retardation effects. The energy transfers involved are small compared with the masses of the mesons exchanged. For this reason alone this approximation should be valid for the σ-, ω-, and ρ-induced interactions, and also, to a lesser extent, for the pion. The role of retardation effects in the exchange terms was investigated in [337] for nuclear matter and in [338] for finite nuclei. It was shown that the total binding energy of a nucleus can be changed by about 5% when retardation due to pions is included, for the other, heavier mesons the contribution of retardation is negligibly small.

The Hamiltonian H in the RHF approximation can be expressed in the second-quantized form as

$$H = T + \sum_i V_i \ , \tag{8.18}$$

where

$$T = \sum_{\alpha_1,\alpha_2} b_{\alpha_1}^\dagger b_{\alpha_2} \int \mathrm{d}^3 r \ \bar{\psi}_{\alpha_1}(\boldsymbol{r})(-\mathrm{i}\boldsymbol{\gamma}\boldsymbol{\nabla} + M)\psi_{\alpha_2}(\boldsymbol{r}) \ , \tag{8.19}$$

8.2 The Relativistic Hartree–Fock Approach for Symmetric Nuclear Matter

$$V_i = \frac{1}{2} \sum_{\alpha_1,\alpha_2,\alpha_3,\alpha_4} b^\dagger_{\alpha_1} b^\dagger_{\alpha_2} b_{\alpha_3} b_{\alpha_4} \quad \text{(isospin)}$$

$$\times \int d^3 r_1 \, d^3 r_2 \, \bar\psi_{\alpha_1}(\boldsymbol{r}_1) \bar\psi_{\alpha_2}(\boldsymbol{r}_2) \left[\Gamma_i(1,2) v(m_i;1,2) \right] \psi_{\alpha_3}(\boldsymbol{r}_2) \psi_{\alpha_4}(\boldsymbol{r}_1) \,. \quad (8.20)$$

Here the v's are Yukawa functions

$$v(m;1,2) = \frac{1}{4\pi} \frac{\exp(-m|\boldsymbol{r}_1 - \boldsymbol{r}_2|)}{|\boldsymbol{r}_1 - \boldsymbol{r}_2|}, \quad (8.21)$$

and the spin and isospin quantum numbers are denoted by α. The isospin factor is 1 for isoscalar mesons and $\boldsymbol{\tau}(1) \cdot \boldsymbol{\tau}(2)$ for isovector ones. The quantities $\Gamma_i(1,2)$ are listed below:

$$\Gamma_\sigma(1,2) = -g_\sigma^2, \quad (8.22)$$

$$\Gamma_\omega(1,2) = g_\omega^2 \gamma_\nu(1) \gamma^\nu(2), \quad (8.23)$$

$$\Gamma_\rho^{VV}(1,2) = g_\rho^2 \gamma_\nu(1) \boldsymbol{\tau}(1) \gamma^\nu(2) \boldsymbol{\tau}(2), \quad (8.24)$$

$$\Gamma_\rho^{VT}(1,2) = -\frac{g_\rho f_\rho}{2M} \gamma_\mu(1) \boldsymbol{\tau}(1) \sigma^{\mu\nu}(2) \partial_\nu(2) \boldsymbol{\tau}(2), \quad (8.25)$$

$$\Gamma_\rho^{TV}(1,2) = -\frac{g_\rho f_\rho}{2M} \sigma^{\mu\nu}(1) \partial_\nu(1) \boldsymbol{\tau}(1) \gamma_\mu(2) \boldsymbol{\tau}(2), \quad (8.26)$$

$$\Gamma_\rho^{TT}(1,2) = -\left[\frac{f_\rho}{2M}\right]^2 \sigma^{\mu\nu}(1) \partial_\nu(1) \boldsymbol{\tau}(1) \sigma_{\mu\lambda}(2) \partial^\lambda(2) \boldsymbol{\tau}(2), \quad (8.27)$$

$$\Gamma_\pi^{PV}(1,2) = -\left[\frac{f_\pi}{m_\pi}\right]^2 \gamma_5(1) \gamma^\mu(1) \partial_\mu(1) \boldsymbol{\tau}(1) \gamma_5(2) \gamma^\nu(2) \partial_\nu(2) \boldsymbol{\tau}(2), \quad (8.28)$$

$$\Gamma_\gamma(1,2) = e^2 \gamma^\nu(1) \frac{1+\tau_3(1)}{2} \gamma_\nu(2) \frac{1+\tau_3(2)}{2}. \quad (8.29)$$

The contribution of the ρ meson is split into four parts here: the vector part (V), the tensor part (T), and the interference term between vector and tensor coupling (VT and TV). In infinite nuclear matter, plane-wave solutions are used:

$$\psi_\alpha(\boldsymbol{r}) = u(\boldsymbol{p},s) \chi_\tau e^{i\boldsymbol{p}\cdot\boldsymbol{r}}. \quad (8.30)$$

The state α actually includes $(\boldsymbol{p}, s, \tau)$, where $\boldsymbol{p}$ is within the Fermi sea of Fermi momentum p_F, and s and τ are the spin and isospin quantum numbers. Following (8.19)–(8.21), we have, in the momentum representation,

$$T = \sum_{\alpha_1,\alpha_2} b^\dagger_{\alpha_1} b_{\alpha_2} \bar u(\boldsymbol{p}_1,s_1) \chi^\dagger_{\tau_1} (\boldsymbol{\gamma}\cdot\boldsymbol{p} + M) u(\boldsymbol{p}_2,s_2) \chi_{\tau_2}, \quad (8.31)$$

$$V_i = \frac{1}{2} \sum_{\alpha_1,\alpha_2,\alpha_3,\alpha_4} b^\dagger_{\alpha_1} b^\dagger_{\alpha_2} b_{\alpha_3} b_{\alpha_4} \text{(isospin)} \times \bar u(\boldsymbol{p_1},s_1) \chi^\dagger_{\tau_1} \bar u(\boldsymbol{p_2},s_2) \chi^\dagger_{\tau_2}$$

$$\times \left[\Gamma_i(1,2) \frac{1}{m_i^2 + \boldsymbol{q}^2} \right] u(\boldsymbol{p}_3,s_3) \chi_{\tau_3} u(\boldsymbol{p}_4,s_4) \chi_{\tau_4}, \quad (8.32)$$

where $p_1 - p_4 = p_3 - p_2 = q$. The operators Γ_i in (8.32) can be obtained using the substitutions $\partial_k(1) \to -iq_k$, $\partial_k(2) \to iq_k$.

Consider the baryon self-energy Σ in the RHF approximation, produced by meson exchanges:

$$\Sigma(p) = \Sigma_S(p) + \gamma_0 \Sigma_0(p) + \boldsymbol{\gamma} \cdot \hat{\boldsymbol{p}} \Sigma_V(p) , \qquad (8.33)$$

where $\Sigma_S(p)$ is the scalar component, and $\Sigma_0(p)$ and $\Sigma_V(p)$ are the time and space components, respectively; $\hat{\boldsymbol{p}}$ is the unit vector along $\boldsymbol{p}$. The tensor piece $\gamma_0 \boldsymbol{\gamma} \cdot \hat{\boldsymbol{p}}$, i.e. $\Sigma_T(p)$, is omitted in (8.33), it does not contribute in the HF approximation for nuclear matter. Equation (8.33) gives the most general form of the baryon self-energy compatible with time-reversal and rotational invariance. The nucleon spinors in an infinite nuclear medium are solutions of the Dirac equation

$$[\boldsymbol{\gamma} \cdot \tilde{\boldsymbol{p}} + M^*] u(\boldsymbol{p}, s) = \gamma_0 \tilde{E} u(\boldsymbol{p}, s) , \qquad (8.34)$$

where the quantities with tildes (the single-particle energy $\tilde{E}$ and momentum $\tilde{\boldsymbol{p}}$) and the scalar effective baryon mass M^* are given by

$$\tilde{\boldsymbol{p}}(p) = \boldsymbol{p} + \hat{\boldsymbol{p}} \Sigma_V(p) , \qquad (8.35)$$
$$M^*(p) = M + \Sigma_S(p) , \qquad (8.36)$$
$$\tilde{E}(p) = E(p) - \Sigma_0(p) , \qquad (8.37)$$
$$\tilde{E}^2 = \tilde{\boldsymbol{p}}^2 + M^{*2} . \qquad (8.38)$$

The Dirac equation has the following general solution corresponding to a positive energy:

$$u(\boldsymbol{p}, s) = \left[\frac{\tilde{E} + M^*}{2\tilde{E}} \right]^{1/2} \begin{pmatrix} 1 \\ \boldsymbol{\sigma} \cdot \tilde{\boldsymbol{p}}/(\tilde{E} + M^*) \end{pmatrix} \chi_s . \qquad (8.39)$$

We introduce the following quantities:

$$\widehat{P} \equiv \frac{\tilde{p}}{\tilde{E}} \equiv \cos\eta(p) , \quad \widehat{M} \equiv \frac{M^*}{\tilde{E}} \equiv \sin\eta(p) , \quad \widehat{P}^2 + \widehat{M}^2 = 1 . \qquad (8.40)$$

The solution (8.39) is normalized as

$$u^\dagger(\boldsymbol{p}, s) u(\boldsymbol{p}, s) = 1 . \qquad (8.41)$$

In symmetric nuclear matter, the HF trial wave function is taken in the following form:

$$|\phi_0\rangle = \prod_{p,s} b^\dagger(\boldsymbol{p}, s) |0\rangle , \qquad (8.42)$$

$|0\rangle$ being the physical vacuum. The total energy density of the system is given by

8.2 The Relativistic Hartree–Fock Approach for Symmetric Nuclear Matter

$$\mathcal{E} = \langle T \rangle + \langle V_\mathrm{D} \rangle + \langle V_\mathrm{E} \rangle . \tag{8.43}$$

The kinetic energy $\langle T \rangle$ and the potential energy $\langle V \rangle$ (separated into the direct and exchange parts, $\langle V_\mathrm{D} \rangle$ and $\langle V_\mathrm{E} \rangle$, shown in Figs. 8.2a,b, respectively) are given by

$$\langle T \rangle = \frac{2}{\pi^2} \int_0^{p_\mathrm{F}} p^2\, \mathrm{d}p\, (p\widehat{P} + M\widehat{M}) , \tag{8.44}$$

$$\langle V_\mathrm{D} \rangle = -\frac{1}{2} \left(\frac{g_\sigma}{m_\sigma} \right)^2 \rho_\mathrm{S}^2 + \frac{1}{2} \left(\frac{g_\omega}{m_\omega} \right)^2 \rho_\mathrm{V}^2 , \tag{8.45}$$

$$\langle V_\mathrm{E} \rangle = \frac{1}{(2\pi)^4} \int_0^{p_\mathrm{F}} p\, \mathrm{d}p\, p'\, \mathrm{d}p' \left[\sum_i A_i(p,p') + \widehat{M}(p)\widehat{M}(p') \sum_i B_i(p,p') \right.$$
$$\left. + \widehat{P}(p)\widehat{P}(p') \sum_i C_i(p,p') + \widehat{P}(p)\widehat{M}(p')D(p,p') \right] , \tag{8.46}$$

where the summation over i corresponds to different possibilities given by (8.22)–(8.29). The scalar and vector densities ρ_S and ρ_V are given by

$$\rho_\mathrm{S} = \frac{2}{\pi^2} \int_0^{p_\mathrm{F}} p^2\, \mathrm{d}p\, \widehat{M}(p) , \tag{8.47}$$

$$\rho_\mathrm{V} = \frac{2}{3\pi^2} p_\mathrm{F}^3 , \tag{8.48}$$

and the coefficients A, B, C, and D are presented in Table 8.1. The term D is connected with the cross vector–tensor ρN coupling.

All the coefficients can be expressed in terms of the functions $\theta(m;p,p')$ and $\phi(m;p,p')$ given by

$$\theta(m;p,p') = \ln \left[\frac{m^2 + (p+p')^2}{m^2 + (p-p')^2} \right] , \tag{8.49}$$

$$\phi(m;p,p') = \frac{p^2 + p'^2 + m^2}{4pp'} \theta(m;p,p') - 1 . \tag{8.50}$$

The HF equations are obtained by requiring the energy per particle be stationary with respect to variations of ρ_V and η with the constraint (8.40). These equations are a basis for studying nuclear matter; they have been solved numerically to obtain $\eta(p)$ and the saturation Fermi momentum p_F^0. Expressions for various components of the baryon self-energy are given in [335].

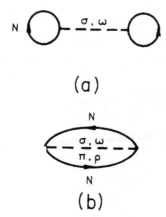

Fig. 8.2. Diagrammatic representation of (**a**) the direct (Hartree) and (**b**) the exchange (Fock) terms in the total binding energy

Table 8.1. The functions A_i, B_i, C_i, and D (see text for the definition of θ (8.49) and ϕ (8.50)) [335]

i	A_i	B_i
σ	$g_\sigma^2 \theta_\sigma$	$g_\sigma^2 \theta_\sigma$
ω	$2g_\omega^2 \theta_\omega$	$-4g_\omega^2 \theta_\omega$
ρ_V	$6g_\rho^2 \theta_\rho$	$-12g_\rho^2 \theta_\rho$
π_{PV}	$-3(g_\pi/2M)^2 \left[m_\pi^2 \theta_\pi\right]$	$-3(g_\pi/2M)^2 \left[m_\pi^2 \theta_\pi\right]$
ρ_T	$-3(f_\rho/2M)^2 \left[m_\rho^2 \theta_\rho\right]$	$-9(f_\rho/2M)^2 \left[m_\rho^2 \theta_\rho\right]$
ρ_{VT}	$\mathcal{D} = 36(f_\rho g_\rho/2M)(p\theta_\rho - 2p'\phi_\rho)$	

i	C_i
σ	$-2g_\sigma^2 \phi_\sigma$
ω	$-4g_\omega^2 \phi_\omega$
ρ_V	$-12g_\rho^2 \phi_\rho$
π	$6(g_\pi/2M)^2 \left[(p^2+p'^2)\phi_\pi - pp'\theta_\pi\right]$
ρ_T	$12(f_\rho/2M)^2 \left[(p^2+p'^2-(m_\rho^2)/2\phi_\rho) - pp'\theta_\rho\right]$
ρ_{VT}	$\mathcal{D} = 36(f_\rho g_\rho/2M)(p\theta_\rho - 2p'\phi_\rho)$

8.3 The Relativistic Hartree–Fock Approach for Finite Nuclei

In this section we consider the case of spherical, closed-subshell nuclei. Just as in the case of nuclear matter, we calculate the HF energy in the tree approximation. The tree approximation corresponds to the case where the ground state of the system is represented by a Slater determinant composed

8.3 The Relativistic Hartree–Fock Approach for Finite Nuclei

of positive-energy spinors. This means, in particular, that the nucleon field is expanded without d and $d^\dagger$ terms (see (8.16), (8.17)). A single-nucleon state with energy E_α is specified by the quantum numbers $\alpha = (q_a, n_a, l_a, j_a, m_a) \equiv (a, m_a)$, where $q_a = -1$ for neutrons and $q_a = +1$ for protons. The nucleon spinors $\psi_\alpha(\mathbf{r})$, given by (4.59), are normalized according to

$$\int d^3 r\, \psi_\alpha^\dagger(\mathbf{r})\psi_\alpha(\mathbf{r}) = \int [G_a^2(r) + F_a^2(r)]\, dr = 1 \,. \tag{8.51}$$

According to the discussion in Appendix A9, the zero-range interactions, contained in Γ_π and Γ_ρ^T, are subtracted from the Hamiltonian. This is done by adding the two terms

$$\delta\left[\Gamma_\pi^{PV} v(m_\pi; 1, 2)\right] = \frac{1}{3}\left[\frac{f_\pi}{m_\pi}\right]^2 \gamma(1)\gamma(2)\gamma_5(1)\gamma_5(2)\delta(\mathbf{r}_1 - \mathbf{r}_2) \,, \tag{8.52}$$

$$\delta\left[\Gamma_\rho^T v(m_\rho; 1, 2)\right] = \frac{1}{3}\left[\frac{f_\rho}{2M}\right]^2 \sigma^{\mu i}(1)\sigma_{\mu i}(2)\delta(\mathbf{r}_1 - \mathbf{r}_2) \,. \tag{8.53}$$

Consider a nucleus with A nucleons. In the HF approximation, the trial ground-state wave function is

$$|\Phi_0\rangle = \prod_{\alpha\ (\text{occupied})} b_\alpha^\dagger |0\rangle \,. \tag{8.54}$$

The HF solution is obtained by requiring the total binding energy

$$E = \langle\Phi_0|\, H_0\, |\Phi_0\rangle - AM \tag{8.55}$$

to be stationary with respect to variations of the spinors ψ_α under the constraint given by the normalization condition (8.51). This is expressed in terms of the following equation:

$$\delta\left[E - \sum_{\alpha\ (\text{occupied})} E_\alpha \int \psi_\alpha^\dagger(\mathbf{r})\psi_\alpha(\mathbf{r})\, d^3r\right] = 0 \,, \tag{8.56}$$

where the E_α are Lagrange multipliers, they appear to be just the Hartree–Fock single-nucleon energies (the nucleon mass included). The calculation of $\langle\Phi_0|\, H_0\, |\Phi_0\rangle$ is involved, but once this value is calculated, it is possible to perform variations of the type (8.56). So one obtains the HF equations for the self-consistent functions (G_a, F_a) and energies E_a. These equations have the following form:

$$\frac{d}{dr}\begin{bmatrix} G_a(r) \\ F_a(r) \end{bmatrix}$$
$$= \begin{bmatrix} -\kappa_a/r - \Sigma_{T,a}^D(r) & M + E_a + \Sigma_{S,a}^D(r) - \Sigma_{0,a}^D(r) \\ M - E_a + \Sigma_{S,a}^D(r) + \Sigma_{0,a}^D(r) & \kappa_a/r + \Sigma_{T,a}^D(r) \end{bmatrix}$$
$$\times \begin{bmatrix} G_a(r) \\ F_a(r) \end{bmatrix} + \begin{bmatrix} -X_a(r) \\ Y_a(r) \end{bmatrix} \,. \tag{8.57}$$

Here, $\Sigma^{\mathcal{D}}_{S,a}$, $\Sigma^{\mathcal{D}}_{0,a}$, and $\Sigma^{\mathcal{D}}_{T,a}$ give the Hartree (direct) contributions to the self-energy, while X_a and Y_a come from the Fock (exchange) terms. The quantity $\kappa_a = \mp(j_a + 1/2)$ for $j_a = l_a \pm 1/2$. The Hartree potentials are local and state-independent. Detailed expressions for all of the potentials can be found in [335].

The Hartree–Fock energy of (8.55) satisfies the relation

$$E = \frac{1}{2} \sum_{\alpha \text{ (occupied)}} (T_\alpha + E_\alpha) - AM , \qquad (8.58)$$

where T_α is the kinetic energy of the orbital α (see (4.127)):

$$T_\alpha = \int d^3 r \, \bar{\psi}_\alpha(\mathbf{r})(-i\boldsymbol{\gamma} \cdot \boldsymbol{\nabla} + M)\psi_\alpha(\mathbf{r}) . \qquad (8.59)$$

The HF equations (8.57) are a set of coupled integro-differential equations that must be solved self-consistently. The basic complications are connected with the exchange contributions, which produce the integral terms, i.e. the HF potentials are totally nonlocal and state-dependent. It is reasonable to try to write down the HF equations in a totally equivalent form of a system of homogeneous differential equations. A possible way to achieve this aim is to represent the inhomogeneous parts as

$$X_a(r) = \frac{G_a(r)X_a(r)}{G_a^2(r) + F_a^2(r)} G_a(r) + \frac{F_a(r)X_a(r)}{G_a^2(r) + F_a^2(r)} F_a(r)$$

$$\equiv P_a(r)G_a(r) + Q_a(r)F_a(r) , \qquad (8.60)$$

$$Y_a(r) \equiv R_a(r)G_a(r) + S_a(r)F_a(r) ; \qquad (8.61)$$

the definition of the functions $R_a(r)$ and $S_a(r)$ is similar to that of $P_a(r)$ and $Q_a(r)$. The HF equations (8.57) now take the following form:

$$\frac{d}{dr}\begin{bmatrix} G_a(r) \\ F_a(r) \end{bmatrix}$$
$$= \begin{bmatrix} -\kappa_a/r - \Sigma^{\mathcal{D}}_{T,a}(r) - P_a(r) & M + E_a + \Sigma^{\mathcal{D}}_{S,a}(r) - \Sigma^{\mathcal{D}}_{0,a}(r) - Q_a(r) \\ M - E_a + \Sigma^{\mathcal{D}}_{S,a}(r) + \Sigma^{\mathcal{D}}_{0,a}(r) + R_a(r) & \kappa_a/r + \Sigma^{\mathcal{D}}_{T,a}(r) + S_a(r) \end{bmatrix}$$
$$\times \begin{bmatrix} G_a(r) \\ F_a(r) \end{bmatrix} . \qquad (8.62)$$

The structure of (8.62) is the same as that which appears in the Hartree approximation, only the potentials are more complicated now. Further computational details can be found in [335].

8.4 Determination of Parameters and Numerical Results

The Lagrangian (8.1)–(8.6) is uniquely determined (in the case $U_{\text{NL}} = 0$) as soon as the four meson masses and six coupling constants are fixed. On the

8.4 Determination of Parameters and Numerical Results

other hand, two main properties of nuclear matter need to be reproduced: the nuclear binding energy and the saturation density. To reduce the number of parameters, it is possible to fix the meson masses at their experimental values, $m_\omega = 783$ MeV, $m_\rho = 770$ MeV, and $m_\pi = 138$ MeV; the bare nucleon mass was taken as $M = 938.9$ MeV. The σ meson mass is not fixed by experiment. In accordance with the OBEP data, it was taken to lie in the region between 400 and 600 MeV. The π–N and ρ–N coupling constants were fixed at their physical values, $f_\pi^2/4\pi = 0.08$ and $g_\rho^2/4\pi = 0.55$. The tensor ρ–N coupling constant is related to the ratio f_ρ/g_ρ. If vector dominance is assumed (see Chap. 5), this ratio f_ρ/g_ρ can be taken to be equal to 3.7. However, to describe the πN scattering data, a bigger value is needed, namely $f_\rho/g_\rho = 6.6$. In accordance with the vector meson dominance model, $f_\omega/g_\omega = -0.12$. This value is very small and was neglected in the RHF numerical calculations for finite nuclei. Let us mention here that the tensor ωN coupling may be quite essential for understanding the nature of the spin–orbit interaction in hypernuclei [347–349].

So only three parameters appear in the $U_{\rm NL} = O$ model: g_σ, g_ω, and m_σ.[3] To obtain the results presented in [335], these values were chosen to reproduce the two nuclear-matter observables $E/A = -15.75$ MeV and $p_F^0 = 1.30$ fm^{-1} ($\rho_V^0 = 0.1484$ fm^{-3}). The σ meson mass was fitted to reproduce the correct charge r.m.s. radius for ^{16}O and was taken to be $m_\sigma = 440$ MeV. The $\delta(\boldsymbol{r})$ part of the π- and ρ-induced interactions was removed in [335] in accordance with the procedure described above, while in [334] these contributions were included in the numerical calculations. The parameters of the model are given in Table 8.2 for various models in the Hartree and HF approximations.

If the model involves isoscalar mesons only (rows (a) and (b)), the coupling constants must be renormalized by 15–20% when going from the Hartree to the Hartree–Fock approximation. Without this modification, one would obtain too little binding in the HF approximation (Fig. 8.3).

The net contribution of π and ρ exchange can be seen well in this figure (dashed–dotted line), the isoscalar meson coupling constants being fixed. One gains in this case about 20 MeV per nucleon (attraction).

Since $f_\pi^2/4\pi = 0.08$ is much weaker than the corresponding isoscalar–scalar and vector coupling constants, g_σ and g_ω need to be only slightly modified to obtain the nuclear-matter saturation point (in the model with the pion included) (see row (c) of Table 8.2).

---

[3] The relativistic HF approximation appears not to be adequate for reproducing nuclear properties if meson parameters (masses and coupling constants) adjusted to fit NN scattering are used (one obtains unbound nuclei in this case). This problem is related to the existence of the strong short-range and tensor terms in realistic NN potentials; these terms must be taken into consideration, including short-range correlations in the nuclear wave function. This may be done by Brueckner–Hartree–Fock methods, which are discussed in the next chapter.

Table 8.2. Isoscalar–scalar (σ) and vector (ω) meson coupling constants derived from fitting the nuclear–matter saturation point in the Hartree (H) and Hartree–Fock (HF) approximations. The isovector ρ meson coupling constants are described in the text. Rows (d) and (e) correspond to $f_\rho/g_\rho = 6.6$ and 3.7, respectively. All results correspond to $m_\sigma = 440$ MeV [335]

		$g_\sigma^2/4\pi$	$g_\omega^2/4\pi$
(a) $(\sigma + \omega)_\text{H}$		6.25	15.16
(b) $(\sigma + \omega)_\text{HF}$		5.54	12.24
(c) $(\sigma + \omega + \pi)_\text{HF}$		5.35	12.42
(d) $(\sigma + \omega + \pi + \rho)_\text{HF}$,	$f_\rho/g_\rho = 6.6$	2.27	10.00
(e) $(\sigma + \omega + \pi + \rho)_\text{HF}$,	$f_\rho/g_\rho = 3.7$	4.16	11.18

Table 8.3. The various components of the baryon self-energy (in MeV), calculated at the saturation density with parameter set (e) of Table 8.2. The values calculated in the Hartree approximation with parameter set (a) are also given

		Σ_s	Σ_0	Σ_V
Direct		−288	261	0
		(σ)	(ω)	
Exchange	σ	24	26	−1
	ω	−105	56	−1
	π	−5	−5	−6
	ρ	−40	−1	11
Total (e)		−414	337	3
Hartree (a)		−431	354	0

The contribution of the ρ meson may be divided into three parts: a vector part proportional to g_ρ^2, a tensor part proportional to f_ρ^2, and a vector–tensor part proportional to $f_\rho g_\rho$. The total ρ meson effect with a tensor coupling $f_\rho/g_\rho = 3.7$ is attractive and is equal to approximately 19 MeV per nucleon in nuclear matter. New values of g_σ and g_ω were worked out for this case (row (e) of Table 8.2).

Let us now discuss the role of the variuos components of the baryon self-energy Σ. In the Hartree approximation, the scalar component Σ_S is determined completely by the scalar σ meson, and the time component Σ_0 comes from the ω meson, while $\Sigma_\text{V}^{(\text{H})} = 0$. This is not the case in the HF approximation, where both σ and ω mesons (time and space components) contribute to the three components of the baryon self-energy. Moreover, the three components (Σ_S, Σ_0, and Σ_V) contain contributions from π and ρ mesons. The various components of Σ, calculated at the saturation density, are given in Table 8.3.

8.4 Determination of Parameters and Numerical Results   151

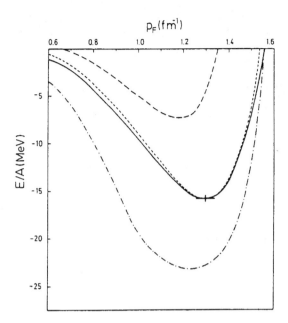

Fig. 8.3. Binding energies per particle in infinite nuclear matter. The *short-dashed curve* corresponds to the Hartree approximation with parameter set (a) of Table 8.2. The *solid curve* was calculated with the parameter set (e) (HF approximation with all mesons included). The *long-dashed curve* corresponds to the HF approximation with the parameters of row (a) and σ, ω mesons only. The *dashed–dotted curve* was calculated in the HF approximation with all mesons included and parameter set (a) (from [335])

The components Σ_S and Σ_0 depend very little on the momentum (note that in the Hartree approximation these functions are momentum-independent). In contrast, the component Σ_V has a stronger momentum dependence; the value of Σ_V is determined almost completely by the contributions of π and ρ mesons, with opposite signs (the mixed vector–tensor term producing the most important contribution). Isoscalar σ and ω mesons give at most (near the saturation point) -1 MeV each. It is worth noting that Σ_V is very small in comparison with Σ_S and Σ_0. From (8.35), it can be seen that the effective momentum $\widetilde{p}$ in nuclear matter is close to the free momentum p, the difference being around 5% at the saturation density. Note that one needs to be careful in comparing the values of Σ_V given in different papers. Indeed, there are different possible ways to perform a relativistic (Lorentz) decomposition of the baryon self-energy Σ. In [335], the authors of that paper made the choice to start from the value given by (8.33) satisfying the Dirac equation (8.34). However, in [350] it was shown that the value of Σ_V can be eliminated in the Dirac equation for nuclear matter, producing new energy-dependent self-energies Σ_S and Σ_0. The values of Σ_S and Σ_0 in [335] are within 10% of

8 The Relativistic Hartree–Fock Approach

Table 8.4. Bulk properties of some finite nuclei obtained with the sets of parameters given in Table 8.2. The total binding energies per particle and the proton spin–orbit splittings for the 1p-state (^{16}O) and 1d-state (^{40}Ca and ^{48}Ca) are presented in MeV; the r.m.s. charge radii are in fm. (From [335])

	^{16}O			^{40}Ca			^{48}Ca		
	$-E/A$	r_c	Δ_{LS}	$-E/A$	r_c	Δ_{LS}	$-E/A$	r_c	Δ_{LS}
(a)	2.04	3.07	4.1	4.06	3.70	5.9	4.61	3.67	6.2
(b)	2.33	2.93	7.1	4.32	3.59	8.6	5.04	3.56	8.9
(c)	3.09	2.91	5.5	4.90	3.59	7.1	5.35	3.57	3.5
(d)	11.94	2.40	9.5	11.15	3.21	8.9	11.42	3.24	1.8
(e)	5.61	2.73	7.3	6.82	3.47	8.0	7.11	3.47	4.1

Table 8.5. Comparison of the RHF results derived with the parameter set (e) (RHF) and experimental values (Expt.) for finite nuclei. The nonrelativistic center-of-mass correction is denoted by c.m. (in MeV). The notation and units are the same as in Table 8.4. (From [335])

	^{16}O			^{40}Ca			^{48}Ca		
	$-E/A$	r_c	Δ_{LS}	$-E/A$	r_c	Δ_{LS}	$-E/A$	r_c	Δ_{LS}
RHF	5.61	2.73	7.3	6.82	3.47	8.0	7.10	3.47	4.1
Expt.	7.98	2.73	6.3	8.55	3.48	7.2	8.67	3.47	4.3
c.m.	0.61			0.20			0.18		

	^{90}Zr		^{208}Pb	
	$-E/A$	r_c	$-E/A$	r_c
RHF	7.40	4.26	6.74	5.47
Expt.	8.71	4.27	7.87	5.50
c.m.	0.08		0.02	

those obtained in [351, 352]. The authors of the latter references begin with the relativistic Brueckner–Hartree–Fock (RBHF) theory of nuclear matter, utilizing one-boson exchange potentials which reproduce the NN data. The self-energies are similar in these two cases, but the coupling constants are very different, for the ω meson in particular.

Finally, we discuss the results obtained in [335] for the ground-state properties of some closed-shell nuclei in the $U_{\rm NL} = 0$ model. These results include the binding energy per particle, the r.m.s. charge radius, and the proton spin–orbit splitting for the 1p state in ^{16}O and the 1d state in ^{40}Ca. These data are given in Tables 8.4 and 8.5, the calculations being performed for the various sets of parameters given in Table 8.2.

8.4 Determination of Parameters and Numerical Results

It was noticed earlier in the discussion of nuclear matter that the bulk properties of finite nuclei depend strongly on the value of m_σ (the other meson masses being fixed). The problem is that m_σ sets the intermediate-range attraction and thus determines the r.m.s. radii. It is reasonable to choose m_σ to correspond to the most realistic case (i.e. the case in which all mesons are included and $f_\rho/g_\rho = 3.7$). It was established by RHF calculations that the appropriate choice is $m_\sigma = 440\,\text{MeV}$. Utilizing the same value of m_σ for all parameter sets, we can see more vividly the effects of including more mesons or modifying the ρ meson tensor coupling constant.

Adding the exchange terms in the simplest (σ, ω) model increases the binding energy slightly but has too strong an effect on the spin–orbit splitting, which appears to be too large in this case, especially for ^{48}Ca.

It is the pion which decreases the spin–orbit splitting, while leaving the binding energy and r.m.s. radius at the level of the (σ, ω) Hartree approximation. This effect has its most striking manifestation in ^{48}Ca. In this case of an $N \neq Z$ nucleus, the role of an isovector meson played by the pion is very pronounced.

The isovector character of the pion is very important for obtaining the proper value of the proton spin–orbit splitting and, in particular, for reproducing in the theory the drastic change of the 1d state spin–orbit splitting when going from ^{40}Ca to ^{48}Ca, the number of protons being the same in both cases [341, 353]. The ρ meson increases the binding energy. It also plays a role in producing the spin–orbit splitting, especially the tensor part of the ρN interaction, the results being quite different for weak and strong tensor coupling.

When the scalar mass m_σ is fitted to obtain the r.m.s. radius of ^{16}O, the radii for all other nuclei agree with experiment to within 1%. The particular value of the radius of the meson-induced interaction ($1/m_\alpha$, m_α being the mass of the respective meson) is very important for describing the surface properties of finite nuclei.

In general, the results obtained in the RHF framework where four mesons are included reproduce the nuclear bulk properties reasonably well. Inclusion of the exchange effects generates a very reasonable dependence of the binding energy on A. A satisfactory description (without nonlinear terms) of the single-particle spectrum, charge densities, and r.m.s. radii is obtained in the RHF framework for doubly magic nuclei. The exchange terms are shown to play an essential role in determining the symmetry energy and spin–orbit interaction. However, as can be seen from Table 8.5, the calculated nuclei are not bound strongly enough. About 1.5 MeV of binding energy is missing over the whole set of nuclei (this defect being more pronounced for light nuclei). This fact is connected with the high compression modulus K of nuclear matter inherent in this model. A better agreement with experiment is expected if meson self-interaction (additional density-dependent) terms are introduced into the Lagrangian. In the next section it will be shown that the $U_{\text{NL}} \neq 0$

model greatly improves the situation in the RHF framework for the binding energies of finite nuclei and for the compression modulus of nuclear matter.

8.5 The Relativistic Hartree–Fock Approach with Meson Self-Coupling Terms

We discuss here the relativistic Hartree–Fock theory for the model including four mesons $(\sigma, \omega, \pi, \rho)$ and self-interaction of the meson fields. We start the discussion from the case where self-interactions of the scalar field only are taken into consideration [343]:

$$U_{\rm NL}(\varphi) = \frac{1}{2}m_\sigma^2 \varphi^2 + \frac{1}{3}g_2 \varphi^3 + \frac{1}{4}g_3 \varphi^4 . \tag{8.63}$$

The general case $U_{\rm NL} = U_{\rm NL}(\varphi, \omega)$ will be considered in the next chapter.

Equation (8.8) enables us to write down an equivalent Hamiltonian density corresponding to the Lagrangian $\mathcal{L}$ in the form $\mathcal{H} = \mathcal{H}_0 + \mathcal{H}_1$, where $\mathcal{H}_0$ is linear in the meson fields, i.e.

$$\mathcal{H}_0 = \bar{\psi}(-i\boldsymbol{\gamma}\cdot\boldsymbol{\nabla} + M)\psi + \frac{1}{2}\bar{\psi}(g_\sigma \varphi + g_\omega \gamma^\mu \omega_\mu + \ldots)\psi , \tag{8.64}$$

whereas $\mathcal{H}_1 = U - (1/2)\varphi(\partial U/\partial \varphi)$ contains only nonlinear terms such as φ^3 and φ^4. The nonlinear character of $\mathcal{H}_1$ makes the calculation of its exchange contribution to the energy rather complicated.

In order to write down the Dirac equation for the nucleons in a Dirac–Hartree–Fock (DHF) approach, one could use (8.9)–(8.13) to eliminate the meson fields in (8.8). In practice, this is difficult because $\hat{m}_\sigma^*$ is an operator. In the RHF framework, the static limit for the meson fields is utilized and the φ field in $\hat{m}_\sigma^*$ and $\mathcal{H}_1$ (but not in $\mathcal{H}_0$) is replaced by its ground-state expectation value φ_0. Then (8.9) reads

$$\partial^\mu \partial_\mu \varphi + m_\sigma^{*2} \varphi + g_\sigma \bar{\psi}\psi = 0 , \tag{8.65}$$

where

$$m_\sigma^{*2} = m_\sigma^2 + g_2 \varphi_0 + g_3 \varphi_0^2 , \tag{8.66}$$

and φ_0 is obtained from the equation

$$(\nabla^2 - m_\sigma^{*2})\varphi_0 = g_\sigma \langle \bar{\psi}\psi \rangle . \tag{8.67}$$

The expectation value is calculated in the ground state. The quantity m_σ^{*2} plays the role of the squared effective mass of the σ meson. Clearly, it becomes the square of the ordinary σ mass in the absence of nonlinearity ($g_2 = g_3 = 0$). Similarly,

$$\mathcal{H}_1 = -g_2 \frac{1}{6}\varphi_0^3 - g_3 \frac{1}{4}\varphi_0^4 . \tag{8.68}$$

8.5 The RHF Approach with Meson Self-Coupling Terms

Table 8.6. Adjusted parameters of the model and properties of symmetric nuclear matter. The dimensionless parameters $\bar{b} = g_2/Mg_\sigma^3$ and $\bar{c} = g_3/g_\sigma^4$ are defined as in [112]. The notations HF(δ), HFSI, and HFSI(δ) are explained in the text. The value of M^* [$\equiv M + \Sigma_s(p)$] at the Fermi surface is given. a_4 is the symmetry energy parameter. The results are taken from [343]

Model	$g_\sigma^2/4\pi$	$g_\omega^2/4\pi$	m_σ (MeV)	$10^3\bar{b}$	$10^3\bar{c}$	K (MeV)	M^*/M	a_4 (MeV)
[335]	4.16	11.18	440	0	0	465	0.56	38.6
HF(δ)	7.19	8.22	525	0	0	399	0.60	35.0
HFSI	4.005	10.4	412	−6.718	−14.61	250	0.61	35.0
HFSI(δ)	6.92	7.5	515	−1.52	−2.62	300	0.63	33.9

We have thus linearized the scalar-meson-field equation with respect to the φ-field operator. Now the φ field can be cast into the following form:

$$\varphi = -g_\sigma \int S_\sigma(x,y)\bar{\psi}(y)\psi(y)\,\mathrm{d}^4 y, \quad (8.69)$$

where the propagator S_σ satisfies the equation

$$(\Box + m_\sigma^{*2})S_\sigma(x,y) = \delta(x-y). \quad (8.70)$$

For the ω_μ, $\boldsymbol{\pi}$, and $\boldsymbol{\rho}_\mu$ fields one can write expressions similar to (8.69) with corresponding propagators S_i, free meson masses m_i, and source terms (see (8.9)–(8.13)). It should be noted that the σ meson effective mass m_σ^* depends on r and consequently, S_σ does not have the simple Yukawa form of S_i ($i = \omega, \pi, \rho$).

Introducing expression (8.69) for the φ field and the corresponding equations for the ω_μ, $\boldsymbol{\pi}$, and $\boldsymbol{\rho}_\mu$ fields into the Dirac equation and $\mathcal{H}_0$, we can write both [343] in a form where only the nucleon fields $\psi(x)$ are present:

$$\left[-i\gamma^\mu \partial_\mu + M + \sum_i \int \bar{\psi}(x_2)\Gamma_i(1,2)S_i(x_1,x_2)\psi(x_2)\,\mathrm{d}^4 x_2 \right]\psi(x_1) = 0 \quad (8.71)$$

$$\mathcal{H}_0 = \bar{\psi}(x_1)(-i\boldsymbol{\gamma}\cdot\boldsymbol{\nabla} + M)\psi(x_1) \\ + \frac{1}{2}\sum_i \int \bar{\psi}(x_1)\bar{\psi}(x_2)\Gamma_i(1,2) \times S_i(x_1,x_2)\psi(x_2)\psi(x_1)\,\mathrm{d}^4 x_2, \quad (8.72)$$

where the operators $\Gamma_i(1,2)$ have been given in the previous section.

It is easy to see that the Dirac equation (8.71) can be obtained from the Hamiltonian density $\mathcal{H}_0$. Thus, in the approach considered here, $\mathcal{H}_1$ plays the role of a perturbation; it does not affect the state of the system but contributes to its energy.

Using the same approximations as in the previous section, we obtain a Dirac equation for the nucleon spinor $\psi_\alpha(\boldsymbol{r}_1)$ in the form

$$[-i\boldsymbol{\alpha} \cdot \boldsymbol{\nabla} + \beta M + \beta \Sigma(\boldsymbol{r}_1)]\psi_\alpha(\boldsymbol{r}_1) = E_\alpha \psi_\alpha(\boldsymbol{r}_1) \,. \tag{8.73}$$

Note that (8.71)–(8.73) are formally identical to the corresponding equations of the previous section; only the σ propagator is different.

To calculate the self-energy Σ entering the Dirac equation (8.73), it is possible to use the expressions developed in the previous section, replacing the Yukawa propagator of the σ meson by the new propagator given by (8.70).

In the case of nuclear matter, it is convenient to use the momentum representation. In this case, the σ propagator takes the form $S_\sigma(q) = (m_\sigma^{*2} + q^2)^{-1}$, where q is the four-momentum of the σ meson exchanged between two nucleons. The self-energy $\Sigma(p)$ is momentum-dependent and is given by (8.33). In the present case, however, the free meson mass m_σ must be replaced by the effective mass m_σ^*. Note that φ_0 and m_σ^* are momentum-independent. On the other hand, they must be calculated self-consistently.

For finite nuclei $S_\sigma(\boldsymbol{r}, \boldsymbol{r}')$ is expanded in the form

$$S_\sigma(\boldsymbol{r}, \boldsymbol{r}') = \sum_{l=0}^{\infty} \sum_{m=-l}^{l} S_l(r, r') Y_{lm}(\Omega) Y_{lm}^*(\Omega') \,, \tag{8.74}$$

where

$$S_l(r, r') = \frac{\eta_l^{(i)}(r_<) \eta_l^{(e)}(r_>)}{W_\sigma} \,, \tag{8.75}$$

$$W_\sigma = \eta_l^{(e)'}(r) \eta_l^{(i)}(r) - \eta_l^{(e)}(r) \eta_l^{(i)'}(r) \,. \tag{8.76}$$

The functions $\eta_l^{(i)}$ and $\eta_l^{(e)}$ are solutions of the differential equation

$$\eta_l''(r) - \frac{l(l+1)}{r^2} \eta_l(r) - m_\sigma^{*2}(r) \eta_l(r) = 0 \,, \tag{8.77}$$

which are regular at the origin and at infinity, respectively.

Now we can see that the replacement of the Yukawa propagator for σ meson by the function $S_\sigma(\boldsymbol{r}, \boldsymbol{r}')$ amounts to the replacements

$$\widetilde{I}_l(m_\sigma r) \to \frac{\eta_l^{(i)}(r)}{r\sqrt{-m_\sigma W_\sigma}} \,, \quad \widetilde{K}_l(m_\sigma r) \to \frac{\eta_l^{(e)}(r)}{r\sqrt{-m_\sigma W_\sigma}} \tag{8.78}$$

in the equations for the case without self-interactions.

The total binding energy in the model with $U_{\mathrm{NL}}(\varphi) \neq 0$ can be obtained by adding to the corresponding part of $\mathcal{H}_0$ the nonlinear contribution that arises from the energy density $\mathcal{H}_1$. We can write the energy density as

8.5 The RHF Approach with Meson Self-Coupling Terms

$$E = \frac{1}{2}\sum_i (T_i + E_i - M) + \int \mathcal{H}_1(\varphi_0)\,\mathrm{d}^3 r\,, \qquad (8.79)$$

where T_i the kinetic energy of the orbital i, which is given by (4.127) and (4.128).

As was mentioned above, it is possible to simulate part of the effect of short-range correlations in the π and ρ contributions by removing spurious δ-force components from the potential part of the nuclear Hamiltonian. To see the importance of these contact interactions, calculations were carried out in [343] for infinite and finite systems, considering three models. In the first model (Hartree–Fock with self-interactions (HFSI) proportional to φ^3 and φ^4), the $\delta(\boldsymbol{r})$ parts of the π- and ρ-induced interactions were removed from the Hamiltonian. The $\delta(\boldsymbol{r})$ components were taken into account, however, in the other two models. Now we distinguish the linear case (where $g_2 = g_3 = 0$, referred to as HF(δ) here) from the nonlinear case (referred to as HFSI(δ)).

The parameters were chosen [343] in the same way as in the previous chapter. It is worth mentioning that the ratio $f_\rho/g_\rho = 3.7$, given by the VDM model, was taken for the HFSI model (this choice corresponds to set (e) in Table 8.2, while $f_\rho/g_\rho = 6.6$, given by πN scattering data, was taken for the HF(δ) and HFSI(δ) models.

We are then left with the following free parameters: m_σ, g_σ, g_ω, g_2, g_3. The values of g_σ, g_2, and g_3 were determined by reproducing the saturation condition of symmetric nuclear matter at $\rho_0 = 0.14$ fm^{-3} for the HFSI model and $\rho_0 = 0.1484$ fm^{-3} for the HF(δ) and HFSI(δ) models, assuming $E/A = -15.75$ MeV and a reasonable value of the compression modulus K. Finally, calculations of finite nuclei put some constraints on the values of m_σ and g_ω (the σ meson mass was adjusted to obtain the correct r.m.s. charge radius of ^{16}O, while the value of g_ω was chosen to obtain reasonable values of spin–orbit splittings).

In the rest of this section we shall refer to the results of case (e) of the previous section for comparison. The values of the parameters used in this section, as well as some calculated nuclear-matter properties, are given in Table 8.6.

By comparing rows 1 and 2 of Table 8.6, one can see that, in order to obtain similar saturation conditions of ρ_0 and E/A, the HF(δ) model needs a larger coupling constant g_σ and a smaller coupling constant g_ω than those used in [335]. This difference would increase if we were to take the same value of the coupling constant f_ρ in both models. This is because of the different m_σ values in the two cases and the strong attraction arising from the tensor component of the ρ meson in [335]. Table 8.6 also shows that models without self-interaction always give values of K somewhat larger than those obtained with models built on nonlinear Lagrangians. If one compares rows 1 and 3 or 2 and 4, one can see that the effects of the self-interactions cause only a small renormalization in the coupling constants g_σ and g_ω, which can justify treating $\mathcal{H}_1$ in a perturbative form.

Table 8.7. Comparison of calculations performed in [343] (HF(δ), HFSI, and HFSI(δ) models) with experimental values and the calculations of [335] for finite nuclei. The total binding energy per particle, nonrelativistic center-of-mass correction (c.m.), and proton spin–orbit splitting Δ_{LS} for the 1p shell (^{16}O) or 1d shell (^{40}Ca and ^{48}Ca) are given in MeV; the r.m.s. charge radii r_c are in fm

	^{16}O			^{40}Ca			^{48}Ca		
	$-E/A$	r_c	Δ_{LS}	$-E/A$	r_c	Δ_{LS}	$-E/A$	r_c	Δ_{LS}
[335]	5.61	2.73	7.3	6.82	3.47	8.0	7.10	3.47	4.1
HF(δ)	6.36	2.74	6.17	7.73	3.46	7.33	7.96	3.47	3.55
HFSI	7.43	2.73	6.4	8.33	3.48	7.05	8.45	3.48	3.27
HFSI(δ)	6.97	2.73	5.62	8.22	3.44	6.88	8.37	3.45	3.12
Exp.	7.98	2.73	6.3	8.55	3.48	7.2	8.67	3.47	4.3
c.m.	0.61			0.20			0.18		

	^{90}Zr		^{208}Pb	
	$-E/A$	r_c	$-E/A$	r_c
[335]	7.40	4.26	6.74	5.47
HF(δ)	8.34	4.24		
HFSI	8.58	4.26	7.78	5.52
HFSI(δ)	8.69	4.21		
Experiment	8.71	4.27	7.87	5.5
c.m.	0.08		0.02	

Let us mention that the Σ_s and Σ_0 components depend very little on the momentum and are very similar in all models, although their absolute values decrease slightly when the scalar self-interactions are present. The Σ_V component has a stronger momentum dependence. Although this dependence is similar in all cases, Σ_V is negative for small momenta in the HFSI model in [335], whereas it is positive in the HF(δ) and HFSI(δ) models. In [343] calculations were performed for spherical doubly closed-shell nuclei: ^{16}O, ^{40}Ca, ^{48}Ca, ^{90}Zr, and ^{208}Pb. The calculated and experimental binding energies, r.m.s. charge radii, and proton spin–orbit splittings are shown in Table 8.7.

The nonrelativistic center-of mass corrections to the total binding energy were taken from [353] and are given as indicative values.

As was mentioned in the previous section, the main problem of the theory with $U_{\mathrm{NL}} = 0$ is the small total binding energy. It is easily seen that all the models considered in this section (HF(δ), HFSI, and HFSI(δ)) improve this property and the spin–orbit splitting in ^{16}O and ^{40}Ca too, especially the HFSI model. For the ^{48}Ca nucleus, all models reproduce rather well the spectacular reduction of the d-state spin–orbit splitting. This is caused, essentially, by

8.5 The RHF Approach with Meson Self-Coupling Terms

the isovector meson contributions. The HFSI(δ) model, however, predicts a smaller r.m.s. charge radius than the experimental value, though this could be improved if we were to choose a slightly smaller value of the saturation density of nuclear matter (for instance, $\rho_0 = 0.14$ fm^{-3} as for the HFSI model).

The single-particle spectra for both protons and neutrons for all the nuclei considered here, together with the charge densities, can be found in [343]. On the whole it is clear that the inclusion of self-interaction effects together with the contributions of π and ρ mesons (both the Hartree and the Fock terms) improves substantially the overall description of nuclear ground-state properties. Although relativistic mean-field models provide a rather good global picture of nuclear properties, it is impossible if one stays in the Hartree approximation to obtain at the same time a good description of the spin–orbit interaction and the total binding energy.

One of the manifestations of the scalar-field self-interactions $U_{\rm NL}(\varphi)$ is the dressing of the scalar meson in the nuclear medium. In this case the σ meson mass is different from the bare mass and becomes density-dependent. Discussion of the most general form of $U_{\rm NL}(\varphi,\omega)$ will be performed in the next chapter. Another important effect of self-interactions is to introduce a strong density dependence in the energy density. This fact is responsible for the better values of the total binding energy of finite nuclei and for lowering the value of the compression modulus below 300 MeV. Furthermore, the exchange (Fock) terms and the isovector mesons are very important for a quantitative description of the symmetry energy in nuclear matter and the spin–orbit interaction in nuclei.

In [354], an exact method to treat the nonlinear self-interaction (NLSI) of the meson fields in the relativistic Hartree–Fock approximation was suggested for nuclear systems. The authors of [354] consider an NLSI constructed from the relativistic scalar nucleon densities, including products of six and eight fermion fields. This type of NLSI corresponds to the zero-range limit of the standard cubic and quartic self-interactions of the scalar field (see Sect. 12.2, on the relativistic point-coupling model). The method for treating the NLSI in the RHF framework is essentially based on utilizing Fierz transformations (see Appendix A12 and references therein), which enable one to express explicitly the exchange (Fock) components in terms of the direct (Hartree) ones. The method is applied to nuclear matter and finite nuclei. It was shown that, in the RHF formalism, the NLSIs, which are explicitly isovector-independent, generate scalar, vector, and tensor nucleon self-energies that are strongly density-dependent. This strong isovector structure of the self-energies is due to the exchange terms that occur in the RHF method. Calculations of nuclear matter and finite nuclei were carried out with a parameterization containing five free parameters. The model allows an adequate description of both types of systems (nuclear matter and finite nuclei). Different forms of NLSI can be treated in the same fashion.

We should mention that in [355] the neutrino cross section and mean free path in neutron stars were calculated in the framework of the DHF approximation.

8.6 Spin–Orbit Interaction

The spin–orbit interaction in nuclei was first introduced in the framework of the shell model [356, 357], its magnitude clearly indicating that relativistic effects play an important role in generating such forces in nuclei. At first, it was natural to try to explain the origin of these forces in nuclei by analogy with the situation in an atom, in which the greater part of the observed spin–orbit coupling is explained as a relativistic correction to the motion of the electron in the field of the nucleus (Thomas coupling). Furry [68] suggested that the different signs of the doublet splittings in atomic and nuclear spectra could be explained by the different type of interaction of the particles. The electrostatic potential in which an electron moves in an atom transforms as the zeroth (time) component of the four-vector of the electromagnetic field. Furry pointed out that if the average field is regarded as a world scalar, then in this case one obtains a Thomas operator of the spin–orbit coupling that differs only in sign from the analogous operator of the atomic problem. However, these early attempts to ascribe the spin–orbit interaction in nuclei to relativity were abandoned after calculations showed that Thomas coupling gives a doublet splitting about 30 times smaller than the experimentally observed splitting.

In [72], a phenomenological model with compensating contributions of two relativistic fields (scalar and vector) leading to a large spin–orbit coupling was considered by Duerr. In [59, 60] it was shown that one can obtain strong spin–orbit coupling of relativistic origin in nuclei if two-particle spin–orbit forces generated by relativistic OBE potentials are used in the framework of Hartree and Hartree–Fock theory.

The relations (3.5), (3.8), and (3.13) determine the Galilean-invariant component of the spin–orbit force of an OBE potential, this component depending only on the relative momentum $\boldsymbol{p}_{12} = (\boldsymbol{p}_1 - \boldsymbol{p}_2)/2$ of two nucleons. The total operator of the two-particle spin–orbit forces, to which only scalar and vector mesons (both isoscalar and isovector) contribute, has a more complicated structure. The two-particle operator of the spin–orbit interaction can be represented as

$$V^{\mathrm{SO}} = V_1^{\mathrm{SO}} + V_2^{\mathrm{SO}}, \qquad (8.80)$$

where

$$V_1^{\mathrm{SO}} = -\frac{1}{8M^2} \boldsymbol{r}_{12} \times \boldsymbol{p}_{12} (\boldsymbol{\sigma}_1 + \boldsymbol{\sigma}_2) \frac{1}{r_{12}} \frac{\mathrm{d}}{\mathrm{d}r_{12}} \Big\{ (J^{\mathrm{S}}(r_{12}) + 3J^{\mathrm{V}}(r_{12})) \\ + \boldsymbol{\tau}_1 \cdot \boldsymbol{\tau}_2 \Big[J^{\mathrm{S}\tau}(r_{12}) + (3 + 4\frac{f_\rho}{g_\rho}) J^{\mathrm{V}\tau}(r_{12}) \Big] \Big\}, \qquad (8.81)$$

$$V_2^{SO} = -\frac{1}{8M^2} \boldsymbol{r}_{12} \times (\boldsymbol{p}_1 + \boldsymbol{p}_2) \cdot (\boldsymbol{\sigma}_1 - \boldsymbol{\sigma}_2) \frac{1}{r_{12}}$$
$$\times \frac{d}{dr_{12}} \left\{ (J^S(r_{12}) - J^V(r_{12})) + \boldsymbol{\tau}_1 \cdot \boldsymbol{\tau}_2 \left[J^{S\tau}(r_{12}) - J^{V\tau}(r_{12}) \right] \right\}, \quad (8.82)$$

i.e. as the sum of a Galilean-invariant term V_1^{SO}, determined by the relative momentum $(\boldsymbol{p}_1 - \boldsymbol{p}_2)$ of the nucleon pair, and a Galilean-noninvariant term V_2^{SO}, which depends on the total momentum of the two nucleons (the functions $J(r_{12})$ are given by (3.1), for example). In a system consisting of two nucleons in free space, Galilean-noninvariant forces are absent by virtue of the relativity principle, whereas in a many-particle system Galilean-noninvariant forces are present, since every pair of particles moves in the field of the other particles, i.e. the relativity principle no longer applies separately to each pair of particles.

In [59], on the basis of the operators (8.81) and (8.82), a single-particle spin–orbit operator was derived for doubly-magic-plus-one-nucleon nuclei, and the mathematical expectation value of the operators (8.81) and (8.82) was calculated in [59] with a nonsymmetrized wave function of the core of the nucleus. The calculation was performed in a short-range approximation, i.e. the action of the operator (8.81) was restricted to P states. As a result, the following single-particle spin–orbit operator was obtained:

$$V_{SO}^H(r) = \frac{\pi}{3} \frac{1}{M^2} \frac{1}{r} \frac{d\rho_V}{dr} \int_0^\infty \frac{d}{dx} \left\{ (J^S(x) + J^V(x)) - \tau_3 \frac{N-Z}{A} \right.$$
$$\left. \times \left[J^{S\tau}(x) + \left(1 + 2\frac{f_\rho}{g_\rho}\right) J^{V\tau}(x) \right] \right\} x^3 \, dx \, \boldsymbol{l} \cdot \boldsymbol{\sigma}, \quad (8.83)$$

where $A - 1 = N + Z$; τ_3 is the nucleon isospin projection, such that $\tau_3 = \pm 1$ (the positive sign corresponds to the proton); and $\rho_V(r)$ is the matter density in the nucleus.

The average nuclear field was calculated in [59] using only the static part of the OBE potential (see (3.6), (3.11), (3.17)). The calculation was performed in the short-range approximation using a Hartree-type wave function for the nucleus ground state. The average field in this case was obtained in the form of a local, velocity-independent potential with a form that repeats that of the density distribution:

$$V_{av}^H(r)$$
$$= 4\pi \rho_V(r) \int_0^\infty \left\{ (J^S(x) - J^V(x)) - \tau_3 \frac{N-Z}{A} (J^{S\tau}(x) - J^{V\tau}(x)) \right\} x^2 \, dx$$
$$= [S(r) + V(r)] - \tau_3 \frac{N-Z}{A} [S^\tau(r) + V^\tau(r)]. \quad (8.84)$$

Note the following important point: the contributions of the vector and scalar mesons to the spin–orbit potential (8.83) have the same signs, while their contributions to the average field (8.84) have opposite signs. One then obtains a situation in which similar contributions compensate each other to a considerable degree in the average field but enhance each another in the spin–orbit potential. It is just this fact that leads to the strong enhancement of spin–orbit coupling in nuclei.[4] We should mention also that the operator V_2^{SO} in (8.82) plays a less important role than the Galilean-invariant operator V_1^{SO} in (8.81), since V_2^{SO} is determined by the difference between the contributions of the scalar and vector mesons. Therefore, the contribution of V_2^{SO} to the spin–orbit splittings is approximately of the same order of magnitude as that of the one-particle Thomas coupling in the old relativistic approach, i.e. it is an order of magnitude smaller than the V_2^{SO} contribution.

Consider a Dirac equation of the following form:

$$i\frac{\partial \Psi}{\partial t} = \Big\{ \boldsymbol{\alpha}\cdot\boldsymbol{p} + \beta\Big[M + S(r) - \tau_3\frac{N-Z}{A}S^\tau(r)\Big] + \Big[V(r) - \tau_3\frac{N-Z}{A}V^\tau(r)\Big] + i\frac{f_\rho}{g_\rho}\frac{1}{2M}\tau_3\frac{N-Z}{A}\beta\boldsymbol{\alpha}\nabla V^\tau(r)\Big\}\Psi \,. \quad (8.85)$$

The equation was introduced in [145].

After performing the Foldy transformation, we obtain the following Thomas-type operator (in the Hartree approximation):

$$\frac{1}{4M^2}\frac{1}{r}\frac{d}{dr}\Big\{[V(r) - S(r)] + \tau_3\frac{N-Z}{A}\Big[S^\tau(r) - \Big(1 + 2\frac{f_\rho}{g_\rho}\Big)V^\tau(r)\Big]\Big\}\boldsymbol{l}\cdot\boldsymbol{\sigma} \,, \quad (8.86)$$

which explicitly coincides with (8.83). So we have obtained a strong single-particle spin–orbit potential of the Thomas form (of relativistic origin) which needs no extra fitting parameters.

Notice that in the Hartree approximation $V(r)$ is determined by the contribution of the ω meson, and $S(r)$ by the contribution of the σ meson. The terms $S^\tau(r)$ and $V^\tau(r)$ are related to the contributions of the scalar–isovector $a_0(\delta)$ meson and the vector–isovector ρ meson, respectively. In nuclear-structure calculations, the scalar–isovector field may be useful[5]

---

[4] Note that there are actually two reasons for the spin–orbit force to be increased in finite nuclei (relative to the old-fashioned relativistic estimates). The first reason is connected with coherent interference of the scalar and vector potentials in the single-particle operator of the spin–orbit force. The second reason is connected with the large reduction of the effective nucleon mass in nuclear media as discussed in Chap. 4.

[5] Effects of the $a_0(\delta)$ meson in the RMF theory of asymmetric nuclear matter are discussed in [358]. The scalar–isovector meson is involved into consideration in the recent versions of the OBEP, see Chap. 3. However, its coupling constant is not determined reliably and the mass is big enough. For this reason at the initial

when one needs to distinguish proton and neutron effective masses in the nuclear medium [359]:

$$M_\tau^* = M + S(r) - \tau_3 \frac{N-Z}{A} S^\tau(r) . \qquad (8.87)$$

From (8.86), it can be seen that in the Hartree approximation the isovector structure of the spin–orbit force is related to $a_0(\delta)$ and ρ mesons, the tensor ρN coupling dominating in this structure. Note also that the isovector dependence in (8.84) is different from that in (8.83).

In [60], an investigation of the influence of the Pauli principle on the spin–orbit splitting and the average nuclear field was performed, the treatment taking into account a wave function of the proper symmetry. The spin–orbit splittings were expressed in terms of the contributions of the spin–orbit operator to the energies of the single-particle states as follows:

$$\Delta E_{nl}^{SO} = \varepsilon_i^{SO}\left(j = l - \frac{1}{2}\right) - \varepsilon_i^{SO}\left(j = l + \frac{1}{2}\right), \qquad (8.88)$$

where, in the Hartree–Fock approximation

$$\varepsilon_i^{SO} = \sum_{k \leq A}\left(\langle ik | V^{SO} | ik \rangle - \langle ik | V^{SO} | ki \rangle\right). \qquad (8.89)$$

If we remember that the operator (8.81) acts only on triplet states and, in the short-range approximation adopted here, only on triplet P states, the exchange matrix elements can be calculated in terms of the direct matrix elements [60]. In this case, for nuclei with one nucleon above a closed shell, we obtain a single-particle spin–orbit operator of the following form (see Appendix A11):

$$V_{SO}^{HF}(i) = \frac{\pi}{4}\frac{1}{M^2}\frac{1}{r}\frac{d\rho_V}{dr}\int_0^\infty \frac{d}{dx}\left\{(J^S(x) + 3J^V(x)) + \left[J^{S\tau}(x)\right.\right.$$
$$\left.\left. + \left(3 + 4\frac{f_\rho}{g_\rho}\right)J^{V\tau}(x)\right]\right\}x^3 dx \left(1 - \tau_3\frac{N-Z}{3A}\right) \boldsymbol{l}\cdot\boldsymbol{\sigma}, \qquad (8.90)$$

where it is also assumed that the neutron density $\rho_{V,n}(r)$ and the proton density $\rho_{V,p}(r)$ are related to the total density by the relations

$$\rho_{V,n}(r) = \frac{N}{A}\rho_V(r), \qquad (8.91)$$

---

stage this meson was not taken into account in the calculations of the nuclear structure at normal density. Mention that vector–isovector and scalar–isovector mesons determine the isovector structure of the relativistic Hartree solutions, the value of the symmetry energy introducing certain restrictions on the relative values of the coupling constants of two mesons.

$$\rho_{\mathrm{V,p}}(r) = \frac{Z}{A}\rho_{\mathrm{V}}(r) \,, \tag{8.92}$$

Comparing (8.83) and (8.90), it is easily seen that the single-particle spin–orbit potential, as a whole, is strongly increased in the HF approximation. The strengthening of the role of the Dirac coupling of the vector mesons (both isoscalar and isovector) and that of the ρ meson tensor coupling in the RHF case should be emphasized, in particular. The calculations carried out in [60] indicate a very important role of the exchange (Fock) terms in calculating the spin–orbit splittings. On the other hand, in the same approximations, the average field is local and has the form

$$V_{\mathrm{av}}^{\mathrm{HF}}(r) = 3\pi\rho_{\mathrm{V}}(r)\int_0^\infty \Big[(J^{\mathrm{S}}(x) - J^{\mathrm{V}}(x)) - (J^{\mathrm{S}\tau}(x) - J^{\mathrm{V}\tau}(x))\Big]x^2\,\mathrm{d}x$$

$$+ \tau_3\pi(\rho_{\mathrm{V,n}}(r) - \rho_{\mathrm{V,p}}(r))\int_0^\infty \Big[(J^{\mathrm{S}}(x) - J^{\mathrm{V}}(x)) - 5(J^{\mathrm{S}\tau}(x) - J^{\mathrm{V}\tau}(x))\Big]x^2\,\mathrm{d}x \,.$$

$$\tag{8.93}$$

From (8.93) it can be seen that the absolute value of the average field in the HF approximation decreases in comparison with the Hartree case.

Up to this point in this section, we have considered only the contribution of the two-particle spin–orbit forces to the one-particle operator of the spin–orbit coupling in nuclei. However, the single-particle spin–orbit operator in nuclei is completely determined by the contribution of two-particle spin–orbit forces only in the case of spin-saturated nuclei, i.e. nuclei for which the two levels of the doublet $j = l \mp 1/2$ are either both filled or both empty. To elucidate completely the problem of the nature of the spin–orbit coupling in nuclei, in [62] the contributions of all components of the OBEP (see (3.2)–(3.19)) to the spin–orbit splittings of single-particle states were calculated for spherical and deformed nuclei. In this case the matrix elements of the finite-range potential (the OBEP) were expanded in Taylor series. If the range of the potential is short compared with p_{F}, then in this expansion we can restrict ourselves to terms quadratic in the relative momenta (of the initial and final states). By applying such an expansion to the one-boson exchange potential, the latter potential was represented as a sum of potentials of zero range depending on the relative momenta. Then the technique developed in [136] was utilized in [62] to derive a single-particle spin–orbit potential generated by the components of the OBEP different from the two-body LS force, and to calculate the corresponding spin–orbit splittings. Two points should be emphasized here:

1. The contributions of different components of the OBEP (especially those related to the pion) to the total spin–orbit splitting are essential.
2. These contributions appear in the Hartree–Fock approximation for spin-unsaturated nuclei only.

These conclusions were confirmed later by calculations for deformed nuclei [62]. This problem has been extensively investigated in the framework of the self-consistent RHF theory in [335, 343] also. The most spectacular example of the role of the exchange effects (in particular, those related to the pion) is demonstrated by the strong (almost two times) reduction of the 1d(3/2)–1d(5/2) proton splitting in ^{48}Ca in comparison with ^{40}Ca. This result of the relativistic HF calculations is excellently confirmed by experiment (see Tables 8.5 and 8.7) and cannot be reproduced theoretically either by nonrelativistic HF calculations or by relativistic Hartree methods, if the pion is excluded from consideration in both cases (see also a discussion of this subject in [360]).

8.7 Pseudospin as a Relativistic Symmetry

The spin–orbit force generates a large spin–orbit splitting. That is why the original $SU(3)$ symmetry had few applications in finite nuclei. However, it gave a concrete form to two directions in nuclear physics, the first one related to pseudospin symmetry, the other one related to the interacting-boson model. In this section we consider the origin of pseudospin symmetry (PSS).

In [361, 362], a quasi-degeneracy of the single-nucleon doublet states with quantum numbers $(n_r, l, j = l + 1/2)$ and $(n_r - 1, l + 2, j = l + 3/2)$ in heavy nuclei was established, n_r, l, and j being the single-particle radial, orbital, and total angular-momentum quantum numbers, respectively. In those papers a "pseudo" orbital angular momentum was introduced in accordance with the prescription $\widetilde{l} = l + 1$. For example, the single-nucleon doublet $(n_r s_{1/2}, (n_r - 1)d_{3/2})$ has $\widetilde{l} = 1$, and for $(n_r p_{3/2}, (n_r - 1)f_{5/2})$ $\widetilde{l} = 2$. The doublets considered here are practically degenerate with regard to the "pseudo" spin $\widetilde{s} = 1/2$, since one has $j = \widetilde{l} \pm \widetilde{s}$ for the doublet states ($\widetilde{j} = j = \widetilde{l} \pm 1/2$, and for the parity $\widetilde{\pi} = -\pi$). For example, the pair of states $(4s_{1/2}, 3d_{3/2})$ and $(3d_{5/2}, 2g_{7/2})$ can be considered as the $(3\widetilde{p}_{1/2}, 3\widetilde{p}_{3/2})$ and $(2\widetilde{f}_{5/2}, 2\widetilde{f}_{7/2})$ pseudospin doublets.

Examples of pairs of states forming pseudospin doublets in spherical nuclei are given in Table 8.8. If the level with $j = \widetilde{l} - 1/2$ is higher than that with $j = \widetilde{l} + 1/2$, we say that the doublet has a "natural ordering"; the term "unnatural ordering" corresponds to the inverted situation.

The terminology used here is taken from experience with the normal spin–orbit splittings, where the natural ordering appears to be the common situation, and this is definitely supported by the experimental data. However, in the case of pseudo-spin–orbit splittings, both possibilities appear to be natural from the experimental and theoretical points of view.

Pseudospin symmetry is widely used in nuclear physics [363–367]. In case of deformed nuclei, the relabeling $[N, n_3, \Lambda](\Omega = \Lambda - 1/2)^\pi \to [\widetilde{N}, \widetilde{n}_3, \widetilde{\Lambda}](\widetilde{\Omega} = \widetilde{\Lambda} + 1/2)^\pi = [N - 1, n_3, \Lambda - 1]\Omega^\pi$ and $[N, n_3, \Lambda'](\Omega' = \Lambda' + 1/2)^\pi \to$

8 The Relativistic Hartree–Fock Approach

Table 8.8. Some pairs of states of pseudospin doublets in spherical nuclei. (Note that there is no pseudospin–orbit pairs in ^{16}O at all, and there are only one pair in ^{40}Ca and ^{48}Ca, $(2s_{1/2}, 1d_{3/2})$, etc.)

n	l	j	$\widetilde{l}$
2 3 4	0	1/2	
1 2 3	2	3/2	$1(\widetilde{p})$
2 3	1	3/2	
1 2	3	5/2	$2(\widetilde{d})$
2 3	2	5/2	
1 2	4	7/2	$3(\widetilde{f})$
2			
1			$4(\widetilde{g})$
2			
1			$5(\widetilde{h})$

$[\widetilde{N}, \widetilde{n}_3, \widetilde{\Lambda}](\widetilde{\Omega} = \widetilde{\Lambda} - 1/2)^\pi = [N-1, n_3, \Lambda'+1]\Omega^\pi$ produces almost degenerate states with the quantum numbers $[\widetilde{N}, \widetilde{n}_3, \widetilde{\Lambda}](\widetilde{\Omega} = \widetilde{\Lambda} \pm 1/2)^\pi$.

Pseudospin symmetry has been utilized to describe various features of deformed nuclei, in particular superdeformation and the existence of identical rotational bands. And it is well established that PSS reveals itself in finite nuclei as a slightly broken symmetry. The procedure described above was considered initially as a relabeling which brings about almost degenerate pseudospin–orbit doublets, which are observed empirically. However, the origin of the pseudospin symmetry has never really been understood in the nonrelativistic framework. Recently [368–371, 373–382] it has been shown that PSS has its origin in a relativistic symmetry of the Dirac Hamiltonian (see also [383] in this connection). Earlier, in [384], a family of external potentials for the Dirac equation was given for which there is an exact $SU(2)$ symmetry, leading to exact pseudo-spin–orbit doublet degeneracy. This family includes an equal superposition of vector and Lorentz scalar potentials. In [369], the limit $S + V = 0$ was investigated. Later, such investigations were carried out within realistic RMF models [370, 371, 373]. In [370] it was shown that in the case $S + V \approx 0$ there exist some similarities in the relativistic single-nucleon wave functions of the pseudospin doublets; these similarities are described by the following equation:

$$F_{\widetilde{l}+1/2}(r) \approx -F_{\widetilde{l}-1/2}(r) , \qquad (8.94)$$

where F is the lower component of the nucleon wave function, $\widetilde{l}$ being (exactly) conserved in the exact pseudospin symmetry limit. This property of the wave function $F(r)$ has been checked by calculations within realistic RMF models [373–380].

There are two possible ways to solve the Dirac equation. If one utilizes (4.86) and (4.87), the relation between the upper and lower components of the total wave function can be obtained from

$$\varphi = \frac{1}{\varepsilon - V - S} \boldsymbol{\sigma} \cdot \boldsymbol{p}\chi , \qquad (8.95)$$

$$\chi = \frac{1}{\varepsilon + 2M^* - V - S} \boldsymbol{\sigma} \cdot \boldsymbol{p}\varphi , \qquad (8.96)$$

where $E = M + \varepsilon$. The following phase convention for vector spherical harmonics is used below:

$$(\boldsymbol{\sigma} \cdot \boldsymbol{n})\Omega_{jlm} = -\Omega_{jl'm} , \qquad (8.97)$$

where

$$l' = (2j - l) = \begin{cases} l+1, & j = l + 1/2 \\ l-1, & j = l - 1/2 \end{cases} . \qquad (8.98)$$

In (8.98) l' is just the pseudo-orbital angular momentum.

From (4.86) and (4.87), one obtains [374, 378]

$$\left[\frac{d^2}{dr^2} + \frac{1}{\varepsilon - V - S} \frac{d}{dr}(S + V) \frac{d}{dr} \right] F_a(r)$$
$$+ \left[\frac{\kappa_a(1 - \kappa_a)}{r^2} - \frac{1}{\varepsilon - V - S} \frac{\kappa}{r} \frac{d}{dr}(S + V) \right] F_a(r)$$
$$= -(\varepsilon + 2M^* - V - S)(\varepsilon - V - S) F_a(r) , \qquad (8.99)$$

$$\left[\frac{d^2}{dr^2} - \frac{1}{\varepsilon + 2M^* - V - S} \frac{d(2M^* - V - S)}{dr} \frac{d}{dr} \right] G_a(r)$$
$$- \left[\frac{\kappa_a(1 + \kappa_a)}{r^2} - \frac{1}{\varepsilon + 2M^* - V - S} \frac{\kappa_a}{r} \frac{d(2M^* - V - S)}{dr} \right] G_a(r)$$
$$= -(\varepsilon + 2M^* - V - S)(\varepsilon - V - S) G_a(r) , \qquad (8.100)$$

where

$$\kappa_a = \begin{cases} -l_a - 1, & j_a = l_a + 1/2 \\ l_a, & j_a = l_a - 1/2 \end{cases} , \qquad (8.101)$$

so one has

$$\kappa_a(\kappa_a - 1) = l'_a(l'_a + 1) \qquad \kappa_a(\kappa_a + 1) = l_a(l_a + 1) . \qquad (8.102)$$

From the equations above, one can see that l'_a physically is "the orbital angular momentum" of the lower component of the Dirac wave function.

To obtain the eigenvalues and the respective eigenfunctions, one can use either (8.99) or (8.100). It is more conventional to use (8.100) for the upper component of the wave function. In this case the spin–orbit splitting is determined by the spin–orbit potential

$$\frac{1}{\varepsilon + 2M^* - V - S} \frac{\kappa_a}{r} \frac{\mathrm{d}(2M^* - V - S)}{\mathrm{d}r} \,.$$

However, it is possible also to use (8.99). If one neglects the term

$$\frac{1}{\varepsilon - V - S} \frac{\kappa_a}{r} \frac{\mathrm{d}(S + V)}{\mathrm{d}r}$$

then the eigenvalues ε for the same l' will be degenerate, i.e. one obtains the phenomenon of pseudospin symmetry discussed in [361–364]. From (8.99), one can see that it is just the term

$$\frac{1}{\varepsilon - V - S} \frac{\kappa_a}{r} \frac{\mathrm{d}(V + S)}{\mathrm{d}r}$$

that spoils pseudospin symmetry, i.e. splits the pseudospin partners. So one obtains the condition [374] $\mathrm{d}(S+V)/\mathrm{d}r = 0$ for the pseudospin symmetry to be realized, while (8.95) and (8.96) give the transformation from the normal spin formalism to the pseudospin formalism and vice versa.

The equation $(\mathrm{d}/\mathrm{d}r)(S+V) = 0$ includes the condition $S+V = 0$, discussed above, as a particular case. However, in spherical nuclei the equation $(\mathrm{d}/\mathrm{d}r)(S+V) = 0$ is satisfied only at the origin ($r = 0$) and, in principle, is not valid for r different from zero. For this reason, the authors of [374] compare the relative magnitudes of the pseudo-spin–orbit potential and the pseudo-centrifugal barrier and show that the degree of degeneracy of the pseudo-spin–orbit splitting becomes smaller if the respective levels are less bound (the energy splitting becomes smaller if the binding energy decreases).

In [381], the properties of the pseudo-spin–orbit potential are investigated in the relativistic Hartree formalism. Let us discuss some of the results obtained in this paper in more detail. Equation (8.99) can also be written in the following form:

$$-F_a'' + \left[\frac{V_{G_a}'}{V_{G_a}} \left(\frac{F_a'}{F_a} - \frac{\kappa_a}{r} \right) + \frac{\tilde{l}_a(\tilde{l}_a + 1)}{r^2} \right.$$
$$\left. + 2M(S+V) + 2\varepsilon_a V + (S^2 - V^2) - \varepsilon_a(\varepsilon_a + 2M) \right] F_a = 0\,, \quad (8.103)$$

where $\tilde{l}$ is the pseudo-orbital angular momentum defined above, $V_{G_a} = -\varepsilon_a + S + V$, and $\kappa_a = (2j_a + 1)(l_a - j_a)$. Further,

$$\kappa_a - 1 = \langle 2\tilde{l} \cdot \tilde{s} \rangle_a = j_a(j_a + 1) - \tilde{l}_a(\tilde{l}_a + 1) - \frac{3}{4}\,, \quad (8.104)$$

where $\langle 2\tilde{l} \cdot \tilde{s} \rangle_a$ is the angular mean value of the operator of the pseudo-spin–orbit interaction in the state a. Bearing this in mind, we introduce the pseudo-spin–orbit operator in the Hartree approximation [381]:

$$\widehat{V}_{\tilde{l}\cdot\tilde{s}}^{\mathrm{H}} = -\frac{2}{r}\frac{V'_G}{V_G}\tilde{\boldsymbol{l}}\cdot\tilde{\boldsymbol{s}} = V_{\tilde{l}\cdot\tilde{s}}^{\mathrm{H}}\tilde{\boldsymbol{l}}\cdot\tilde{\boldsymbol{s}}. \qquad (8.105)$$

The extra factor of 2 in the right-hand side of (8.105) has appeared because of (8.104), which relates κ_a and $\langle\tilde{l}\cdot\tilde{s}\rangle_a$.

Thus, (8.103) has the structure of a Schrödinger-like equation in which the central potential and the pseudo-spin–orbit potential (PSOP) are strongly state-dependent (via ε_a and F_a). Equation (8.103) also includes the pseudo-centrifugal barrier $\widehat{V}_{\mathrm{cf}} = \tilde{l}_a(\tilde{l}_a+1)/r^2$.

The PSOP is determined, in particular, by the derivative of the central potential, given mainly by $S'+V'$, rather than by $S'-V'$ as in the case of the normal spin–orbit operator. Although $S'+V'$ is much smaller than $S'-V'$, this fact by itself does not mean that the pseudo-spin–orbit potential is small enough for our purposes in comparison with the ordinary spin–orbit potential. The basic feature of (8.105), which determines the properties of the pseudo-spin–orbit potential, is its energy dependence,[6] which causes a singularity in this equation. For each energy value ε_a, there is a certain point r_0 at which the value of V_{G_a} becomes equal to zero; V_{G_a} is negative for $r < r_0$ and positive for $r > r_0$ (note that ε_a is negative). One should bear in mind also that ε_a is negative, V'_{G_a} is mainly positive (it would be always positive for self-energies S and V of the Woods-Saxon type), and V_{G_a} is negative in the central part of the nucleus.

Formally, to eliminate terms containing F'_a in (8.103), one needs to introduce a new function $\tilde{F}_a = V_{G_a}^{-1/2}F_a$. However, using this function appears not to be reasonable, since V_{G_a} changes sign at r_0 and, consequently, this possibility should be discarded.

Let us remember that there is no problem of this type when one tries to obtain a Schrödinger-like equation for the large component of the Dirac spinor. In this case one needs to make the transformation $\widetilde{G}_a = V_{F_a}^{-1/2}G_a$, where $V_{F_a} = 2M+\varepsilon_a+S-V$ is a positive function everywhere. The property mentioned above is related only to $\tilde{F}_a$ (see also Appendix A6).

Let us note that although $V'_{G_a}/V_{G_a} \to \infty$ for $r \to r_0$, the term $(V'_{G_a}/V_{G_a})(F'_a/F_a - \kappa_a/r)$ in (8.103) is not singular in this point, because $(F'_a/F_a - \kappa_a/r) \to 0$ for $r \to r_0$, and the entire term remains finite. This behavior can be understood by noting that the other terms in (8.103) are clearly not singular. The addend, proportional to κ_a, which is responsible for the pseudo-spin–orbit splitting, is singular at the point $r = r_0$ when considered alone. However, this singularity changes sign when one goes from $r = r_0 - \varepsilon$ to $r = r_0 + \varepsilon$ (where $\varepsilon \to 0$), and for this reason it gives a finite contribution to the single-particle binding energies.

---

[6] Notice that the ordinary spin–orbit operator (appearing in the equation for the large component of the wave function $G(r)$) is also energy-dependent. However, in contrast to the PSOP, this energy dependence is very weak and can be neglected in practice.

The potential determined by (8.105) is singular at $r = r_0$. However, it is possible to work out a corrected PSOP satisfying the following two conditions: (a) it is a continuous function for all values of r, and (b) it has the same mean value as the potential given by (8.105). Such calculations have been done in [381] for ^{40}Ca which has only a single pseudospin doublet and for Zr isotopes, with a more developed pseudospin doublet structure. The following conclusions have been drawn:

1. The important role of the destructive interference of the two large contributions (to the single-particle energy) arising from the PSOP near the singularity point must be stressed, this interference being essential for obtaining a small value of the splitting in the pseudospin doublet.
2. A corrected PSOP has been defined that enables one to make a detailed comparison of its contribution to the single-particle energy with the corresponding contribution of the pseudo-centrifugal barrier.
3. Perturbation theory suggests that the effect of the PSOP on the single-particle binding energies is about a factor of five smaller than that of the pseudo-centrifugal potential. However, the effect of the PSOP is not only to produce a splitting between the two pseudospin partners but also to shift their energy (so as to increase it), so that this potential cannot be treated by perturbation theory.
4. A strong correlation has been established between the compressibility modulus K and the scalar meson mass values of the RMF models (which mainly affect the smoothness of the nuclear surface) and the energy ordering of the partner levels, smaller K and m_σ values favoring unnatural ordering. On the other hand, it is also found that the more strongly bound levels always have a natural ordering.
5. PSS cannot be explained only by the size of the pseudo-spin–orbit potential in comparison with the pseudo-centrifugal barrier; it is also necessary to consider the actual nuclear surface smoothness (which depends strongly on the range of the nuclear interaction), which indirectly favors the accomplishment of this symmetry.
6. The $S + V = 0$ limit is not reasonable in finite nuclei, since there are no nuclear bound states in this limit; PSS cannot be justified by the fact that $S \approx -V$ either.
7. Exact PSS does not need an exact (up to a phase) similarity of the small components of the Dirac spinor in the pseudospin doublet.
8. The diffuseness of the nuclear surface prevents the exact realization of PSS even in the hypothetical case where $\kappa = 0$.
9. Although the pseudo-spin–orbit splittings are usually smaller than those produced by the spin–orbit interaction, the κ term plays a much more important role in the equation for $F(r)$ than does the corresponding term in the equation for the upper component of the nucleon Dirac spinor responsible for the spin–orbit splitting. If this latter term is switched off, the single-particle levels and their corresponding wave functions are

considerably modified, but not in a crucial way as it happens if the κ term in the equation for $F(r)$ is switched off.

10. The (slightly) broken PSS of the pseudospin doublets in real nuclei can be explained as a result of a nonperturbative transformation from nonphysical solutions of the Dirac equation, which satisfy almost exact PSS, to the physical solutions.

11. Self-consistent effects are very important in generating the total pseudospin–orbit splittings. The total pseudo-spin–orbit splitting is determined dynamically by the general structure of the Dirac equation through a delicate compensation of different contributions to the single-particle energy.

The pseudospin doublet structure of the single-particle levels is revealed via the equations for both the upper and the lower components of the nucleon Dirac spinor. However, in the first case it is revealed only implicitly, i.e. only via the respective single-particle energies, while in the second case it is manifested explicitly via the PSS-breaking term (via the pseudo-spin–orbit potential). This means that PSS emerges naturally in the relativistic formalism. A very strong indication of the relativistic origin of PSS is given by the fact that the angular momentum of the lower component of the Dirac spinor is identical to the nucleon pseudo-orbital angular momentum.

Let us mention also [382]; the main purpose of that paper is to derive conditions for PSS in the relativistic Hartree–Fock (RHF) framework and to check the conclusions obtained earlier in the relativistic Hartree approximation. The RHF equation for the lower component $F(r)$ of the nucleon wave function has been considered in this chapter (see (8.57), (8.62)); it contains the direct (Hartree) components and the inhomogeneous parts, arising from the Fock terms, which include nonlocal (state-dependent) potentials. It was shown that the RHF results satisfy the criterion of quasi-degeneracy of the pair of levels of the pseudospin doublet imposed by PSS, though the symmetry appears to be slightly broken even in the $S + V = 0$ limit. The validity of the criterion for PSS of the small components of the Dirac wave function of the pseudo-spin–orbit doublet was checked. It was shown that PSS operates reasonably well (as a broken symmetry) in the RHF approximation.

Another aspect that is worth noting is that for the states of a pseudospin doublet one has $j_a = l_a + 1/2$, whereas $j_b = l_b - 1/2$. Thus, in a simple shell-model approximation in which the nucleons move in a central and an LS potential, the effect of the LS term shifts the two single-particle energies of the pseudospin doublet states in opposite directions. This means that the pseudo-spin–orbit splitting depends crucially on the strength of the LS interaction (see also [368, 381]). In this picture that is not self-consistent, the two states of a pseudospin doublet can be forced to be degenerate if the LS interaction is suitably chosen. For self-consistent models, the relation between the LS scheme and pseudo-LS formalism is qualitatively similar, although the situation is more complicated in this case: there are extra contributions to the LS splittings of the terms in the equation for G that are different from the LS term, as a result of the self-consistent procedure. It is worth noting also that a large LS force increases the relative size of the function F in comparison with the component G favoring a similarity between F_a and F_b. Thus, PSS and the LS scheme are strongly related, and PSS cannot be understood independently from the LS interaction.

Let us mention that the relativistic framework enables one to investigate also the special status of the intruder states in finite nuclei [372] .

In [373] it has been realised also that the spin symmetry in the antinucleon spectrum has the same origin as the PSS in the particle spectrum.

The pseudospin symmetry is an approximate symmetry in finite nuclei. Its accuracy has been related, in particular, to the number of nodes of F: the bigger is the number of nodes, the higher is the accuracy of the symmetry [381]. We may conclude that the PSS is an important feature of the nuclear structure, it is a dynamical symmetry [381] directly related to the spin symmetry, both of them obtained natural explanation in the framework of the relativistic formalism.

9 Brueckner–Hartree–Fock Methods for Nuclear Matter and Finite Nuclei

9.1 The Brueckner–Hartree–Fock Approach

In the present chapter we consider some methods developed during recent years for considering correlations in the relativistic description of the properties of nuclear matter and finite nuclei. First of all, we define more precisely what is meant by correlations. At the present stage, a detailed investigation has been carried out of interactions of free nucleons, both in the case of nucleon–nucleon scattering and in the case of the description of the properties of the deuteron in the relativistic framework on the basis of meson exchange. Quite reasonable agreement with experiment has been obtained [34, 40, 50, 78, 275] in such a framework. The main product of this investigation is the one-boson exchange potentials, embodied in a specific meson exchange model. The next natural step is to describe, on the same footing, complicated systems containing more than two nucleons. The simplest tool to handle the problem in this case is the concept of the mean field generated by the nucleon–nucleon potential. Such a philosophy is a copy of the Hartree and Hartree–Fock methods successfully applied to problems related to the electronic shells in atoms. Although, as was shown in the previous chapters, a quantitative description of nuclear bulk properties has been achieved in the framework of the mean-field approximation, in general this method cannot be considered as a final stage of the investigation of nuclear systems. The commonly accepted reason for the deficiencies of this method is the important modification of the nucleon–nucleon interaction in the nuclear medium in comparison with the interaction of free nucleons. It is conventional to treat the effects generating these modifications as correlations [30, 385].

Investigations of the nucleon–nucleon correlations started when the nucleon–nucleon interaction was described purely phenomenologically without addressing the meson exchange concept. The phenomenological interaction contains a strong repulsive core. Taking this core into consideration leads inevitably to the nonsensical results. The solution to the problem of the proper treatment of this feature of the NN potential has been given by Brueckner (see the review papers [386]). The Brueckner method has also become an obligatory attribute of practically all calculations of nuclear systems utilizing OBE potentials. The source of the correlations described by the Brueckner method is two-particle–two-hole excitations. Considering these excitations leads effectively to a modification of the nucleon–nucleon interaction amplitude.

The relativistic Brueckner–Hartree–Fock theory has been developed and reviewed in [34, 387–394]. It should be noticed that the Brueckner-type correlations form only a part of the correlations that appear in the nuclear medium. The Brueckner method does not take into account modifications of the Green's function of the mesons propagating in the medium. As shown by recent investigations, these modifications may be quite important, a density dependence of the meson mass being the basic effect. From this point of view, the question of the role of the correlations of the Brueckner type remains open.

From the point of view of applications, there exist two mostly efficient approaches that enable us to describe nuclear systems properly. The first of these two approaches is based on the proper treatment of both the Brueckner-type correlations and the influence of the medium on the processes involving the Δ-isobar. This approach includes also three-particle forces, the respective method being described in [34, 50, 78]. The influence of the nuclear medium on the meson propagator has also been investigated in the framework of this approach. A method was suggested to take this influence into account properly, the binding energy and equilibrium density of nuclear matter being calculated by this method.

In the second approach, the main attention is paid to the mostly complete solution of the self-consistency problem, with the interaction obtained via the Brueckner method starting from the bare nucleon interaction. More complete self-consistency is achieved in this case because the nucleon wave functions of the components with negative energy and the average field of the components which couple states with positive and negative energy are included, and also the average field of the components which couple different states with negative energy to each other is included. These components originate from the amplitudes of the single-boson exchange, which do not reveal themselves in problems containing two nucleons, but can be constructed in a parameter-free way similarly to the one-boson exchange potentials. In the framework of the second approach, the saturation properties of nuclear matter have been reproduced, the effective nucleon interaction in the nuclear medium has been investigated, a comparison with the phenomenological description has been carried out, and the optical potential has been derived. In spite of the simple character of the model, the results obtained on its basis, are, on the whole, quite satisfactory. It is still not completely clear how these results may be influenced if the effects of the medium on meson exchange, the Δ-isobar degrees of freedom, and many-particle forces are considered. The second approach has been reviewed in [79, 387].

From the theoretical point of view, some very fundamental investigations have been carried out in [75, 391], the field-theoretical description of a system of interacting mesons and nucleons being the basis of these investigations. The starting point in this case is the Lagrangian of the Walecka model [74] including neutral scalar and vector mesons, or generalized Lagrangians with other mesons included. However, these methods are too complicated even in the case of nuclear matter. Moreover, it is not completely clear to what extent the notions of mesons and nucleons are applicable in nuclear matter, since these particles are really composite quark–gluon objects. For this reason, a truly fundamental theory probably should be constructed on the quark–gluon basis. Such a theory, even if realizable in principle, will be developed only in the future. From this point of view, simple models producing good agreement with experiment and enabling us to understand the basic properties of many-nucleon systems are highly desirable.

9.2 The Brueckner–Bethe–Goldstone Method for Internucleon Correlations

The Brueckner–Bethe–Goldstone method is an approximate method to treat the many-nucleon problem (infinite nuclear matter), the two-particle potential being fixed. The method is based on the Brueckner–Goldstone series in numbers of independent hole states, i.e. states with a momentum chosen arbitrarily inside the Fermi sphere without breaking the momentum conservation law [386]. Estimates show that such an expansion, in the case of nuclear matter of normal density, demonstrates good convergence. For this reason, in calculations of nuclear-matter properties it is sufficient to consider only the first-order diagrams in the effective internucleon interaction, which is calculated in the theory. This follows from the fact that the processes of higher order contain no fewer than three independent hole states and, for this reason, give a comparatively small contribution (the processes of the second order in the effective interaction are absent in the Brueckner–Goldstone theory).

Note that the criterion of convergence with respect to the number of independent hole states coincides with the criterion of convergence with respect to the powers of the density of the nuclear matter. In fact, n independent holes may be created as a result of the interaction of no fewer than n particles, which is possible if the particles come within a sufficiently short distance of each other. As a measure of this distance, one should take the radius of the repulsive (in the case of OBE potentials, the soft) core rather than the radius of nuclear forces, since it is the strong core interaction that leads mainly to considerable correlation effects in nuclear matter. The probability for n particles to be found in a region with the dimensions of the core is proportional to ρ^n, where ρ is the nuclear-matter density.

Let us discuss briefly the nature of the effective interaction. The physical reason for introducing this interaction into the theory is connected with con-

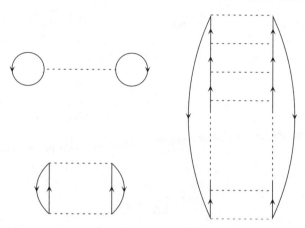

Fig. 9.1. Two-hole nonexchange diagrams. A *continuous line* stands for a particle if the arrow points upward and for holes if it points downward; a *dashed line* represents interaction through the original two-nucleon forces

sidering the "ladder" diagrams related to the process of scattering[1] of a pair of particles (above the Fermi surface) interacting via the initial two-nucleon potential. It is known that a ladder diagram may be a composite part of various diagrams [34].

Sequences of diagrams of two types containing ladder diagrams are shown in Figs. 9.1 and 9.2. Diagrams which are sums of sequences of the diagrams shown in Figs. 9.1 and 9.2 are shown in Fig. 9.3. These two diagrams are the only first-order diagrams in the effective interaction. It can easily be seen that they are none other than the Hartree–Fock direct and exchange diagrams, in which the effective nucleon–nucleon interaction is taken as the interaction.

We write the energy of the nuclear matter in the form

$$E = \sum_{m<k_F} \left\{ \langle m | \widehat{T} | m \rangle + \frac{1}{2} \sum_{n<k_F} [\langle mn | G | mn \rangle - \langle mn | G | nm \rangle] \right\}, \quad (9.1)$$

where m and n are states of the nucleon (its momentum, and spin and isospin projections), the summations are restricted to states within the Fermi surface, and the Fermi momentum p_F is related to the density ρ of the nuclear matter by $\rho = 2p_F^3/3\pi^2$. In (9.1), G is the effective nucleon–nucleon interaction, its matrix elements satisfying the Bethe–Goldstone equation

$$\langle q' | G(P) | q \rangle = \langle q' | V | q \rangle - \int \frac{Q(k,P) \langle q' | V | k \rangle \langle k | G(P) | q \rangle}{E(P+k) + E(P-k) - W(q,P)} d^3k . \quad (9.2)$$

---

[1] By scattering in nuclear matter we mean any change of the state of the particles in which there is no phase shift when the particles move apart to an infinite distance between them. Because of the Pauli exclusion principle, a phase shift resulting from an interaction is impossible in nuclear matter [386].

9.2 The Brueckner–Bethe–Goldstone Method for Internucleon Correlations

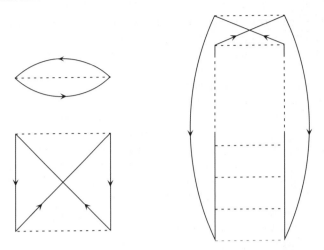

Fig. 9.2. Two-hole exchange diagrams. The notation is the same as in Fig. 9.1

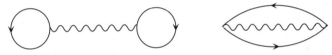

Fig. 9.3. Direct and exchange diagrams of first order in the effective interaction. The *wavy line* is the interaction through the effective nucleon–nucleon forces, which include the corresponding ladder processes

Here V is the original internucleon potential; q is the relative momentum of the interacting nucleons; $\boldsymbol{P} = (\boldsymbol{k}_1 + \boldsymbol{k}_2)/2$ is their mean momentum; $E(\boldsymbol{P} \pm \boldsymbol{k})$ are the energies of the two nucleons of a pair under consideration with momenta outside the Fermi sphere (we start this section from a discussion of the general features of the nonrelativistic theory), such that

$$E(\boldsymbol{k}_m) = \frac{k_m^2}{2M} \ ; \tag{9.3}$$

$Q(\boldsymbol{k}, \boldsymbol{P})$, the Pauli operator, is equal to 1 if both of the momenta $\boldsymbol{k} + \boldsymbol{P}$ and $\boldsymbol{k} - \boldsymbol{P}$ lie outside the Fermi sphere and 0 otherwise; and $W(\boldsymbol{q}, \boldsymbol{P})$ is the so-called starting energy. For the processes shown in Figs. 9.2 and 9.3,

$$W(\boldsymbol{q}, \boldsymbol{P}) = \widetilde{E}(\boldsymbol{P} + \boldsymbol{q}) + \widetilde{E}(\boldsymbol{P} - \boldsymbol{q}) \ , \tag{9.4}$$

where $\widetilde{E}(\boldsymbol{P} \pm \boldsymbol{q})$ are the Hartree–Fock energies of the nucleons below the Fermi boundary:

$$\widetilde{E}(\boldsymbol{k}_m) = \frac{k_m^2}{2M} + \sum_{n<p_F} [\langle mn|\, G(\boldsymbol{P})\, |mn\rangle - \langle mn|\, G(\boldsymbol{P})\, |nm\rangle]\ . \tag{9.5}$$

As can be seen from (9.2), the effective interaction depends on the mean momentum $\boldsymbol{P}$ of the pair of nucleons and, therefore, is not Galilean-invariant, in contrast to the original nucleon interaction in the nonrelativistic limit ($v/c \to 0$). This is explained by the presence of the medium, which distinguishes that system which is at rest with respect to the medium; in this system, the total momentum of all the particles of the medium is zero. The dependence on the total momentum of the particles has the consequence that the effective interaction also depends on the angle between the relative momentum and the vector $\boldsymbol{P}$. This means there is violation of isotropy in the space of the relative momentum, resulting in nonconservation of the relative angular momentum of a pair of nucleons interacting with effective forces. However, the dependence of G on $\boldsymbol{P}$ is comparatively weak, so that this violation of symmetry is slight and is usually ignored (see below).

Note that the relations (9.2), (9.4), (9.5) form a system of coupled equations – in accordance with (9.5), the single-particle energies $\widetilde{E}(\boldsymbol{k}_m)$ depend on the effective interaction G, which, in turn, depends on the single-particle energies through the starting energy (9.4). Therefore, to find the effective interaction, it is necessary to solve a self-consistency problem. Note that in this case there is no need to find the single-particle wave functions, since they can only, because of the symmetry of the problem (translational invariance), be plane waves. Nevertheless, a self-consistency problem arises because the interaction is not taken to be the fixed vacuum internucleon interaction but rather an effective interaction, which itself depends on the state of the nucleons.

The Bethe–Goldstone equation is usually solved approximately under the two following simplifying assumptions:

1. The Pauli operator $Q(\boldsymbol{k}, \boldsymbol{P})$ is averaged over the angle between the vectors $\boldsymbol{k}$ and $\boldsymbol{P}$:

$$\bar{Q}(\boldsymbol{k}, \boldsymbol{P}) = \frac{1}{2} \int_0^\pi Q(\boldsymbol{k}, \boldsymbol{P}) \sin\theta \, d\theta , \qquad (9.6)$$

where θ is the angle between $\boldsymbol{k}$ and $\boldsymbol{P}$. Bearing in mind that only the angles θ for which

$$|\cos\theta| < (P^2 + k^2 - p_F^2)/2kP \qquad (9.7)$$

contribute to the integral, we obtain

$$\bar{Q}(\boldsymbol{k}, \boldsymbol{P}) = \begin{cases} 0, & \text{if } P^2 + k^2 \leq p_F^2 \\ 1, & \text{if } k - P \geq p_F \\ (P^2 + k^2 - p_F^2)/2kP, & \text{otherwise} \end{cases}, \qquad (9.8)$$

where we have also taken into account the fact that the average momentum satisfies $P \leq p_F$ because of the conservation of the momentum of the nucleon pair in the homogeneous nuclear matter.

9.2 The Brueckner–Bethe–Goldstone Method for Internucleon Correlations

2. The Hartree–Fock energies are taken in the form

$$\widetilde{E}(\boldsymbol{k_m}) = \frac{k_m^2}{2M_{\rm NR}^*} + A_0 , \qquad (9.9)$$

where $M_{\rm NR}^*$ and A_0 are adjustable parameters, $M_{\rm NR}^*$ playing the role of the (nonrelativistic) effective mass. The values of $M_{\rm NR}^*$ and A_0 are determined by the self-consistency condition: In accordance with the initial approximation for $M_{\rm NR}^*$ and A_0, the value of G is calculated, and then, using (9.5), the values of $\widetilde{E}(\boldsymbol{k_m})$ ($k_m < p_{\rm F}$) and the new values of $M_{\rm NR}^*$ and A_0 that give the best approximation of $\widetilde{E}(\boldsymbol{k_m})$ are found; such iterations are repeated until the process converges. Both approximations lead to only slight errors in the calculation of the energy of nuclear matter (see [34]).

With the approximations adopted, the Bethe–Goldstone equation (9.2) can be solved in the partial-wave representation. In this representation, one can write the matrices of the interaction V and the effective interaction G in the following form (see, for example, [395]):

$$\langle \boldsymbol{q'} | G(\boldsymbol{P}) | \boldsymbol{q} \rangle = \frac{2}{\pi} \sum_{all\, l'M} i^{l-l'} G_{ll'}^{\alpha}(P, q', q) \mathcal{J}_{lS}^{JM}(\hat{q}') \mathcal{J}_{l'S}^{JM+}(\hat{q}) , \qquad (9.10)$$

$$\langle \boldsymbol{q'} | V | \boldsymbol{q} \rangle = \frac{2}{\pi} \sum_{all\, l'M} i^{l-l'} V_{ll'}^{\alpha}(q', q) \mathcal{J}_{lS}^{JM}(\hat{q}') \mathcal{J}_{l'S}^{JM+}(\hat{q}) , \qquad (9.11)$$

where α denotes the set of quantum numbers J, S, T, which are the angular momentum, the spin, and the isospin of the pair of nucleons in the given partial wave; and $\mathcal{J}_{lS}^{JM}(\hat{q})$ are spherical spinors:

$$\mathcal{J}_{lS}^{JM}(\hat{q}) = \sum_{m_l, m_S} \langle lSm_l m_S | JM \rangle Y_{lm_l}(\hat{q}) | Sm_S \rangle . \qquad (9.12)$$

Substituting the expressions (9.10) and (9.11) into (9.2), we obtain the following integral equation for the matrices of the effective interaction in the partial-wave representation:

$$G_{ll'}^{\alpha}(P, q', q)$$
$$= V_{ll'}^{\alpha}(q', q) - \frac{2}{\pi} \sum_{l''} \int_0^{\infty} \frac{V_{ll''}^{\alpha}(q', k) \bar{Q}(k, P) G_{l''l'}^{\alpha}(P, k, q)}{(P^2 + k^2)/2M - W(q, P)} k^2 \, dk , \qquad (9.13)$$

where we have taken into account the fact that, in accordance with (9.4) and the approximation (9.9), the starting energy does not depend on the angle between the vectors $\boldsymbol{k}$ and $\boldsymbol{P}$.

Let us consider the results of numerical calculations based on (9.13), using the OBE potentials given in [34]. Equation (9.13) may be simplified by replacing the mean momentum $\boldsymbol{P}$, which can vary in the interval $[0; p_{\rm F}]$, by an average value P_{av}. This does not strongly distort the solution of the equation, since the relative momentum k of the pair of nucleons appreciably exceeds, on average, the limiting momentum p_F, and therefore P as well. This last is due to the fact that the short-range repulsive core knocks nucleon pairs into states high above the Fermi sphere (the mean value of the relative momentum is approximately 1/(core radius), which corresponds to an energy greater than or equal to 300 MeV, which is appreciably higher than the Fermi energy of approximately 50 MeV).

Taking this remark into account, (9.13) can be solved by a matrix inversion method, which entails replacing the integral by a sum over a finite number of momenta k; this transforms the integral equation into a system of linear algebraic equations for $G^{\alpha}_{ll'}(P, q', q)$, where q' belongs to the set of momentum values over which the summation is performed.

The potential used in the Bethe–Goldstone equation has the following form in the momentum representation:

$$V(\boldsymbol{p}, \boldsymbol{q}) = \sqrt{\frac{M}{E_p}} V_{\text{OBE}}(\boldsymbol{p}, \boldsymbol{q}) \sqrt{\frac{M}{E_q}}, \qquad (9.14)$$

where $V_{\text{OBE}}(\boldsymbol{p}, \boldsymbol{q})$ is the one-boson exchange potential; the factors $\sqrt{M/E_{p,q}}$ are the "minimal relativity" correction. Let us discuss the meaning of this correction, following [396]. The scattering amplitude $\widetilde{R}(\boldsymbol{k}, \boldsymbol{k}')$ of free relativistic nucleons satisfies the following equation [397, 398]:

$$\widetilde{R}(\boldsymbol{k}, \boldsymbol{k}') = V_{\text{OBE}}(\boldsymbol{k}, \boldsymbol{k}') - \frac{2M^2}{(2\pi)^3} \int \frac{V_{\text{OBE}}(\boldsymbol{k}, \boldsymbol{q}) \widetilde{R}(\boldsymbol{q}, \boldsymbol{k}')}{(q^2 - k'^2 + i\varepsilon)(M^2 + q^2)^{1/2}} d^3q. \qquad (9.15)$$

The nonrelativistic Lippman–Schwinger equation for the scattering amplitude has the form

$$R(\boldsymbol{k}, \boldsymbol{k}') = V_{\text{OBE}}(\boldsymbol{k}, \boldsymbol{k}') - \frac{2M}{(2\pi)^3} \int \frac{V_{\text{OBE}}(\boldsymbol{k}, \boldsymbol{q}) R(\boldsymbol{q}, \boldsymbol{k}')}{q^2 - k'^2 + i\varepsilon} d^3q. \qquad (9.16)$$

The amplitude $R(\boldsymbol{k}, \boldsymbol{k}')$ obtained from (9.16) does not satisfy the relativistic unitarity condition, which entails violation of the general relativistic relation between the forward scattering amplitude and the total scattering cross section. However, if the potential (9.14) is used in (9.16) instead of the OBEP, and the relativistic amplitude is taken to be

$$\widetilde{R}(\boldsymbol{p}, \boldsymbol{q}) = \sqrt{\frac{M}{E_p}} R(\boldsymbol{p}, \boldsymbol{q}) \sqrt{\frac{M}{E_q}}, \qquad (9.17)$$

then (9.15) for $\widetilde{R}(\boldsymbol{p}, \boldsymbol{q})$ will be satisfied and relativistic unitarity reestablished. Therefore, we may assume that if we use the nonrelativistic equation then we must use in it not the OBE potential itself but a potential of the form

(9.14). This has just been demonstrated for the Lippman–Schwinger equation describing two free particles. This conclusion can also be extrapolated to the Bethe–Goldstone equation (9.2), which describes a pair of particles in the nuclear medium.

Note that besides taking into account relativity by changing the potential in accordance with (9.14), there is a further way, which consists of replacing the nonrelativistic kinetic-energy operator (9.3) in the denominator of (9.2) by the relativistic form

$$E(\boldsymbol{k_m}) = \left(M^2 + k_m^2\right)^{1/2} - M \; . \tag{9.18}$$

This way of taking relativity into account was proposed by Lee and Tabakin [395].

The following qualitative argument indicates that relativistic corrections, in particular those for minimal relativity, are important: the nucleon–nucleon correlations due to the core lead to states with a relative momentum corresponding to energies of 300 MeV, at which relativistic effects appear perfectly naturally.

There are two nuclear-matter observables related to the saturation of nuclear matter: the binding energy per nucleon and the Fermi momentum. These values correspond to equilibrium in nuclear matter with respect to density variations, i.e. to the minimum of the total energy as a function of density. The empirical value of the density of nuclear matter is taken as equal to the nucleon density in the center of heavy nuclei. The binding energy per nucleon is derived from the Weizsäcker formula extrapolated to $A = 2N = 2Z \to \infty$ with neglect of the Coulomb interaction between the protons. The empirical values of these nuclear-matter observables are 15.7 MeV and 1.4 fm^{-1}, respectively.[2] In fact, there is a window in the vicinity of these numbers which determines phenomenologically acceptable values for the binding energy and saturation density. Initially, several different sets of realistic OBE potentials were developed, both in momentum and coordinate space, which accurately reproduce the two-nucleon data (phase-shift equivalent potentials). However, the same OBE potentials generate strongly different results for the nuclear-matter observables when they are utilized in the framework of the BHF theory. The very important role of the tensor force in nuclear saturation has been demonstrated by calculations. In particular, it was shown that the value of the nuclear-matter binding energy calculated with different realistic OBE potentials in momentum space strongly correlates with the D-state admixture

---

[2] In principle, the following can also be treated as nuclear-matter observables: the compression modulus K=200–300 MeV; the symmetry energy coefficient a_4 (≈ 35 MeV); the nucleon effective mass ($\approx 0.6M$ at normal nuclear density), related directly to the scalar component of the nucleon self-energy, or, in the general case, three components of the nucleon self-energy namely the scalar, timelike, and spacelike components; and the skewing parameter (related to the third derivative of the binding energy with respect to the density).

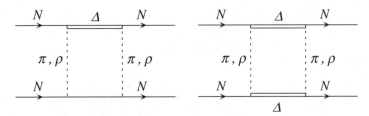

Fig. 9.4. Direct diagrams for two-meson exchange

in the deuteron: the smaller the value of this admixture is, the bigger is the binding energy (we should mention also that the D-state admixture shows strong sensitivity to the choice of the form factor, being noticeably smaller for the eikonal form factor; see Chap. 3 and [34]). Calculations performed on the basis of the nonrelativistic BHF theory described above have shown that none of the existing OBE potentials produce results falling inside the window of empirical values for the binding energy and saturation density. Various modifications of the theory have been introduced to improve the situation. The most important modifications were connected with considering two effects: the contribution of the $\Delta(1236)$ isobar and that of many-body (first of all, three-body) forces. These modifications appeared to be essential, and for this reason we shall consider them qualitatively in the next section. However, looking ahead, we should mention that the solution of the problem of nuclear-matter saturation (taking the correlations involved into consideration) has been obtained only in the framework of the relativistic BHF theory, relativity being taken into account explicitly in a dynamical way, rather than only kinematically; in this case the empirical saturation of nuclear matter can be explained quantitatively [387–394].

9.3 Δ-Isobar for Nuclear Interactions and Nuclear Structure

Let us now discuss the contribution of the Δ resonance to the nuclear interactions of free nucleons and of nucleus embedded in the nuclear medium. This problem has been extensively investigated recently (see [34] and references therein). The processes involved in which the $\Delta(1236)$ resonance participates are shown in Fig. 9.4.

Together with the diagram shown in Fig. 9.5, these diagrams represent all irreducible processes of two-meson exchange between two nucleons, if only one nucleon resonance is taken into consideration (the Δ-isobar). The only possible exchange mesons are the π and ρ mesons, because of the isotopic invariance of strong interactions (the isospin of the Δ-isobar is equal to 3/2). The restriction to the Δ-isobar can be justified by the strong excitation mostly

9.3 Δ-Isobar for Nuclear Interactions and Nuclear Structure

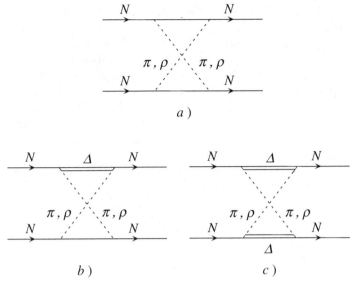

Fig. 9.5. Crossed diagrams for two-meson exchange: (**a**) no Δ-isobar, (**b**) with one Δ-isobar, (**c**) with two Δ-isobars

of the $\Delta(1236)$ resonance in the processes of scattering of mesons by nucleons. Thus, taking the processes shown in Fig. 9.4 into consideration means that the 2π exchange is partially taken into account. On the other hand, the fictitious σ mesons are introduced in the theory of OBE potentials just to allow effective consideration of the 2π meson exchange. So it follows that a part of the σ interaction is taken into account directly now. However, there remains a part, corresponding to the processes shown in Fig. 9.5, which is still not taken into consideration. Corresponding calculations show that from one-third to one-half of the σ interaction is described by the diagrams given in Fig. 9.4. To obtain an interaction including two-meson processes, we write down the equation for the R-matrix in the following form [399]:

$$R_\alpha^\beta = V_\alpha^\beta + \sum_\gamma V_\gamma^\beta \left(\frac{P}{E - H_0}\right)^\gamma R_\alpha^\gamma, \qquad (9.19)$$

where α, β, γ denote the channel number (NN, NΔ, and $\Delta\Delta$ correspond to the channel numbers 1, 2, 3, respectively); V_α^β stands for the interaction in the channels α and β (the lower index corresponds to the incoming state, and the upper index to the outgoing state); H_0 is the Hamiltonian describing free motion of nucleons and nucleon resonances (kinetic energy plus the Δ excitation energy); P is the principal value; and E is the starting energy.

Calculation of the processes shown in Figs. 9.4 and 9.5 is extremely complicated and has not been carried out up to now. However, the procedure is greatly simplified if one neglects the crossed two-meson exchange diagram

(Fig. 9.5a) and interaction in the channels 2 and 3 and between those channels, setting $V_3^2 = V_2^3 = V_2^2 = V_3^3 = 0$. This means that in this case one considers only transitions NN ↔ NN, NN ↔ NΔ, and NN ↔ $\Delta\Delta$, while the vertex $\Delta\Delta$ and processes NΔ ↔ ΔN are not taken into account. In the language of Feynman diagrams, this procedure corresponds to neglecting the diagrams given in Fig. 9.5 and to considering the diagrams shown in Fig. 9.4 only. Notice that also ω mesons contribute to the vertex part $\Delta\Delta$, along with π and ρ mesons. The contribution of the vertex part $\Delta\Delta$ is poorly known. However, it leads to repulsion and can be taken into account effectively by energy-cutting the NΔ vertex. In the approximation formulated above one can easily derive from (9.19) the following equation for the projection of the R-matrix onto the NN channel:

$$R_1^1 = V_{\text{eff}} + V_{\text{eff}} \left(\frac{P}{E - H_0} \right)_1^1 R_1^1 , \qquad (9.20)$$

where

$$V_{\text{eff}} = V_1^1 + V_2^1 \left(\frac{P}{E - H_0} \right)_2^2 V_1^2 + V_3^1 \left(\frac{P}{E - H_0} \right)_3^3 V_1^3 \qquad (9.21)$$

plays the role of the effective interaction in the NN channel which takes into account (in the approximation considered here) the processes of creation of virtual Δ resonances. V_1^1 is the interaction in the NN channel, i.e. the conventional OBE interaction, and V_2^1 and V_3^1 are the transition potentials into the second and third channels.

Equation (9.21) is valid for two-channel systems (note that for the deuteron the transition to the channel NΔ is forbidden by isospin). In the case of the interaction of two nucleons embedded in the nuclear medium, we obtain the following formula for the effective interaction:

$$V'_{\text{eff}} = V_1^1 + V_2^1 \left(\frac{PQ}{E - H_0} \right)_2^2 V_1^2 + V_3^1 \left(\frac{P}{E - H_0} \right)_3^3 V_1^3 , \qquad (9.22)$$

where Q is the projection operator of the nucleon states onto the Hilbert space above the Fermi sea. The presence of the operator Q leads to a strong density dependence of V'_{eff}. Looking ahead, we should mention that the second and third terms in (9.21) and (9.22) lead to an additional attraction, approximately equal to one-half of the σ interaction. With increasing density the operator Q reduces the second term to a greater and greater extent, decreasing the attraction. Another factor operates in the same direction – the starting energy E of the nucleon pair becomes smaller with increasing nuclear density.

Since in the fitting procedure of the experimental scattering phases the total attraction of free nucleons remains approximately the same both in the approach where the Δ-isobar is considered and in the approach where it

9.3 Δ-Isobar for Nuclear Interactions and Nuclear Structure

is not (it is only redistributed between the σ interaction and the Δ-isobar term), the decrease of the attraction is manifested by a repulsive interaction generated in the nuclear medium. This is an essential feature of the nuclear interaction (when the Δ-isobar contribution is taken into account), which is not characteristic of the conventional OBE potentials.

The transition potentials between channels 1 and 2 and between 1 and 3 may be derived using the interaction Lagrangian of the nucleon field, the Δ-isobar field, and the π and ρ meson fields. We have

$$\mathcal{L}_{NN\pi} = ig_\pi \bar{\psi}\boldsymbol{\tau}\gamma_5\psi\boldsymbol{\pi} \,, \tag{9.23}$$

$$\mathcal{L}_{N\Delta\pi} = \frac{f_\pi^*}{m_\pi} \bar{\psi}\boldsymbol{T}\Delta^\mu \partial_\mu \boldsymbol{\pi} + \text{h.c.} \,, \tag{9.24}$$

where ψ, Δ^μ, and $\boldsymbol{\pi}$ are, respectively, the nucleon field, the Δ-isobar field, and the pion field; $\boldsymbol{\tau}$ and $\boldsymbol{T}$ are the isospin operators; and g_π and f_π^* are the coupling constants (see Chap. 7).

The calculations carried out in [399] include two versions: with the N$\Delta\rho$ vertex and without the N$\Delta\rho$ vertex. Since the N$\Delta\rho$ contribution has a sign opposite to that of the N$\Delta\pi$ contribution, in the second version cutoff dipole form factor of the type

$$\frac{\Lambda^2 - m_\pi^2}{\Lambda^2 - \Delta^2} \tag{9.25}$$

is introduced for N$\Delta\pi$, where Λ is the cutoff mass, and Δ^2 is the squared transfer momentum. This form factor for the transition potential was introduced in [400]. Using the eikonal form factor in the transition potential is not acceptable, since its radius is much shorter than that of the ρ meson exchange. For this reason one cannot obtain sufficient cutoff with the eikonal form factor.

In [399] the results of fitting the NN scattering phases for the effective interaction V_{eff} (9.21) are given, V_1^1 being taken in the form of the HM1 potential [401] and the HM2 potential [402]. The meson coupling constants, the fictitious σ meson mass, the meson cutoff masses (in the case where the HM1 potential was utilized for V_1^1), the parameter γ of the eikonal form factor (for the HM2 potential), and the cutoff mass Λ in (9.25) for the N$\Delta\pi$ vertex were used as fitting parameters.

For the meson parameters, the modifications concern mostly the σ meson, a reduction of the σ meson contribution being the main effect. This effect is naturally explained by the additional attraction generated by 2π exchange via the Δ resonance in the channel with the spin equal to zero. A small reduction of the ρ meson contribution also takes place, which is explained by the interaction produced by 2π exchange in the channel with the spin equal to one. From the calculations, the conclusion was drawn, that the cutoff parameter of the N$\Delta\pi$ vertex Λ is less than or equal to $5m_\pi$, while a value $\Lambda > 1\,\text{GeV}$ leads to an obvious overestimate of the Δ resonance contribution.

The agreement with experiment of the scattering phases calculated with the potentials HM1 + Δ and HM2 + Δ is, in general, almost the same as that of the calculations without the Δ resonance. Phases with isospin equal to zero are described a little better when the Δ-isobar is taken into account, since in this case only the $\Delta\Delta$ channel makes a contribution; the cutoff factor (9.25) enters twice in the interaction via this channel. For this reason, for large values of L this interaction is weak, i.e. the scattering phases decrease and the agreement with experiment becomes better on the whole. On the other hand, for the same reason, in the case of the $^3D_2$ wave the agreement becomes worse. The authors of [403] argue that this discrepancy can be eliminated by introducing the eikonal form factor for the π meson exchange.

Notice that the explicit inclusion of the N$\Delta\rho$ vertex leads to a considerably smaller sensitivity of the interaction to the N$\Delta\pi$ vertex cutoff. The vertex N$\Delta\rho$ produces the same effect as a potential that contains only the N$\Delta\pi$ vertex, with $f^{*2}_{N\Delta\pi} = 4\pi \times 0.36$ and the cutoff parameter equal to 800 MeV. The calculations show that this cutoff is not sufficient, the best agreement being achieved at $\Lambda \simeq 650$ MeV. The authors of [399] argue that a solution to the problem can be obtained by a more appropriate field-theoretical description of the transition potential. Taking into account certain diagrams corresponding to the processes of channel coupling could increase, to the extent needed, the strength of the N$\Delta\pi$ vertex at the expense of the N$\Delta\rho$ vertex (the cutoff appears to have the same effect as for $\Lambda \simeq 550$ MeV).

In the field of nucleon–nucleon scattering, consideration of the Δ resonance does not lead to any significant improvement in the ability to reproduce the experimental data, though the introduction of Δ-isobar into the theory is an essential piece of progress, since the description of the NN interaction becomes more appropriate from the theoretical point of view.

Let us discuss briefly some effects of the Δ-isobar on nuclear-structure calculations (for more details, see the exhaustive reviews [34]). In this case the Δ resonance manifests itself first of all by generating extra repulsion with increasing density of the nuclear medium. The reason for the appearance of this extra repulsion is the density dependence of the effective interaction V'_{eff} given by (9.22), this density dependence originating both from the presence of the projection operator Q onto the space of free nucleon states lying above the Fermi surface (the Pauli correction), and from the dependence of the starting energy E, which must be determined self-consistently for each pair of nucleons (the dispersive correction). The calculations carried out in [399] show that both corrections are comparable in magnitude.

It is evident that the presence of a repulsion increasing with density leads to a shift of the saturation point to smaller values of density and binding energy. The three-body force has been taken into account in calculations of nuclear matter and light nuclei by many authors, using different mechanisms to generate this force and different methods to handle it. In [34], the appropriate mechanisms for producing three-body forces are discussed, and it is argued that the Δ-isobar plays an essential role in this case also. We should mention that in the scheme discussed above in which OBE potentials are used, the influence of the medium on meson exchange is not taken into account. This influence is reduced to only nucleon–nucleon correlations in the framework of the Brueckner theory. However, it is clear that the meson propagators in the medium and in the vacuum are quite different. This problem is important from the theoretical point of view [34] The role of the Δ-isobar has been shown to be important in this case also.

We may conclude this section by saying that in the nuclear medium, the Δ-isobar manifests itself via the contribution of

(a) medium effects,
(b) many-body forces.

Both effects are large but cancel each other, having opposite signs. In the framework of a theory that has a regular pattern, both effects should be taken into account simultaneously [34].

9.4 Relativistic Extension of the BHF Theory for Nuclear Matter

The relativistic Brueckner–Hartree–Fock model was introduced, developed, and utilized in [53, 79, 351, 352, 387–394, 404, 405]. In this model (which takes the internucleon correlations into account), one considers the admixture in the nucleon states of the Dirac spinors with negative energy. This admixture arises from the self-consistent field acting on the nucleons, which contains large scalar and vector components S and V. For states with positive energy, these two components of the average field cancel each other to a considerable extent. In contrast, in the case of states with negative energy, they add coherently; the resultant value of the mass operator appears not to be small in comparison with the nucleon rest mass.

In the RBHF method, the internucleon interactions are constructed starting from the interaction of free nucleons, the medium corrections being taken into account using the Brueckner method. The basic equations of the RBHF model can be obtained by various methods, the method of unitary transformations [404] of single-particle states being one of them.

In the method of unitary transformations, the starting point is a Hamiltonian containing effective two-body forces which is derived from the OBE potentials by the Brueckner method, the terms corresponding to the states with negative energy being retained in the Hamiltonian. The single-particle states are transformed by a unitary transformation that mixes nucleon states with positive and negative energy, the Hamiltonian remaining invariant with respect to this transformation. The coefficients of the transformation for each nucleon state are chosen to make diagonal the single-particle part of the Hamiltonian, so that the residual interaction, not taken into account in the model, leads to the admixtures of the type two-particles–two-holes, the hole states including both the states with positive energy, below the Fermi level, and the states with negative energy. Owing to the extension of the number of basis states by including the nucleon states with negative energy, in the RBHF model a more complete self-consistency is achieved in comparison with the nonrelativistic theories; from this point of view, the latter theories can be treated as theories that are not self-consistent.

The RBHF model enables one to obtain an acceptable description of the saturation property of nuclear matter. In the diagram of binding energy versus nuclear density, the point corresponding to equilibrium appears to be shifted from the Coester band to lower density, i.e. in the proper direction. The improvement of the description of the saturation property of nuclear matter in the RBHF model is explicitly related to taking states with negative energy into consideration.

We discuss here some results obtained in the framework of the RBHF model (see [34, 53, 405]). The nucleon spinors in an infinite nuclear medium are solutions of the Dirac equation (8.34). The following approximation is used below. The scalar and vector self-energies are, in general, momentum-dependent; however, as was shown in Chap. 8, this momentum dependence is weak (see, however, [394] and references therein) and in the RBHF framework it is usually completely ignored. In what follows, the constant (momentum-independent) components of the nucleon self-energy will be denoted by S and V (for the scalar attractive and vector repulsive components, respectively). The momentum dependence is more important for the space component of the nucleon self-energy. However, the value of this component is itself small, and for this reason this component is not taken into account at all in the RBHF scheme (in this case $\widetilde{\boldsymbol{p}}$, defined by (8.35), coincides with $\boldsymbol{p}$).

As in the standard BHF approach, the basic value in the Dirac–Brueckner–Hartree–Fock (DBHF) theory is a G-matrix which is a solution of an integral equation. In [53], the Thompson equation is chosen (meson retardation being ignored in this case). For this equation, one has in nuclear matter

$$\widetilde{G}(\boldsymbol{q}',\boldsymbol{q}|\boldsymbol{P},\widetilde{z})
= \widetilde{V}(\boldsymbol{q}',\boldsymbol{q}) + \int \frac{\mathrm{d}^3 k}{(2\pi)^3} \widetilde{V}(\boldsymbol{q}',\boldsymbol{k}) \frac{M^{*2}}{\widetilde{E}^2_{\boldsymbol{P}+\boldsymbol{k}}} \frac{Q(\boldsymbol{k},\boldsymbol{P})}{\widetilde{z}-2\widetilde{E}_{\boldsymbol{P}+\boldsymbol{k}}} \widetilde{G}(\boldsymbol{k},\boldsymbol{q}|\boldsymbol{P},\widetilde{z}) , \quad (9.26)$$

where

$$\widetilde{z} = 2\widetilde{E}_{\boldsymbol{P}+\boldsymbol{q}} ; \quad (9.27)$$

$\boldsymbol{P}$ is one-half of the center-of-mass momentum; M^* is the nucleon effective mass; and $\boldsymbol{q}$, $\boldsymbol{k}$, and $\boldsymbol{k}'$ are the relative momenta of the interacting nucleons. In (9.26), the p_F dependence has been suppressed and the spin and isospin indices have been omitted. For $|\boldsymbol{P}\pm\boldsymbol{q}|$ and $|\boldsymbol{P}\pm\boldsymbol{k}|$, the angle-averaged values are used. Equation (9.26) may be treated in the same way as in the conventional Brueckner approach and can be solved by the standard methods described above.

The main difference from the BHF approach is the use of the potentials $\widetilde{V}$ in (9.26). The tilde indicates that this potential is obtained by using solutions of (8.34) (with the momentum dependence of the self-energies omitted) rather than free spinors, which are used in the scattering problem and standard BHF theory. The potential $\widetilde{V}$ is strongly density-dependent owing to the density dependence of S and M^*. We should mention that the suppression of the attractive (scalar) exchange is an important relativistic effect generated in the RBHF method; this effect increases with density, supplying some extra saturation. This effect appears to be very strongly density-dependent, so that the nuclear-matter observables (binding energy per nucleon, saturation density, and compression modulus) can be reproduced; the value of the Landau parameter f_0 is greatly improved, while the other parameters do not become worse.

In the theory considered here, the Dirac equation includes a potential U given by the sum of an attractive scalar field S and a repulsive vector field V,

$$U = S + \beta V \,, \tag{9.28}$$

so that the single-particle potential $U(m)$ can be defined as

$$U(m) = \frac{M^*}{\widetilde{E}_m} \langle m | \, U \, | m \rangle = \frac{M^*}{\widetilde{E}_m} \langle m | \, S + \beta V \, | m \rangle = \frac{M^*}{\widetilde{E}_m} S + V \,, \tag{9.29}$$

which is expressed via the G-matrix as follows:

$$U(m) = \sum_{n \le p_F} \frac{M^{*2}}{\widetilde{E}_n \widetilde{E}_m} \langle mn | \, \widetilde{G}(\widetilde{z}) \, | mn - nm \rangle \,. \tag{9.30}$$

The constant fields S and V are obtained from (9.30), where m corresponds to a state below or above the Fermi surface. The energy per particle of nuclear matter in the lowest order is given by

$$E/A = \frac{1}{A} \sum_{m \le p_F} \frac{M^*}{\widetilde{E}_m} \langle m | \, \boldsymbol{\gamma} \cdot \boldsymbol{p}_m + M \, | m \rangle$$

$$+ \frac{1}{2A} \sum_{m,n \le p_F} \frac{M^{*2}}{\widetilde{E}_m \widetilde{E}_n} \langle mn | \, \widetilde{G}(\widetilde{z}) \, | mn - nm \rangle - M \,. \tag{9.31}$$

We have utilized here the following equation:

$$\widetilde{z} = \widetilde{E}_m + \widetilde{E}_n \,. \tag{9.32}$$

Equation (9.31) corresponds to just the DBHF approximation. If M is used instead of M^*, one obtains the BHF approximation corresponding to the standard Brueckner method (these two cases will be referred to, in what follows, as "relativistic" and "nonrelativistic" calculations, respectively).

As was mentioned above, a repulsive relativistic many-body effect is the common feature of all DBHF calculations. In most calculations, OBE potentials have been used for the nuclear force.[3] Three different OBE potentials BM–A, BM–B, and BM–C have been developed in [53] for use in nuclear-structure calculations. The parameters of these potentials can be found in Chap. 3, and the predictions for a two-nucleon system are given in [53]. Some one-boson exchange potentials developed by the Bonn group earlier utilized πNN vertices with pseudoscalar coupling, which leads to an extremely large attractive contribution in a relativistic approach. The potentials obtained in [53] use a pseudovector πNN interaction which does not contain this effect. The three OBE potentials BM–A, BM–B, and BM–C differ mainly in the strength of the tensor force, which determines the D-state probability of the deuteron, P_D. Potential BM–A generates the weakest tensor force and predicts that P_D is equal to 4.5%, while for potentials BM–B and BM–C this value is equal to 5.1% and 5.5%, respectively. The strength of the tensor force is known also to fix the position of the saturation point for nuclear matter on the Coester band.

An interesting observation was made by Banerjee and Tjon (see [391]). These authors studied the pion contribution to the nuclear-matter ground state energy using a relativistic Dirac–Brueckner analysis. It was shown that the role of the tensor force in the relativistic analysis of the saturation mechanism is strongly reduced in comparison with its role in a conventional non-relativistic consideration. The reduction of the role of the pion in the nuclear medium is related to many-body effects involved in the relativistic consideration. Banerjee and Tjon showed that the damping of the pseudovector-coupled OPEP is connected with the decrease of M^*/M with increasing density. This effect is also discussed in [39] in connection with the problem of pion condensation and is not manifested for pseudoscalar coupling.

In Fig. 9.6, the repulsive relativistic many-body effect generated in nuclear matter in the DBHF calculation is illustrated.

In Table 9.1, the following as a function of the Fermi momentum p_F: the energy per particle E/A, M^*/M, the scalar and vector potentials S and V, and the wound integral k (see [407]).

From the results presented in Table 9.1, one can see that the contribution of the relativistic effect to the energy per particle, $\Delta(E/A)_{\text{rel}}$, can be well reproduced by the following formula:

---

[3] Note that in [406] a more realistic model to describe the NN force is utilized, the fictitious σ meson being avoided in this model. The intermediate-range attraction is generated in [406] by a 2π exchange involving the Δ-isobar. It was established there that the conclusion that a relativistic effect provides saturation remains valid in a more realistic case also. This calculation appears to support the use of the OBE model for the nucleon–nucleon force.

9.4 Relativistic Extension of the BHF Theory for Nuclear Matter

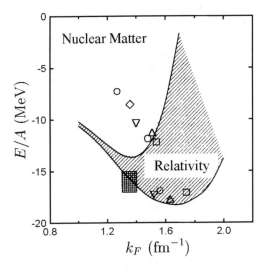

Fig. 9.6. Repulsive relativistic many-body effect generated in nuclear matter in the DBHF calculation with potential BM–B. Saturation points obtained from standard calculations are given in the background. The *checked rectangle* represents the acceptable values for the nuclear-matter observables (from [53])

$$\Delta(E/A)_{\rm rel} \approx 2 \text{ MeV} \times \left(\frac{\rho}{\rho_0}\right)^{8/3}, \qquad (9.33)$$

where $\Delta(E/A)_{\rm rel}$ is the difference between the relativistic and the nonrelativistic calculation. This result is supported by the estimates obtained in [408–410].

Representing the nucleon by a relativistic spinor with an effective mass M^* (see (8.39)) may be treated as an effective consideration of virtual nucleon–antinucleon excitations in the nuclear medium (these are called manybody Z graphs; see Fig. 9.7). This can be shown by decomposing the in-medium spinor (see (8.39)) into series in terms of free Dirac spinors, negative-energy states included.

In Fig. 9.8, a comparison is made between relativistic and nonrelativistic BHF calculations for the three potentials BM–A, BM–B, and BM–C. In the nonrelativistic case, all the saturation points are on the Coester band, while in the relativistic case these points appear to be on a new band, shifted to lower densities and intersecting the empirical values. This shift is caused by an extra strongly density-dependent repulsion introduced in the relativistic treatment. In Table 9.2, the nuclear-matter observables are given (for the three OBE potentials) as obtained by the relativistic and nonrelativistic approaches.

In [53], a relativistic extension of the Brueckner theory was obtained using the Dirac equation to describe single-particle motion in nuclear matter.

Table 9.1. The energy per particle E/A, the ratio M^*/M, the scalar and vector single-particle potentials S and V, and the wound integral k, as a function of the Fermi momentum p_F. Results obtained by the RBHF method are compared with the nonrelativistic calculations for the BM–B potential. (From [53])

	Relativistic					Nonrelativistic			
p_F (fm^{-1})	E/A (MeV)	M^*/M	S (MeV)	V (MeV)	k (%)	E/A (MeV)	M^*/M	U_0[a] (MeV)	k (%)
0.8	−7.02	0.855	−136.2	104.0	23.1	−7.40	0.876	−33.0	26.5
0.9	−8.58	0.814	−174.2	134.1	18.8	−9.02	0.836	−41.0	21.6
1.0	−10.06	0.774	−212.2	164.2	16.1	−10.49	0.797	−49.0	18.5
1.1	−11.18	0.732	−251.3	195.5	12.7	−11.69	0.760	−58.1	14.2
1.2	−12.35	0.691	−290.4	225.8	11.9	−13.21	0.725	−68.5	12.9
1.3	−13.35	0.646	−332.7	259.3	12.5	−14.91	0.687	−80.5	13.1
1.35	−13.55	0.621	−355.9	278.4	13.0	−15.58	0.664	−86.8	13.2
1.4	−13.53	0.601	−374.3	293.4	13.8	−16.43	0.651	−93.2	13.5
1.5	−12.15	0.559	−413.6	328.4	14.4	−17.61	0.618	−106.1	13.0
1.6	−8.46	0.515	−455.2	371.0	15.8	−18.14	0.579	−119.4	12.7
1.7	−1.61	0.477	−491.5	415.1	18.4	−18.25	0.545	−133.2	13.2
1.8	9.42	0.443	−523.4	463.6	21.9	−17.65	0.489	−147.2	14.3
1.9	25.26	0.418	−546.7	513.5	25.2	−16.41	0.480	−160.7	15.0
2.0	47.56	0.400	−563.6	568.6	27.5	−13.82	0.449	−173.6	15.3
2.1	77.40	0.381	−581.3	640.9	30.2	−9.70	0.411	−186.3	15.7
2.2	114.28	0.370	−591.2	723.5	33.3	−3.82	0.373	−198.1	16.3

[a] U_0 is to be compared with $S + V$

Table 9.2. BHF results for nuclear matter: the binding energy per nucleon E/A; Fermi momentum p_F and the incompressibility K at saturation, with and without relativistic effects [53]

	Relativistic			Nonrelativistic		
Potential	E/A (MeV)	p_F (fm^{-1})	K (MeV)	E/A (MeV)	p_F (fm^{-1})	K (MeV)
BM-A	−15.59	1.40	290	−23.55	1.85	204
BM-B	−13.60	1.37	249	−18.30	1.66	160
BM-C	−12.26	1.32	185	−15.75	1.54	143

As mentioned above, three relativistic OBE potential models were developed (BM–A, BM–B, and BM–C). The two-nucleon system is treated in this approach using a relativistic three-dimensional reduction of the Bethe–Salpeter equation, introduced by Thompson (see [42] and references therein). The

9.4 Relativistic Extension of the BHF Theory for Nuclear Matter

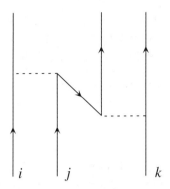

Fig. 9.7. Z diagram for three-body interaction

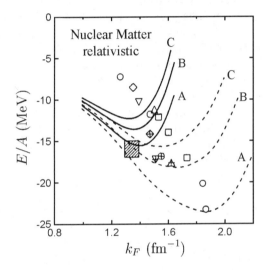

Fig. 9.8. Results of calculations with a family of relativistic potentials (BM–A, BM–B, and BM–C) reveal a new Coester band which intersects the nuclear matter experimental area. *Continuous lines*, relativistic calculations; *dashed lines*, nonrelativistic calculations. Saturation points from conventional calculations are given in the background. The *shaded square* corresponds to the empirical nuclear-matter saturation window [53]

properties of nuclear matter were calculated in this framework, the nucleon self-energies and the single-particle wave functions being calculated fully self-consistently. The main feature of this approach is a suppression of the attractive scalar meson exchange.

This relativistic effect (strongly density-dependent) reproduces the saturation properties of nuclear matter correctly (the calculations in [53] include also the saturation mechanisms of the conventional nonrelativistic theory, i.e.

the Pauli and dispersive effects). We may conclude this section by saying that the RBHF method provides a parameter-free description of the nuclear many-body system. RBHF calculations contain certain approximations, which need to be checked. However, they yield a unified reproduction of NN scattering data and the saturation properties of nuclear matter.

9.5 Finite Nuclei

The application of the relativistic procedure developed in [53] and discussed in the previous section to finite nuclei was studied in [340, 405].[4] However, the authors of [405] restricted themselves to consideration of the ground-state properties of ^{16}O only, the RBHF description of a finite structure being a complicated problem, a much more complicated one than its nonrelativistic counterpart. It should be emphasized also that the Dirac BHF method yields substantial improvements in reproducing the ground-state properties of finite structures. However, it does not provide a completely satisfactory agreement with the experimental data, and the reasons for this discrepancy are not explicitly understood at present.

For this reason, different approximate methods to treat finite structures have been developed. Most of these methods introduce effective interactions that, when they are considered within the Dirac mean–field or Dirac–Hartree–Fock approximation, reproduce the RBHF results for nuclear matter (the nucleon self-energy and the binding energies). The effective interactions obtained in this way are then utilized without introducing new parameters to calculate the properties of finite nuclei. A major consequence of this method is that the meson–nucleon coupling constants or the meson masses become density-dependent, reflecting the genuine density dependence of the Dirac–Brueckner NN G-matrix in the nuclear medium. In [152] it was suggested that the relativistic mean-field theory should be interpreted as a phenomenological approach to the RBHF theory.

The method introduced in [152] will be referred to, in what follows, as the relativistic density-dependent Hartree approach. It is based on the standard σ–ω model Lagrangian with density-dependent coupling constants

$$\mathcal{L}_{\mathrm{RDDH}} = \bar{\psi}\left[i\gamma_\mu \partial^\mu - M - g_\sigma(\rho)\varphi - g_\omega(\rho)\gamma_\mu \omega^\mu\right]\psi$$
$$+ \frac{1}{2}(\partial^\mu \varphi)^2 - \frac{1}{2}m_\sigma^2 \varphi^2 - \frac{1}{4}(\partial_\mu \omega_\nu - \partial_\nu \omega_\mu)^2 + \frac{1}{2}m_\omega^2 \omega_\mu^2 \ . \quad (9.34)$$

It should be underlined that (9.34) does not contain meson self-interaction terms.

---

[4] The authors of [340, 405] took into account the dependence on the medium of the Dirac spinors in the interaction and the kinetic energy, using the effective-density method together with the RBHF results.

In the Hartree approximation, one finds that the nucleon self-energy in nuclear matter is

$$\Sigma_{\text{RDDH}}(\rho) = S(\rho) + V(\rho)\gamma_0 \ . \tag{9.35}$$

Here the scalar and vector potentials are expressed in terms of the coupling constants $g_\sigma(\rho)$ and $g_\omega(\rho)$ through

$$S(\rho) = -\frac{g_\sigma^2(\rho)}{m_\sigma^2}\rho_\text{S} \ , \qquad V(\rho) = -\frac{g_\omega^2(\rho)}{m_\omega^2}\rho_\text{V} \ , \tag{9.36}$$

where ρ_S and ρ_V are the scalar and vector densities (introduced earlier), respectively. The connection of the RDDH approach to the RBHF theory is made through the nucleon self-energy in nuclear matter, (9.35). In fact, one can express the RBHF self-energy in nuclear matter in the form of (9.35) (see 9.29). The density-dependent coupling constants can then be obtained via (9.36), where $S(\rho)$ and $V(\rho)$ are the results of RBHF calculations for nuclear matter.

We write the equations of motion for a finite nuclei in the following form:

$$[-i\boldsymbol{\alpha}\cdot\boldsymbol{\nabla} + \beta M^*(r) + V(r)]\psi_i(\boldsymbol{r}) = E_i\psi_i(\boldsymbol{r}) \ , \tag{9.37}$$

where $M^*(r) = M + g_\sigma(r)\varphi(r)$ and $V(r) = g_\omega(r)\omega^{(0)}(r) + e(1/2)(1+\tau_3)A^{(0)}(r)$. The Klein–Gordon equations for φ, ω^0, and A^0 are

$$(-\Delta + m_\sigma^2)\varphi(r) = -g_\sigma(r)\rho_\text{S}(r) \ , \tag{9.38}$$

$$(-\Delta + m_\omega^2)\omega^{(0)}(r) = g_\omega(r)\rho_\text{V}(r) \ , \tag{9.39}$$

$$-\Delta A^{(0)}(r) = e\rho_\text{p}(r) \ . \tag{9.40}$$

The scalar, vector, and proton densities ρ_S, ρ_V, and ρ_p are obtained via the Dirac wave functions. The coupled differential equations (9.37)–(9.40) are solved self-consistently.

The basic features of the RDDH method are shown in Fig. 9.9. In the upper part of Fig. 9.9, the energy per particle E/A is shown as a function of the Fermi momentum p_F. The saturation point is shown by the square. The calculated results [53] pass through the saturation point, and the energy minimum seems to appear at a slightly larger density. This result was obtained with the parameter set BM–A for the OBEP, which has a relatively small tensor force (see Sect. 9.3).

In the lower part of Fig. 9.9, the vector and scalar potentials V and S are shown as a function of p_F. At the normal density of nuclear matter, $p_\text{F} = 1.35$ fm^{-1}, $V = 274.7$ MeV, and $S = -355.7$ MeV. As a comparison, the vector and scalar mean-field potentials obtained using constant (density-independent couplings) are shown; that is, the parameters of the scalar and vector potential are those of potentials BM–A [53]. This comparison clearly indicates that one needs density-dependent coupling constants to reproduce the nuclear-matter results within the relativistic Hartree framework. In this

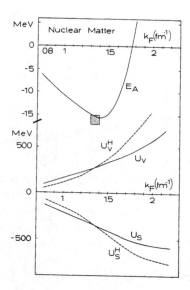

Fig. 9.9. *Upper part*: energy per particle E/A in the RBHF approach as a function of p_F [53]. *Lower part*: vector and scalar potentials for the relativistic density-dependent Hartree approximation (*solid lines*) and for the relativistic Hartree approximation only (*dashed lines*), as a function of p_F

case the RDDH self-energies were made identical in magnitude, sign, and density dependence to those obtained by the RBHF method.

In [152] the coupling constants $g_\sigma(\rho)$ and $g_\omega(\rho)$ were fitted to the self-energies S and V with the choice that m_σ and m_ω were the masses in potential BM–A, as a standard choice: $m_\sigma = 550$ MeV and $m_\omega = 782.6$ MeV [53]. As can be seen from (9.36), the relevant quantities are g_σ^2 and g_ω^2. At $p_F = 0.8$, 1.1, 1.5 fm^{-1} one obtains [152] $g_\sigma^2/4\pi = 12.3$, 8.91, 6.23 and $g_\omega^2/4\pi = 18.63$, 13.48, 9.06, respectively. This shows that the coupling constants at the nuclear surface are more than 40% bigger than in the interior. At the nuclear-matter density ($p_F = 1.35$ fm^{-1}), the coupling constants are not modified, as one can see from Fig. 9.7. But at smaller densities, both coupling constants are growing. The momentum dependence of the nucleon self-energies is not taken into account in [152]; the reasons for the approximations made here have been discussed earlier.

Results for the two nuclei ^{16}O and ^{40}Ca were obtained as examples within the RDDH approach in [152]. The calculated results for the binding energies, single-particle energies, and charge radii (using potential BM–A) are tabulated in Table 9.3.

As a comparison, some results of nonrelativistic BHF calculations for ^{16}O with the same potential set (BM–A) are shown in the column N-BHF [405]. As can be seen from Table 9.3, the RDDH results are very close to the exper-

Table 9.3. The binding energy per nucleon, r.m.s. charge radius, and single-particle energies obtained by the RDDH approach, the nonrelativistic Brueckner–Hartree–Fock (N-BHF) method, and experiment. Potential BM–A [53] was used. The upper part of the table is for ^{16}O, and the lower part for ^{40}Ca. (From [152])

	RDDH	N-BHF	Experiment
^{16}O			
E/A (MeV)	−7.5	−6.6	−7.98
r_c (fm)	2.66	2.29	2.70 ± 0.05
$\varepsilon(1s_{1/2})$ (MeV)	−43.5	−50.5	−40 ± 8
$\varepsilon(1p_{3/2})$ (MeV)	−21.8	−22.9	−18.4
$\varepsilon(1p_{1/2})$ (MeV)	−16.5	−15.5	−12.1
^{40}Ca			
E/A (MeV)	−8.0	–	−8.5
r_c (fm)	3.36	–	3.5
$\varepsilon(1s_{1/2})$ (MeV)	−53.3	–	−50 ± 8
$\varepsilon(1p_{3/2})$ (MeV)	−36.0	–	−34 ± 5
$\varepsilon(1p_{1/2})$ (MeV)	−32.5	–	−34 ± 5
$\varepsilon(1d_{5/2})$ (MeV)	−19.3	–	−14 ± 2
$\varepsilon(2s_{1/2})$ (MeV)	−14.3	–	−10 ± 1
$\varepsilon(1d_{3/2})$ (MeV)	−13.6	–	−7 ± 1

imental results, the improvements on the N-BHF results being remarkable. If we compare the relativistic and nonrelativistic results, we find that the radius is larger for the relativistic calculation. This is natural, since the lower component of the relativistic $1p_{3/2}$ wave function looks like a nonrelativistic $1d_{3/2}$ wave function. This shifts part of the density to the surface and leads to a larger radius in the relativistic case. This has consequences for the binding energy per nucleon. Using the Bethe–Weizsäcker mass formula, one can estimate that surface and Coulomb effects lead to 2 to 3 MeV less repulsion for the relativistic calculation, since the radius is larger. Although the volume effect is 1 to 2 MeV more repulsive in the relativistic case, as determined for nuclear matter in [53], the relativistic description of finite nuclei yields more binding energy per nucleon altogether. This shows that relativistic effects lead off the Coester band, which exists for finite nuclei too. In Fig. 9.10, a comparison is made [152] of the experimental charge densities and the charge densities obtained by the RDDH method for ^{16}O and ^{40}Ca.

As can be seen from this figure, the central density (for the potential BM–A; dash-dotted curve) turns out to be higher than the experimental value and the density falls off slightly faster than in the experiment. This observation about the density, and the slightly smaller binding energy discussed above seem to be a reflection of the equation of state of nuclear matter, as shown

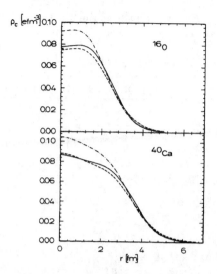

Fig. 9.10. Comparison of the densities obtained by the RDDH method and experiment for ^{16}O and ^{40}Ca, respectively. Results obtained with potential BM–A are shown by the *dash–dotted curves*, while the corresponding results for potential BM–C are shown by the *dashed curves* [152]

in Fig. 9.8. In fact, when we take the RBHF results with potential BM–C, whose saturation density is almost perfect but which gives a saturation energy about 4 MeV above the experimental value, the charge density distributions are found to be very good, as can be seen in Fig. 9.9 (dashed curve). In this case, however, the binding energies of finite nuclei are found to be too small, i.e. $E/A = -5.9$ MeV for ^{16}O and $E/A = -6.0$ MeV for ^{40}Ca.

Let us discuss now the density dependence of the meson mass. In the extraction of the density-dependent coupling constants $g_\sigma(\rho)$ and $g_\omega(\rho)$ by fitting to S and V obtained from the RBHF results, the meson masses for the free potential were chosen in [152]. Actually, this is not a unique choice. The contribution of the exchange term and the iteration of the interactions to obtain the G-matrix would lead to the mass being different from the free case. When the σ meson mass is decreased by 10% (and the coupling constant is changed by the same amount in order not to change the nuclear-matter results), the absolute value of the binding energy is decreased by 15% and the radius is increased by 4%. These changes are in accordance with previous findings [96, 411], but the amount of change is about a factor of two smaller than in the previous mean-field results, owing to the density dependence of the coupling constants. When the ω mass is decreased by 10%, we find changes that are similar but a factor of two smaller and opposite in sign.

It is necessary to emphasize that the procedure of [152] considered above is parameter-free in the following sense: no parameters in the nucleon–nucleon interaction are adjusted to reproduce the properties of nuclei. The calculated binding energies per nucleon and r.m.s. radii are found to be close to experiment.

The calculation carried out in [152] solves a long-standing problem of nuclear physics, since in other calculations the binding energy comes out too small (often by a factor of two) if the r.m.s. radius of the nucleus is reproduced, and, vice versa, the r.m.s. radius is too large by a factor of two if the binding energy is correct.

The approach containing a density dependence of the coupling constants has been investigated further in [360, 412, 413]. The philosophy of density-dependent coupling constants was used also in [414] to examine the possibility of describing nuclear deformation properties in a parameter-free framework.

The problem of obtaining simultaneously a realistic description of the observed saturation densities, charge radii, and binding energies for finite nuclei has been investigated in [415, 416]. In those papers, an effective hadron field theory with density-dependent meson–baryon vertices is suggested, the dynamical structure of in-medium vertices and the use of the RBHF results for nuclear matter in an effective quantum field theory being considered. The treatment of medium effects [415, 416] leads to important changes in the field equations for baryons. In the Euler–Lagrange equations, variations of the vertex functionals with respect to the baryon fields are involved also, giving rise to rearrangement self-energies. In this case the quasiparticles acquire extra dressing by high-momentum excitations of the background nuclear matter from the short-range nucleon–nucleon repulsion. Calculations carried out in [415, 416] show that rearrangement improves the agreement with experiment.

Another approximation was suggested in [417], this approximation is based on an effective Lagrangian including σ and ω meson fields and self-interactions of those fields, the parameters of the Lagrangian being fitted in the framework of the mean-field (Hartree), approximation by reproducing the "nuclear-matter observables" obtained by the RBHF method using various OBE models for the NN potential.

In [117, 340–342], the method of the effective Lagrangian was used in a relativistic Hartree–Fock scheme, explicitly taking into account the contributions of π and ρ mesons. In particular, in [117] an effective interaction for RHF calculations of nuclear structure was constructed. This includes a nonlinear functional of the simplest general form that takes into account interactions and self-interactions of isoscalar meson fields (see Chap. 8). The parameters were determined so as to reproduce DBHF nuclear-matter results obtained from various types of one-boson exchange potential that fit NN scattering data. The effective interaction was then used to calculate ground-state properties of finite nuclei. The results for some specific OBEPs (the BM–B version) are in good agreement with experiment. The symmetry energy coefficient, in particular, is well reproduced. It was found that the $\sigma^2\omega^2$ and ω^4 components of the effective interaction play an important role. The approach developed in [117] can be considered as a method for avoiding complete DBHF calculations for finite structures. The inclusion of self-interactions and interactions between different meson fields leads to meson dressing, i.e. to a density dependence of the meson effective masses. Meson dressing in the nuclear medium is one of the key ingredients of the theory developed in [117]. Thus the effective interaction introduced in that paper may be considered as a convenient tool for investigating the influence of medium effects on the meson propagators when properties of nuclear structures are calculated within the RHF framework.

We should mention also that in [418], modifications of meson masses by the medium were included directly. This investigation demonstrated that the results obtained by use of the DBHF method do not change essentially, the saturation properties being still very good.

It has been known for many years that the equation of state of nuclear matter contains a symmetry energy term. Studies of the symmetry energy and its density dependence present an acute problem; this has become more important because of the recent development of radioactive-ion-beam facilities [156]. The symmetry energy and its density dependence also have a strong effect on the properties of neutron stars. In [419], a systematic investigation of the nuclear symmetry energy in the framework of the RBHF theory using an OBE potential was performed. The authors of [419] found the symmetry energy coefficient to be equal to 31 MeV (at the saturation point), in good agreement with the empirical value 30 ± 4 MeV (see [117]). These authors also found that the symmetry energy increases practically linearly with the density, in contrast with the results of nonrelativistic approaches.

In the present chapter, we have discussed the role of the effects of the medium in the context of nuclear structure, in the framework of the RBHF theory. For nuclear reactions, similar medium effects are present and change the results obtained in the relativistic impulse approximation. Some of the medium effects in nuclear reactions, the Pauli blocking effect in particular, were investigated in [420], and were shown to be important at low incident energies. The authors of [421] carried out a complete study of nuclear medium effects in the framework of the Dirac–Brueckner (DB) method and investigated their role in elastic and quasi-elastic scattering. Three types of medium effects were considered in [421]:

(a) the Pauli exclusion principle,
(b) the dispersive effect,
(c) modification of the nucleon spinor (M^* dependence).

The authors of [421] calculated proton–nucleus elastic scattering. It was established that the DB optical potential reproduces the data reasonably well at medium and low energies. They also carried out an eikonal-type calculation of the quasi-elastic scattering, this calculation producing the right order of the reduction effects in the analyzing power at the quasi-elastic peak.

10 Excited Nuclear States in the Relativistic RPA Method

10.1 RRPA Method for Nuclear Matter and Finite Nuclei

The relativistic theory provides a natural basis to study the properties of the nuclear ground state. With a small number of fitting parameters, all of which have clear physical meanings, the relativistic mean-field approximation reproduces the nuclear bulk properties reasonably well. A version of this approximation extended by including the nonlinear meson self-interaction terms is now used as a standard model for the description of finite nuclei [103, 116]. The small component of the single-particle wave function is greatly enhanced in this approximation owing to the small value of the nucleon effective mass in the nuclear medium. The success of the relativistic Hartree and Hartree–Fock approximations ensured a good basis for the study of nuclear excitations in relativistic models. The RMF theory has been extended, so far, to the description of excited states in two directions. The first direction is related to the description of the yrast states of rotating nuclei [422, 423]. The second direction is related to consideration of giant resonances (and other excitations) in the framework of the relativistic version of the RPA theory the RRPA theory [82] or of the time-dependent Hartree theory [84].

Elementary excitations may occur because of the displacement of nucleons from occupied states to unoccupied states, and the random-phase approximation provides a method for constructing collective states which can arise as a result of the residual particle–hole interaction.

RPA calculations for nuclear matter were first carried out in [75], following similar calculations for relativistic electron gas. In [75] it is shown that in the high-density limit, vector meson exchange dominates in the σ–ω model. The excitation spectrum of nuclear matter corresponds to zero sound, and the sound velocity aproaches the speed of light c (from below) with increasing density. Further contributions to this field have been made in [200, 424–427]. Note that the calculations performed in [75] give an increasing mass of the ω meson in nuclear matter, related to a process similar to Compton scattering. In [75] only the particle–hole excitations, i.e. only the polarization of the Fermi sea, were taken into account. An analogous, but more complicated, calculation for the polarization of nuclear matter by the ρ meson [428, 429] also generated a small in-medium increase of the mass of the ρ meson.

Further investigations in this field were motivated by arguments, based on the scale invariance of QCD, which suggest that all the hadron masses, except that of the pion, decrease in the nuclear medium [430–432]. A lot of investigations have been performed in this field [433–449]. In particular, it has been realized [433, 434, 437, 440, 443] that the mass of the vector mesons is strongly influenced also by the vacuum polarization of the nucleons in the medium. The vacuum polarization dominates over the Fermi sea polarization; the total effect generates the effect that the ω and ρ meson masses decrease with density. However, the theoretical situation in relation to the density dependence of the vector mesons is, to a certain extent, controversial at present (cf. [444, 445, 450]); further investigations in this field are needed.

The nuclear-matter ground state is stable against small-amplitude oscillations of the meson fields which form the ground state (excitations of normal parity). A problem may arise for excitations of fields with a symmetry different from that of the ground state (excitations of abnormal parity, for example, with the symmetry of the pion field). In [39], the problem of the stability of relativistic nuclear matter against pion condensation has been treated in the framework of the RRPA method (see also Chap. 6). It was shown that a relativistic treatment of the problem of pion condensation provides new, essential aspects of the solution of the problem. In particular, in the case of pseudovector pion–nucleon coupling, it was shown that for a particular choice of the parameters (which determine the nuclear-matter ground state) the system remains stable at all densities. Such behavior is directly related to the relativistic treatment of the problem: the polarization integral appears in this case with a factor (the squared ratio of the effective nucleon mass to the bare mass) that essentially suppresses the contribution of the polarization integral. This effect brings stability to the system. In [9], the stability problem has been investigated for pseudoscalar pion–nucleon coupling, relativistic effects (of a different character) are manifested in this case also. New relativistic effects appear in the problem of pion condensation because the transition operator (containing the γ_5 matrix) in this case mixes the upper and lower components of the nucleon wave function, the latter component being greatly enhanced, as was mentioned above.

For this reason, there is a particular motivation to perform calculations for RPA excitations of abnormal parity[1] (0^- and 2^-) with isospin $T = 0$ and $T = 1$ for light nuclei (^{4}He and ^{16}O, for example). The properties of these

---

[1] Note that states of normal and abnormal parity differ only by quantum numbers (spins and parities). However, they are generated by the same RPA excitation mechanism. States with normal parity have an even spin value and positive parity or an odd spin value and negative parity, and vice versa for states of abnormal parity (excitations involving the quantum numbers of the pion field, for example). In the first case (normal parity), the following meson fields must be taken into consideration: the scalar–isoscalar field, and the time components of the vector

states have been suggested as signatures of precursors to pion condensation. Because of the reasons mentioned above, a relativistic treatment of these excitations would lead to new manifestations of relativity in nuclear structure.

Relativistic RPA methods have been developed within certain approximations by several groups and applied to finite nuclei [424, 425, 434, 451–461] to describe different types of excitations (with different spins and parities, both isoscalar and isovector excitations); the theoretical status of these approaches is discussed in [82, 85]. All these contributions[2] involve calculations (with or without Dirac sea effects) of the relativistic nuclear response in models without nonlinear σ and ω meson self-interaction terms.

10.2 RRPA with Nonlinear Interactions for Giant Resonances

Models with self-interactions of mesonic fields [103, 116, 147] are now considered as standard models for the description of the ground state of a finite nucleus. Therefore it is desirable to apply the RRPA in nonlinear models, and this program was realized in [462–464].

The meson propagators in the presence of self-interaction terms are no longer simple Yukawa functions. In fact, the key point in solving the RRPA equation with nonlinear interactions is to work out the meson propagators. From field theory, the meson propagators can be obtained in principle from the second variation of the action of the system. This can be done more easily in momentum space, where the meson propagators in nonlinear models can be calculated numerically. This idea was realized in the RRPA framework in a number of papers [462–464].

In this section, we consider the description of the giant resonances of finite nuclei in the RRPA by using relativistic effective Lagrangians with nonlinear interactions, such as TM1 [116] and NL-SH [147], which have been recently proposed for use in a quantitative description, in the framework of the RMF approach, of ground-state properties of stable and unstable nuclei up to the nucleon drip lines. Following [462–464], we discuss to what extent the corresponding parameter sets can also reproduce the most well-known collective

---

fields (both vector and tensor couplings). In the second case (abnormal parity), a contribution is made by the pseudoscalar field, the space components of the vector fields (vector and tensor couplings), and nonstatic terms. It should be emphasized that this statement concerns excitations described by the RPA on the basis of RMF theory.

[2] We consider here two methods to introduce the basic equations of the RRPA for finite nuclei. One of these methods, developed in [456, 457, 462–464], is discussed in the present chapte; the other method was developed by Fomenko, Savushkin, and Toki [465] (this method allows a straightforward extension to include nonlinear self-interactions of mesonic fields and to consider RRPA oscillations of the Hartree–Fock basis).

excitations. The giant resonances in nuclear matter have been estimated in a relativistic approach using a local Lorentz boost and scaling model [434], which directly connects the properties of giant resonances with the bulk properties described by the effective Lagrangian. The question may be raised as to whether this connection is still valid for finite nuclei. We shall consider mainly the isoscalar monopole and quadrupole, and isovector monopole and dipole giant resonances in spherical nuclei, because they are well established experimentally and have been extensively investigated in the nonrelativistic approaches.

The Hartree single-particle Green's function is defined by

$$G_H(x,y) = -i \langle 0 | T(\Psi_H(x)\bar{\Psi}_H(y)) | 0 \rangle \, , \tag{10.1}$$

where $x \equiv (t, \boldsymbol{r})$, $\Psi_H(x)$ is the Hartree field operator, and $|0\rangle$ is the uncorrelated Hartree ground state. The single-particle energies and wave functions are obtained from the Dirac equation

$$\{\gamma^0(E_\lambda - \Sigma_0) + i\boldsymbol{\gamma}\boldsymbol{\nabla} - (M + \Sigma_S)\} \Phi_\lambda(\boldsymbol{r}) = 0 \, , \tag{10.2}$$

where the scalar potential Σ_S and vector potential Σ_V, due to meson–nucleon couplings, are obtained by solving the meson-field equations self-consistently. The Hartree Green's function, in a standard spectral representation, is expressed as

$$G_H(\boldsymbol{r_1}, \boldsymbol{r_2}; E) = \sum_\alpha \frac{\Phi_\alpha(\boldsymbol{r_1}) \cdot \bar{\Phi}_\alpha(\boldsymbol{r_2})}{E - E_\alpha + i\eta} + \sum_{\bar{\alpha}} \frac{\Phi_{\bar{\alpha}}(\boldsymbol{r_1}) \cdot \bar{\Phi}_{\bar{\alpha}}(\boldsymbol{r_2})}{E - E_{\bar{\alpha}} - i\eta}$$
$$+ \sum_{a < E_F} \Phi_a(\boldsymbol{r_1}) \cdot \bar{\Phi}_a(\boldsymbol{r_2}) \left(\frac{1}{E - E_a - i\eta} - \frac{1}{E - E_a + i\eta} \right) \equiv G_H^F + G_H^D \, ,$$
$$\tag{10.3}$$

where we denote by α all positive-energy states ($E_\alpha > 0$), $\bar{\alpha}$ all negative-energy states ($E_{\bar{\alpha}} < 0$), and a the states in the Fermi sea. The unoccupied positive-energy states will be denoted by A in the following expressions. The superscripts F and D refer to the Feynman and density-dependent parts, respectively, of the Green's function.

To study particle-hole excitations, we define the unperturbed, or Hartree, polarization operator

$$\Pi_0(P,Q;x_1,x_2) = i \langle 0 | T[\bar{\Psi}_H(x_1)P\Psi_H(x_1)\bar{\Psi}_H(x_2)Q\Psi_H(x_2)] | 0 \rangle$$
$$= i \operatorname{Tr}[P G_H(x_1, x_2) Q G_H(x_2, x_1)] \, , \tag{10.4}$$

where P and Q are general 4×4 matrices. The polarization operator in (10.4) contains terms of the type $G^D G^F$, $G^F G^D$, and $G^D G^D$, which are finite, and of the type $G^F G^F$, which are divergent. The divergence can be removed in principle by a renormalization procedure [75]. However, in [462] the no-sea

10.2 RRPA with Nonlinear Interactions for Giant Resonances

approximation is adopted, which is consistent with the RMF approximation, and therefore the $G^F G^F$ terms are neglected there. The contributions from $G^D G^F$ and $G^F G^D$ in (10.4) contain excitations of a particle from an occupied Fermi sea state to another occupied Fermi sea state (h–h), or from a state in the Dirac sea to one in the Fermi sea (N̄–h), which obviously violate the Pauli principle. Those Pauli-blocked excitations would be exactly cancelled by terms from $G^D G^D$ and $G^F G^F$, and therefore the neglect of $G^F G^F$ contributions introduces some Pauli violations [462].

In the relativistic approach, the residual particle–hole interactions are produced by σ, ω, and ρ meson exchanges as specified by the effective Lagrangian. To work out the classical meson propagators, it is convenient to work in momentum space and then solve the RRPA equation in momentum space. The Fourier transform of the Hartree polarization operator is defined by

$$\Pi_0(P,Q;x_1,x_2) = \frac{1}{(2\pi)^4} \int dE\, d\boldsymbol{k}_1\, d\boldsymbol{k}_2\, \exp(-iE(t_1-t_2))\exp(i\boldsymbol{k}_1\cdot\boldsymbol{r}_1)$$
$$\times \Pi_0(P,Q;\boldsymbol{k}_1,\boldsymbol{k}_2;E)\exp(-i\boldsymbol{k}_2\cdot\boldsymbol{r}_2)\,. \quad (10.5)$$

Since the unperturbed ground state is treated in the RMF approximation, the RRPA corresponds to the ring approximation. The correlated polarization operator Π, therefore, is a solution of the following RRPA integral equation:

$$\Pi(P,Q;\boldsymbol{k},\boldsymbol{r}',E)$$
$$= \Pi_0(P,Q;\boldsymbol{k},\boldsymbol{r}',E) - \sum_i g_i^2 \int d^3k_1\, d^3k_2\, \Pi_0(P,\Gamma^i;\boldsymbol{k},\boldsymbol{k_1},E)$$
$$\times D_i(\boldsymbol{k_1}-\boldsymbol{k_2},E)\Pi(\Gamma_i,Q;\boldsymbol{k_2},\boldsymbol{r}',E)\,. \quad (10.6)$$

In this equation, the index i runs over σ, ω, and ρ mesons; g_i and D_i are the corresponding coupling constants and meson propagators; and $\Gamma^i = 1$, γ^μ, and $\gamma^\mu\boldsymbol{\tau}$ for $i = \sigma$, ω, and ρ, respectively. The explicit forms of the meson propagators $D_i(\boldsymbol{k},E)$ will be given below.

In practice, (10.6) is solved by making multipole expansions of Π and Π_0,

$$\Pi(P,Q;\boldsymbol{k};\boldsymbol{k}';E) = \sum_{LM} Y_{LM}^+(\widehat{\boldsymbol{k}})\Pi^{(L)}(P,Q;\boldsymbol{k},\boldsymbol{k}';E)Y_{LM}(\widehat{\boldsymbol{k}'}))\,, \quad (10.7)$$

and transforming the RRPA integral (10.6) into a set of uncoupled equations for different multipoles. In the no-sea approximation the multipole components of the unperturbed polarization operator can be explicitly expressed in terms of particle–hole configurations.

We have
$$\Pi_0^{(L)} = \Pi_0^{(L)+} + \Pi_0^{(L)-}\,, \quad (10.8)$$

where

$$\Pi_0^{(L)+}(P,Q;k_1,k_2;E) = \frac{(4\pi)^2}{2L+1} \sum_{a,A} (-)^{j_a+j_A}$$

$$\times \left(\frac{\langle \bar{\Phi}_a | \, |P_L| \, |\Phi_A\rangle \langle \bar{\Phi}_A | \, |Q_L| \, |\Phi_a\rangle}{E_a - E_A + E + i\eta} \right.$$

$$\left. + \frac{\langle \bar{\Phi}_A | \, |P_L| \, |\Phi_a\rangle \langle \bar{\Phi}_a | \, |Q_L| \, |\Phi_A\rangle}{E_a - E_A - E + i\eta} \right). \tag{10.9}$$

Here the sum over a and A runs over all possible particle–hole excitations from the Fermi sea to the unoccupied positive-energy states. The expression for $\Pi_0^{(L)-}$ is similar to (10.9) with the replacement of Φ_A, E_A, and $i\eta$ by $\Phi_{\bar{a}}$, $E_{\bar{a}}$, and $-i\eta$, respectively. $\Pi_0^{(L)-}$ is a Pauli blocking term, which is not included in the calculations in [462]. In (10.9) the quantities P_L and Q_L are the L-multipoles of the one-body operators $\bar{\Psi}_H(x) P \Psi_H(x)$ and $\bar{\Psi}_H(x) Q \Psi_H(x)$. The reduced particle–hole matrix elements are calculated for the operators $r^L Y_L$, $r^L Y_L \tau_3$, $j_L(kr) Y_L$, $\gamma^\mu j_L(kr) Y_L$, and $\gamma^\mu \tau_3 j_L(kr) Y_L$, which correspond to the isoscalar and isovector electric multipoles and Γ^σ, Γ^ω, and Γ^ρ, respectively. Finally, the response function of the nucleus to the external field is given by the imaginary part of the RPA polarization operator,

$$R(P,Q;\bm{k},E) = \frac{1}{\pi} \operatorname{Im} \Pi(P,Q;\bm{k},\bm{k};E) \,. \tag{10.10}$$

10.3 Construction of the Meson Propagators

The relativistic RPA calculations in [462–464] are based on the effective Lagrangian with nonlinear meson self-interactions given by (4.2). The coupling constants g_σ, g_ω, g_ρ, the self-interaction terms of the φ and ω fields, with coupling constants g_2, g_3, c_3, and the σ meson mass are adjusted to reproduce the ground-state properties of finite nuclei and the bulk properties of nuclear matter. Various parameter sets, including Walecka's linear σ–ω model as a special case, have been established and can be found in [116, 147, 237].

In field theory, the equations of motion for fermion and boson fields are obtained by variation of the action of the system with respect to the corresponding fields. In order to work out the meson propagators, we write the action in the usual manner, as the integral of the Lagrangian density:

$$S[\varphi] \equiv \int d^4x \, \mathcal{L}(\varphi(x), \partial^\mu \varphi(x)) \equiv \int d^4x \, \mathcal{L}(x) \,, \tag{10.11}$$

where only the dependence on the scalar meson field is given explicitly, for notational convenience. Consider variations of the meson field operator around its classical value:

$$\varphi(\bm{r},t) = \varphi^{(0)}(\bm{r}) + \widetilde{\varphi}(\bm{r},t) \,. \tag{10.12}$$

10.3 Construction of the Meson Propagators

The effective action can be expanded around the value $\varphi^{(0)}$ which makes the action stationary. The first variation of the action with respect to the meson field gives the field equation (Klein–Gordon equation) satisfied by the classical field $\varphi^{(0)}$,

$$\partial^\mu \partial_\mu \varphi^{(0)} = j - \left.\frac{\partial U_\varphi(\varphi)}{\partial \varphi}\right|_{\varphi^{(0)}}, \qquad (10.13)$$

where

$$U = U_\varphi + U_\omega, \qquad (10.14)$$

$$U_\varphi = \frac{1}{2}m_\sigma^2 \varphi^2 + \frac{1}{3}g_2 \varphi^3 + \frac{1}{4}g_3 \varphi^4, \qquad (10.15)$$

$$U_\omega = \frac{1}{2}m_\omega^2 \omega_\mu \omega^\mu + \frac{1}{4}c_3(\omega_\mu \omega^\mu)^2, \qquad (10.16)$$

and j is a source function produced by the nucleon field in the effective Lagrangian (4.2). The second-order variation of the action gives the inverse of the meson propagator [76]:

$$\left(\partial^\mu \partial_\mu + \left.\frac{\partial^2 U_\varphi(\varphi)}{\partial \varphi^2}\right|_{\varphi_0}\right) D_\varphi(x, y) = -\delta^4(x-y). \qquad (10.17)$$

It is clear that the meson propagators with the nonlinear self-interaction in the effective Lagrangian (4.2) are no longer simple Yukawa forms. Taking the Fourier transform of (10.17), we arrive at an expression for the propagator in momentum space,

$$(E^2 - \boldsymbol{k}^2)D_\varphi(\boldsymbol{k} - \boldsymbol{k}', E)$$
$$- \frac{1}{(2\pi)^3}\int S_\varphi(\boldsymbol{k} - \boldsymbol{k_1})D_\varphi(\boldsymbol{k_1} - \boldsymbol{k}', E)\,\mathrm{d}^3 k_1 = (2\pi)^3 \delta(\boldsymbol{k} - \boldsymbol{k}'), \qquad (10.18)$$

where $S_\varphi(\boldsymbol{k} - \boldsymbol{k}')$ is the Fourier transform of $\partial^2 U_\varphi(\varphi)/\partial\varphi^2$. For the effective Lagrangian (4.2), the function S for the σ and ω mesons can be expressed as

$$S_\varphi(\boldsymbol{k} - \boldsymbol{k}') = \int \exp(-\mathrm{i}(\boldsymbol{k} - \boldsymbol{k}')\cdot \boldsymbol{r})\left(m_\sigma^2 + 2g_2\varphi(r) + 3g_3\varphi^2(r)\right)\mathrm{d}^3 r, \qquad (10.19)$$

$$S_\omega(\boldsymbol{k} - \boldsymbol{k}') = \int \exp(-\mathrm{i}(\boldsymbol{k} - \boldsymbol{k}')\cdot \boldsymbol{r})\left(m_\omega^2 + 2c_3\omega_0^2(r)\right)\mathrm{d}^3 r, \qquad (10.20)$$

$\varphi(r)$ and $\omega_0(r)$ being the classical values of the φ and ω fields at the point r. They can be obtained by a self-consistent calculation in the RMF approximation. In the limit of the linear model, one recovers the usual meson propagator, which is a local function in momentum space.

Table 10.1. Properties of nuclear matter calculated in the RMF approximation with the TM1, NL-SH and HS parameter sets: the incompressibility K, effective mass M^*/M, and symmetry energy a_{sym} [462]

	TM 1	NL-SH	HS
K (MeV)	281	355	545
M^*/M	0.634	0.597	0.541
a_{sym} (MeV)	36.9	36.1	35.0

10.4 Results for Collective States and Giant Resonances

The linear response function calculated in the limit of small momentum transfer was used in [462] to study the predictions of the nonlinear models for collective excitations in closed-shell nuclei. The bulk properties of nuclear matter calculated with the three parameter sets TM1, NL-SH, and HS are given in Table 10.1; these bulk properties are expected to affect the collective states, and especially the giant resonances, in finite nuclei. It is well known that the linear model HS [237] provides a considerably larger incompressibility and smaller effective mass than what are expected in the nonrelativistic approach. Therefore, this model should predict the isoscalar giant monopole resonance (ISGMR) and isoscalar giant quadrupole resonance (ISGQR) to be at energies that are too high. The incompressibility is greatly reduced and the effective mass is slightly increased in the nonlinear models. Thus one may expect some improvement in the predictions of the ISGMR and ISGQR when using the TM1 and NL-SH models. The three parameter sets produce more or less the same symmetry energy in nuclear matter and, consequently, the corresponding results for isovector excitations should not differ very much. We shall examine the predictions of the isovector giant monopole resonance (IVGMR) and isovector giant dipole resonance (IVGDR). The closed-shell nuclei ^{208}Pb, ^{90}Zr, ^{40}Ca, and ^{16}O were studied in [462].

We shall not discuss here the calculational procedure used to obtain the collective excitations, it is discussed in detail in [462].

The residual particle–hole interactions in the relativistic approach are produced by meson exchanges. In the ring approximation for the RRPA, only direct particle–hole matrix elements appear. Thus, in the isoscalar channel, only the isoscalar σ and ω mesons can contribute, whereas in the isovector channel, the only contributions in the effective Lagrangian (4.2) come from the isovector ρ meson. However, this statement is not valid for the states of abnormal parity 0^-, 1^+, 2^-, 3^+, 4^-, ...), even for excitations described by the RPA on the basis of RMF theory. Although the nuclear states are not pure isospin states, this has little effect on the particle–hole matrix elements, and the calculations in [462] were performed with σ and ω exchanges for isoscalar modes and ρ exchange for isovector modes.

10.4 Results for Collective States and Giant Resonances

The calculated results for the peak energies and centroid energies of the ISGMR, ISGQR, IVGMR, and IVGDR modes are summarized in Table 10.2, where they are compared with the corresponding data. The ISGMR energies are of great interest because they are the only source of experimental information that can be used to extract the incompressibility of nuclear matter. The theoretical results for the ISGMR shown in Table 10.2 were obtained in [463]. Good agreement with the experimental data is found for the nonlinear models TM1 and NL-SH, although their predictions of the incompressibilities of nuclear matter are different (see Table 10.1). Excessively large monopole energies and less collectivity are produced by the linear model, an expected result in view of the much larger incompressibility. However, the discrepancies of the results of the linear model for finite nuclei are less than a simple $\sqrt{K}$ law would predict. This shows that there are important contributions from surface, symmetry and Coulomb effects and that the ISGMR is far from being just a volume mode, a fact already well known from other analyses [466].

The response functions of the ISGQR mode have been calculated in [462] with the three parameter sets for the nuclei ^{208}Pb, ^{90}Zr, ^{40}Ca, and ^{16}O. In ^{90}Zr, a low-lying collective state below the particle emission threshold is found in all cases. Its energy is around 5–6 MeV; the results of the calculations are presented in Fig. 10.1 for ^{90}Zr and ^{208}Pb. Corresponding results for ^{16}O and ^{40}Ca can be founded in [462]. Experimentally, several low-lying 2^+ states between 2.2 MeV and 6.3 MeV are known, exhausting about 8% of the energy-weighted sum rule [467]. Similarly, the calculations predict low-lying collective 2^+ states between 4 MeV and 6 MeV in ^{208}Pb, while the data show a strong state at 4.09 MeV with 13.4% of the energy-weighted strength [467]. The strength of the low-lying states can be subtracted out by using a pure Lorentzian distribution with $\Gamma = 2\,\text{MeV}$. The corresponding RRPA ISGQR response functions show well-marked peaks after this low-lying strength has been subtracted. The results show that the main features of the ISGQR can be well described by the RRPA. The peak energies of the giant quadrupole resonance produced in both the nonlinear models and the linear model are in reasonable agreement with the experimental data (Table 10.2) [468], though larger deviations for the linear model are observed, which might be attributed to the small effective mass $M^*/M = 0.541$. In general, the calculated giant quadrupole resonance energies in the RRPA are about 2 MeV higher than the experimental data. From a study of nuclear matter, large quadrupole energies were expected owing to a small effective mass [434].

The IVGMR response functions calculated with the three parameter sets are shown in Fig. 10.2 for ^{208}Pb and ^{90}Zr. For this isovector mode, in the effective-Lagrangian equation (4.2) only the ρ meson is effective in the residual particle–hole interaction (see, however, the remark made above concerning excitations of abnormal parity). This residual interaction is weakly repulsive and, consequently, the RRPA strength distribution is shifted upwards compared with the unperturbed Hartree distribution, but the isovector

Table 10.2. Peak energies and centroid energies of isoscalar monopole (ISGMR), isoscalar quadrupole (ISGQR), isovector monopole (IVGMR), and isovector dipole (IVGDR) giant resonances calculated in the RRPA with the parameters TM1, NL-SH, and HS. The experimental energies shown in the last column are taken from [466] (ISGMR), [468] (ISGQR), [469, 470] (IVGMR), and [471] (IVGDR). All energies are in MeV. From [462]

	TM1		NL-SH		HS		Experiment
	E_{peak}	$\bar{E}$	E_{peak}	$\bar{E}$	E_{peak}	$\bar{E}$	
ISGMR							
^{208}Pb	12.8	13.3	13.0	13.7	16.0	16.4	13.7 ± 0.40
^{90}Zr	17.4	18.2	17.4	18.9	21.2	21.4	16.2 ± 0.50
^{40}Ca	21.5	22.8	22.0	23.4	24.5	24.6	
^{16}O	23.0	25.3	24.0	25.9	23.0	24.9	
ISGQR							
^{208}Pb	12.0	9.56	12.6	9.55	13.4	9.24	10.9 ± 0.1
^{90}Zr	15.6	13.5	15.6	13.7	16.4	13.9	14.1 ± 0.1
^{40}Ca	19.4	20.2	19.4	20.5	19.2	20.8	17.8 ± 0.3
^{16}O	23.6	25.3	23.6	25.8		25.2	20.7
IVGMR							
^{208}Pb		26.7		27.5		27.8	26.0 ± 3.0
^{90}Zr		32.8		33.4		32.5	28.5 ± 2.6
^{40}Ca		33.4		33.5		31.3	31.1 ± 2.2
IVGDR							
^{208}Pb	12.6	13.2	13.0	13.6	13.4	13.8	13.5 ± 0.2
^{90}Zr	16.0	17.2	16.4	17.6	16.4	17.5	16.5 ± 0.2
^{40}Ca	18.0	19.9	18.0	20.1	17.2	19.3	19.8 ± 0.5
^{16}O	20.6	23.3	20.8	23.5	19.4	21.8	$22.3 - 24$

monopole mode does not exhibit a strong collectivity. As a result, the distribution is considerably Landau damped and the RRPA response function is as broadly distributed as the unperturbed response function. The results obtained with $\Delta = 2\,\text{MeV}$ [3] show many structures and therefore, the curves

---

[3] The procedure used in [462] of discretizing the single-particle continuum introduces discrete structures into the RRPA spectrum above the particle emission threshold. In order to smooth out and diminish some unphysical structures an appropriate energy averaging of the response function was performed in [462]. This was done by replacing all real energies E in the polarization operators (10.5)–(10.9) and in the meson propagators (10.18) by complex $E + i\Delta/2$, with the value $\Delta = 2\,\text{MeV}$ being chosen.

10.4 Results for Collective States and Giant Resonances

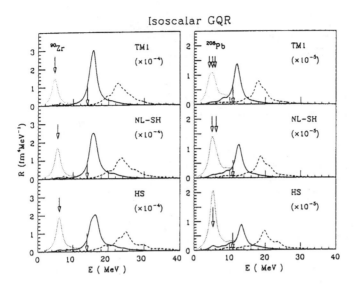

Fig. 10.1. ISGQR response functions $R_0(E)$ (RMF, *dashed curves*), $R(E)$ (RRPA after subtracting out the strength of low-lying states, *solid curves*) and the full strength (*dotted curves*) calculated for ^{90}Zr and ^{208}Pb using the HS, NL-SH, and TM1 models. The *arrows* on the horisontal axes indicate the experimental energies of the ISGQR, while the *arrows* pointing to the dotted peaks show the positions of the calculated low-lying states [462]

in Fig. 10.2 were calculated with $\Delta = 5$ MeV to display more clearly the main features of the IVGMR strength. The results obtained with the three parameter sets do not differ much from one another, and they slightly overestimate the experimental energies [469, 470].

The IVGDR strengths are plotted in Fig. 10.3 for ^{208}Pb and ^{90}Zr (corresponding results for ^{16}O and ^{40}Ca can be found in [462]). In contrast to the IVGMR case, the dipole strength is much less Landau damped. The unperturbed strength exhibits a well-marked peak, and the residual interaction just shifts this peak to higher energies without much affecting its shape. The peak energies of the giant dipole resonance for the three parameter sets are in good agreement with the experimental data, especially for heavy nuclei (see Table 10.2) [471]. This is consistent with the fact that the three models give values of the symmetry energy coefficient rather than close to one another (see Table 10.1). The calculated strength of the IVGDR in light nuclei is more fragmented, which is consistent with the experimental findings.

In [463], a systematic and detailed investigation of isoscalar monopole modes (i.e. ISGMR) in finite nuclei was carried out in the same framework. It was found that for some models, such as TM1 [116] and NL-SH [147], the ISGMR energies in medium and heavy nuclei are correctly reproduced,

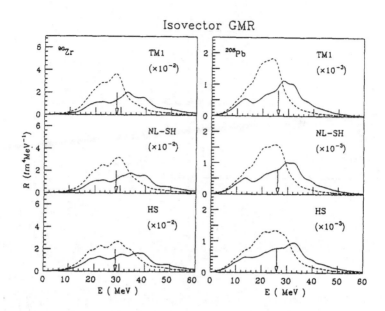

Fig. 10.2. IVGMR response functions $R_0(E)$ (RMF, *dashed curves*) and $R(E)$ (RRPA, *solid curves*), calculated for ^{90}Zr and ^{208}Pb using the HS, NL-SH, and TM1 models. The *arrows* indicate the experimental energies of the IVGMR. The averaging parameter is $\Delta = 5$ MeV in this figure [462]

although the calculated bulk incompressibilities may differ by more than 20% (see Table 10.1).

The general conclusion drawn in [462, 463] is that the effective Lagrangians recently developed can describe well not only the nuclear ground-state properties of finite nuclei, both stable and unstable nuclei up to the nucleon drip lines, but also the collective excited states and the giant resonances in doubly closed-shell nuclei. This should be an incentive to apply the RRPA method with nonlinear effective Lagrangians to the study of other systems, such as unstable nuclei near drip lines.

The recent development of radioactive nuclear beams has initiated a new era of study of unstable nuclei and of exploration of new physical phenomena. One of the most interesting discoveries made with radioactive beams is the existence of a neutron halo in nuclei containing neutrons with a very low separation energy, having a far-extending neutron distribution. The experiments by Tanihata et al. [155, 156] showed quite large reaction cross sections in collisions of ^{11}Li with high-Z targets, an indication that the matter radius of ^{11}Li is much larger than what is predicted by the usual $r_0 A^{-1/3}$ law. New types of excitation modes are expected owing to the excess of neutrons at the surface. A schematic picture of the new modes has been proposed by various authors [472]. The distribution of weakly bound nucleons extends beyond the

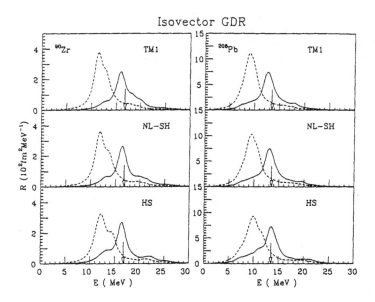

Fig. 10.3. IVGDR response functions $R_0(E)$ (RMF, *dashed curves*) and $R(E)$ (RRPA, *solid curves*), calculated for ^{90}Zr and ^{208}Pb using the HS, NL-SH, and TM1 models. The *arrows* indicate the experimental energies of the IVGDR. The averaging parameter is $\Delta = 5$ MeV in this figure [462]

usual nuclear radius and is decoupled from the nuclear core. The collective motion is separated into two parts, one being produced by the vibration of the core and the other originating from the motion of the weakly bound nucleons against the core. This second type of excitation is predicted to be at a much lower energy than the core mode. Recently, radioactive-beam facilities have become available, and they have allowed the possibility to experimentally measure such new excitation modes. It is therefore desirable to investigate the new modes theoretically.

This type of investigation has been carried out in [464]. In that paper, the response of Ar nuclei (from $A = 30$ to $A = 52$) to the isovector dipole operator was calculating by solving the RRPA equation (10.6) using the nonlinear model TM1. From the static Dirac–Hartree spectrum it can already be seen that, near the proton and neutron drip lines, there are particle–hole configurations at low energy produced by the jump of a loosely bound nucleon into the nearby continuum. These configurations are well separated from the usual $1\hbar\omega$ configurations which carry the major part of the dipole strength. The two types of configurations do not couple to each other and should lead to separated dipole peaks. This is confirmed by the RRPA calculations, which show that the strength of the giant dipole resonance is split into two parts for nuclei near the neutron and proton drip lines. One peak is at the normal excitation energy of the giant dipole resonance, and this is reproduced well by an empirical two-parameter formula. The other peak is located at lower energy and is due to unperturbed particle–hole excitations involving loosely bound occupied states. The lower dipole strength becomes stronger as the isotopes move towards the drip lines. One may expect that a manifestation of this would be an enhancement of the observed cross sections in Coulomb excitation experiments involving these nuclei.

We should mention that in [84] a review is given of applications of time-dependent RMF theory for the description of oscillations with small amplitudes, as well as large-amplitude collective motion, in connection with relativistic Coulomb excitations. We shall not discuss this approach in detail here, but instead make reference to the respective papers [473].

The relativistic RPA theory needs to be developed further along the following lines:

1. Take into account contributions from the Dirac sea (see, for example, [460]).[4]
2. Include the tensor interaction of the ρ meson for excitations of normal parity.
3. Investigate the role of the spatial components of the vector fields in excitations of the abnormal-parity states.
4. Consider the nonstatic terms in connection with excitations of abnormal-parity states.
5. Investigate oscillations of the RHF basis in a model with nonlinear self-interactions of mesonic fields.

Essential progress has been achieved in [477] in applying the RRPA for calculation of the monopole and dipole compression modes in nuclei. In that paper the emphasis is put on the effects of Dirac sea states, which are generally neglected in relativistic RPA calculations. It is found that these effects

---

[4] In [83], it is said in summary that the inclusion of the effects of the Dirac sea in RPA calculations of nuclear excitations and of ground-state properties of nuclei via extension to the relativistic RPA has three effects, which deserve to be mentioned here:
(a) It solves the magnetic-moment problem. The isoscalar convection current is improved from $\boldsymbol{P}/M^*$ to $\boldsymbol{P}/M$ (see Sect. 5.2); this correction gives the Schmidt values for the single-particle magnetic moments [185–191].
(b) If a sufficient number of particle–hole configurations is admixed, this brings the spurious state related to pure center-of-mass motion down to zero excitation energy [458].
(c) It ensures conservation of the electromagnetic current [458]. It should be mentioned also that both the scalar and the vector meson propagators actually have poles at $q_0 = 0$ and $|\boldsymbol{q}| \neq 0$ at finite baryon density in the RRPA [77, 474], these poles indicating an instability of the RMF ground state against density oscillations. This problem needs further investigation. The structure of the vertex in the relativistic theory is considered in [475, 476].

10.4 Results for Collective States and Giant Resonances

can be quite important for the isoscalar monopole mode. The main contributions from the particle–hole pairs constructed from Fermi and Dirac sea states arise through the exchange of the scalar meson, while the vector mesons play a negligible role. The large contribution to the RRPA strengths from the Dirac states can be understood in the following way. In the relativistic approach, the nucleon potential is the result of a strong cancellation between the isoscalar–scalar and vector potentials, and so are the particle–hole residual interactions in the isoscalar channel. Owing to the structure of the γ-matrix, the matrix elements of the residual interaction between particle–hole pairs through the exchange of scalar and vector mesons are largely cancelled, to two orders of magnitude. However, this is not true for pairs built from Fermi and Dirac states, because the large component of the wave function is the upper component for a Fermi state and the lower component for a Dirac state. The orthogonality between positive- and negative-energy states with the same quantum numbers is mainly due to the cancellation between the upper and lower components. Again, owing to the structure of the γ-matrix, the matrix elements between Fermi and Dirac states corresponding to the exchange of vector mesons (ω or ρ) are small, whereas those corresponding to a scalar meson are large. These scalar meson matrix elements are repulsive and can overcome the large energy gap between the Dirac and Fermi states.

In [477] it is concluded that the inclusion of only the positive-energy particle–hole pairs in the isoscalar modes, where the isoscalar mesons play an important role, would provide too strong an attraction and therefore, too low an isoscalar giant resonance energy. This will be reflected in too large a polarizability. On the other hand, the contributions from Dirac states can be neglected in the case of isovector modes where the ρ meson dominates.

In [477], the RRPA monopole strengths with and without Dirac states were calculated for different effective Lagrangians. The strengths are shifted down strongly to lower energies when only the positive-energy particle–hole pairs are included. In some parameterizations, such as NL1, the strong attractive residual interaction produced by the positive-energy particle–hole pairs would bring the system beyond a critical point, where the system would not be stable. In complete RRPA calculations with the Dirac states included, the strengths are shifted to higher energies and the instabilities disappear. In a recent publication [478], such an effect due to the Dirac states has been demonstrated analytically in the framework of nuclear matter and Walecka's linear model.

In [477], good agreement between the monopole energies calculated in the RRPA and in the time-dependent relativistic mean-field approximation (TDRMFA) is achieved when the Dirac state contributions are included in the calculation in the RRPA (in the TDRMFA, the density change $\delta\rho(t)$ at time t necessarily has matrix elements not only between positive-energy particle–hole pairs but also between pairs formed from Dirac sea states and occupied Fermi sea states of the static solution).

In [477, 479], it is emphasized also that a large discrepancy remains between theory and experiment in the case of the dipole compression mode.

Recently, a complete relativistic calculation of isoscalar giant resonances has been performed also in [480]. In that paper, the longitudinal response of ^{208}Pb was calculated using the RRPA for different parameterizations of the Walecka model with scalar self-interactions. From a nonspectral calculation of the response [427, 481] – which automatically includes the mixing between positive- and negative-energy states – the distributions of strength for the isoscalar monopole, dipole, and high-energy octupole resonances were obtained. In [480], a consistent formalism is considered that uses the same interaction in the calculation of the ground state and in the calculation of the response. As a result, the conservation of the vector current is strictly maintained throughout the calculation. Further, the spurious dipole strength associated with the uniform translation of the center of mass is shifted to zero excitation energy and is cleanly separated from the physical excitations.

The further development of the RRPA theory and of its applications is considered in [482–485].

10.5 Cranked Relativistic Mean-Field Theory

In 1989 the Munich group [423] made the first effort to consider rotating nuclei by combining RMF theory and the cranking assumption (see also [486]). The rotation of a nucleus with an angular velocity Ω_x about the x-axis is considered using a single-dimensional cranking approach. The value of Ω_x is derived from the following condition (cranking prescription):

$$J(\Omega_x) = \langle \Phi_\Omega | \hat{J}_x | \Phi_\Omega \rangle = \sqrt{I(I+1)} \; ; \qquad (10.21)$$

this means that the average value of the total angular momentum for a spin I has a definite value. The rotation of a nucleus can be considered by transforming the coordinate system to a frame rotating with a constant angular velocity Ω_x around a fixed axis in space. In this case the variational procedure generates a stationary Dirac equation for the nucleons in the rotating frame,

$$\left\{ \boldsymbol{\alpha} \left(-i \boldsymbol{\nabla} - g_\omega \boldsymbol{\omega}(\boldsymbol{r}) \right) + g_\omega \omega_0(\boldsymbol{r}) + \beta \left(M + g_\sigma \varphi(\boldsymbol{r}) \right) - \Omega_x \hat{J}_x \right\} \psi_i = E_i \psi_i \; , \qquad (10.22)$$

and time-independent inhomogeneous equations for the mesonic fields.

$$\left\{ -\Delta - \left(\Omega_x \hat{L}_x \right)^2 + m_\sigma^2 \right\} \varphi(\boldsymbol{r}) = -g_\sigma \rho_S(\boldsymbol{r}) - g_2 \varphi^2(\boldsymbol{r}) - g_3 \varphi^3(\boldsymbol{r}) \; , \qquad (10.23)$$

$$\left\{ -\Delta - \left(\Omega_x \hat{L}_x \right)^2 + m_\omega^2 \right\} \omega_0(\boldsymbol{r}) = g_\omega \rho_V(\boldsymbol{r}) \; , \qquad (10.24)$$

10.5 Cranked Relativistic Mean-Field Theory

$$\left\{-\Delta - \left(\Omega_x(\hat{L}_x + \hat{S}_x)\right)^2 + m_\omega^2\right\}\boldsymbol{\omega}(\boldsymbol{r}) = g_\omega \boldsymbol{j}(\boldsymbol{r}), \tag{10.25}$$

where $\boldsymbol{j}(\boldsymbol{r})$ is the baryonic current, and the term

$$\Omega_x \hat{J}_x = \Omega_x \left(\hat{L}_x + \frac{1}{2}\hat{\Sigma}_x\right) \tag{10.26}$$

corresponds to the Coriolis field. Note that for brevity we have not written down the equations for the (uncharged) components of the ρ meson field or for the electromagnetic field and for the same reason have not included the respective fields in (10.22) (however, they have been taken into account in the calculations performed in practice). We should mention also that ω^4 terms have not yet been included in calculations in the framework of the cranked RMF theory.

The main features of the equations (10.22)–(10.25) are the presence of the Coriolis term $\Omega_x \hat{J}_x$ in the Dirac equation (10.22) and the squared Coriolis terms $(\Omega_x \hat{L}_x)^2$ and $(\Omega_x(\hat{L}_x + \hat{S}_x))^2$ in the Klein–Gordon equations (10.23)–(10.25). The presence of the term $g_\omega \boldsymbol{\omega}(\boldsymbol{r})$ together with the space components of the ω meson field (as well as the presence of the similar terms for the ρ meson) is related to the breaking of time-reversal invariance by the Coriolis term. The effects determined by the space components of the ω field are usually treated as nuclear magnetism and are essential, for example, for a proper understanding of the moments of inertia [84]. For this reason, the space components of the vector fields should be taken into consideration self-consistently (their role in reproducing the magnetic moments in odd-mass nuclei has been discussed earlier).

This approach was applied by the Munich group to the yrast states of ^{20}Ne.[5] Later it was shown to reproduce the moments of inertia of medium-heavy and heavy superdeformed nuclei (pairing correlations were neglected because of the large angular momentum in this region).

The authors of [487] formulated a general relativistic mean-field theory for rotating nuclei in which they adopted the tetrad formalism (i.e. they took into account the fact that the rotating frame is an accelerated frame), and also performed calculations for three zinc isotopes, including the newly discovered superdeformed band in ^{62}Zn, which is the first experimental observation in this mass region. Further discussion of the relationship between these two approaches can be found in [487], while [84, 486] contain an exhaustive discussion of investigations of nuclear rotations up to very high angular momenta in the framework of the cranking model.[6]

---

[5] The calculational procedure for obtaining self-consistent solutions of (10.22)–(10.25) is described in [84, 92]. It is based on the expansion of the Dirac spinors and the meson fields in terms of eigenfunctions of a deformed oscillator.

[6] It is remarkable that the cranked RMF theory (with only six or seven parameters) is able to generate a microscopic description of rotational bands in superdeformed nuclei.

11 The Equation of State of Nuclear Matter for Supernovas and Neutron Stars

11.1 Thomas–Fermi Method for Nonuniform Matter

A supernova is a spectacular stellar phenomenon that happens when a massive star ($M \geq 8M_\odot$) uses up its nuclear fuel. Owing to the gravitational collapse of the core, a supernova explosion may lead to the formation of a neutron star or black hole in the central part of the massive star. In order to clarify the mechanism of the explosion and the whole field of the phenomena of supernova explosions, it is necessary perform numerical simulations. For this purpose, we need an equation of state of nuclear matter that covers the wide density and temperature range, with various proton fractions, which occurs during the collapse and the explosion.

There have been several EOSs worked out so far for supernova study [488–490]. Most of them are based on the nonrelativistic Skyrme–Hartree–Fock (SHF) framework. Although some of these EOSs have been carefully constructed in a form suitable for supernova simulations, there are many places where the EOSs are discontinuous or not available, and the supernova simulations stop there owing to the incompleteness of the EOS table. Sometimes these places are not even documented, and it is difficult to fix the problem. Lattimer and Swesty worked out an EOS in the compressible liquid drop model on the basis of a nonrelativistic framework [490]. This EOS table is almost unique, and many simulations are performed with it at present.

Now that success has been achieved in describing nuclear properties using the relativistic approach, it is extremely important to calculate an EOS for astrophysical purposes. One has to perform consistent calculations for high-density nuclear matter, inhomogeneous nuclear matter, and finite nuclei for the wide range of density, proton fraction, and temperature which occurs inside neutron stars and supernovas. For this purpose one has to study the properties of dense matter with both homogeneous and inhomogeneous distributions in the RMF framework. This program has been realized in [491], this chapter being essentially based on the results obtained in [491]. So, along the lines of [491], the RMF theory with the parameter set TM1 has been adopted in the present chapter. The RMF theory with the TM1 parameter set as was discussed in Chap. 4, is known to provide excellent results for the properties of the ground states of heavy nuclei, including unstable nuclei [116, 492]. As was shown in the Chap. 10, TM1 can also be used to describe the giant resonances within the RPA formalism and has been demonstrated to provide good descriptions of giant resonances[462–464].

For the description of inhomogeneous nuclear matter, which appears below $\rho \sim \rho_0/3$, where ρ_0 is the normal nuclear-matter density, the Thomas–Fermi approximation[1] is used [493]. We assume [491] that the inhomogeneous matter is composed of a lattice of spherical nuclei. At finite temperatures, a proton gas, as well as a neutron gas, may be present outside the nuclei. The free energy of the system is minimized with respect to the cell volume, the proton and neutron radii, the surface diffuseness of the nucleus, and the densities of the outside neutron and proton gases. To determine the phase of the dense matter, the calculated free energy is compared with that of homogeneous matter. This will be done for many values of the density, temperature, and proton fraction covering the wide range necessary for astrophysical purposes.

The relativistic mean-field theory was described earlier (see Chap. 4). In [491] a Lagrangian with nonlinear σ and ω terms is adopted (see (4.2)). The parameter set TM1 was introduced and studied in [116], this parameter set being determined to be the best one for reproducing the properties of finite nuclei over a wide mass range in the periodic table, including neutron-rich nuclei. The authors of [116] performed a least-squares fit to the experimental data for the binding energies and charge radii of proton-magic nuclei from Ca to Pb, including unstable Pb isotopes. The RMF theory with the TM1 parameter set was also shown to produce satisfactory agreement with experimental data in studies of nuclei with a deformed configuration [460] and of giant resonances within the RPA formalism [462–464]. Details of the parameter set TM1 can be found in Chaps. 4 and 10 (see also [94, 506]).

The derivation of the Euler–Lagrange equations and the following treatment of the EOS of homogeneous nuclear matter at finite temperature follows the same procedure as described in Chap. 4 and [94, 506]. The expressions for physical quantities such as the energy density and entropy, which are used in the Thomas–Fermi calculations, can be found in Chap. 4 and [94]. The properties of homogeneous nuclear matter and supernova matter, which also contains leptons, at high densities (above ρ_0) at finite temperature in the RMF theory have been discussed in the same references.

For a density below 10^{14} g/cm^3, where heavy nuclei can be formed together with a free-nucleon gas in order to lower the free energy, the authors of [491] performed a Thomas–Fermi calculation based on work done by Oyamatsu [493]. In this case the system is inhomogeneous and is modeled by a mixture of electrons, muons, free nucleons, and a single species of heavy nuclei. The leptons can be treated as uniform noninteracting relativistic particles. They play no role in the minimization of the free energy for a system with fixed proton fraction. So the main attention in [491] is paid to the baryon

---
[1] Various aspects of recent extensions of the Thomas–Fermi method in the nuclear-structure context are discussed in [494–505].

11.1 Thomas–Fermi Method for Nonuniform Matter

contribution. As in [493], it is assumed that each heavy spherical nucleus is in the center of a charge-neutral cell consisting of a vapor of neutrons, protons and leptons. The nuclei form a body-centered cubic (BCC) lattice to minimize the Coulomb lattice energy.

It is useful to introduce the Wigner–Seitz cell to simplify the calculation of energy of the unit cell. The Wigner–Seitz cell is a sphere whose volume is the same as that of the unit cell in the BCC lattice. The lattice constant a is defined as the cube root of the cell volume. Then, one has

$$V_{\text{cell}} = a^3 = N_B/\rho_B \,, \tag{11.1}$$

where N_B and ρ_B are the baryon number per cell and the average baryon number density, respectively. The Coulomb energy is calculated using this Wigner–Seitz approximation, and a correction energy for the BCC lattice [493] is added. This correction energy is negligible unless the size of the nucleus is comparable to the cell size.

The nucleon distribution functions $\rho_i(r)$ for neutrons (i =n) and protons (i =p) are assumed to be

$$\rho_i(r) = \begin{cases} (\rho_i^{\text{in}} - \rho_i^{\text{out}}) \left[1 - \left(\dfrac{r}{R_i}\right)^{t_i}\right]^3 + \rho_i^{\text{out}}, & 0 \le r \le R_i \\ \rho_i^{\text{out}}, & R_i \le r \le R_{\text{cell}} \end{cases}, \tag{11.2}$$

where r represents the distance from the center of the nucleus and R_{cell} is defined by the relation

$$V_{\text{cell}} = \frac{4\pi}{3} R_{\text{cell}}^3 \,. \tag{11.3}$$

The density parameters ρ_i^{in} and ρ_i^{out} are the densities at $r = 0$ and $r \ge R_i$, respectively. The parameters R_i and t_i determine the boundary and the relative surface thickness of the nucleus. In order to illustrate these parameters, the neutron and the proton distributions are shown in Fig. 11.1 for the case $T = 10 \,\text{MeV}$, $Y_p = 0.3$, and $D_B = 10^{13.5} \,\text{g/cm}^3$. The baryon mass density is defined here as $D_B = m_u \rho_B$, m_u being the atomic mass unit. For a system with fixed temperature T, proton fraction Y_p, and baryon mass density D_B, there are seven independent parameters in the nine variables a, ρ_n^{in}, ρ_n^{out}, R_n, t_n, ρ_p^{in}, ρ_p^{out}, R_p, t_p. The thermodynamically favorable state will be the one that minimizes the free-energy density with respect to these seven parameters. In zero temperature case, there are no free protons outside the nucleus, so the number of independent parameters is reduced to six.

In this model, the free-energy density contributed by baryons is given by

$$f = (E - TS)/a^3 \,, \tag{11.4}$$

where the energy per cell E can be written as

$$E = E_{\text{bulk}} + E_{\text{s}} + E_{\text{c}} \,. \tag{11.5}$$

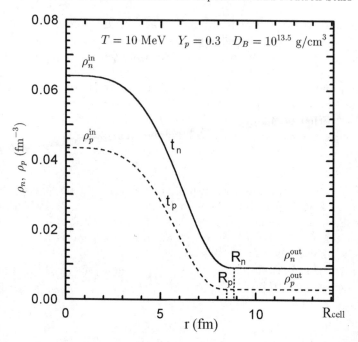

Fig. 11.1. The neutron distribution and the proton distribution in the Wigner–Seitz cell at $T = 10\,\text{MeV}$, $Y_p = 0.3$ and $D_B = 10^{13.5}\,\text{g/cm}^3$ are shown by the *solid curve* and *dashed curve*, respectively. The parameters ρ_i^{in}, ρ_i^{out}, R_i, and t_i of the density distributions for protons and neutrons are defined in (11.2), and R_{cell} represents the radius of the spherical Wigner–Seitz cell [491]

The bulk energy E_{bulk} and the entropy per cell S are calculated from

$$E_{\text{bulk}} = \int_{\text{cell}} \varepsilon_{\text{RMF}}(\rho_n(r), \rho_p(r))\,\mathrm{d}^3 r\,, \tag{11.6}$$

$$S = \int_{\text{cell}} s_{\text{RMF}}(\rho_n(r), \rho_p(r))\,\mathrm{d}^3 r\,, \tag{11.7}$$

where ε_{RMF} and s_{RMF} are the energy density and the entropy density in the RMF theory, expressed as functionals of the neutron density ρ_n and the proton density ρ_p. Note that ε_{RMF} and s_{RMF} are calculated at each radius in the RMF theory for uniform nuclear matter with the corresponding densities ρ_p and ρ_n. For the surface energy term E_s due to the inhomogeneity of the nucleon distribution, a simple form is taken,

$$E_s = \int_{\text{cell}} F_0\,|\nabla(\rho_n(r) + \rho_p(r))|^2\,\mathrm{d}^3 r\,. \tag{11.8}$$

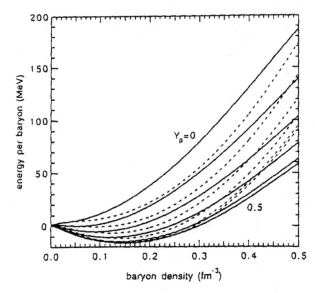

Fig. 11.2. The energy per baryon as a function of the baryon density for homogeneous nuclear matter at zero temperature, with proton fractions $Y_p = 0$, 0.1, 0.2, 0.3, 0.4, and 0.5. The *solid curves* represent the results of the RMF theory, while nonrelativistic results taken from [490] are shown by *dashed curves* for comparison [491]

The parameter $F_0 = 70$ MeV fm^5 was determined by performing the Thomas–Fermi calculations of finite nuclei so as to reproduce the gross properties of the nuclear mass, the charge radii, and the beta stability line as described in the appendix of [493]. Here the same expression for the Coulomb energy term E_c is adopted as in (3.8) of [491].

The minimization of the free energy was performed for many values of the density, temperature, and proton fraction covering the wide range necessary for astrophysical purposes. The phase transition from nonuniform matter to uniform matter occurs when the free-energy density cannot be reduced by making heavy nuclei.

11.2 Equation of State of Nuclear Matter

In [491], the EOS of nuclear matter is constructed for a wide range of the baryon mass density ($D_B = 10^5$–$10^{15.5}$ g/cm^3), temperature ($T = 0$–50 MeV), and proton fraction ($Y_p = 0.01$–0.5). Uniform matter and nonuniform matter are treated consistently, using the same RMF theory. Hence, all the resulting thermodynamic quantities are consistent and smooth over the whole range of this EOS. The EOS of nuclear matter with $Y_p = 0.0$ (i.e.

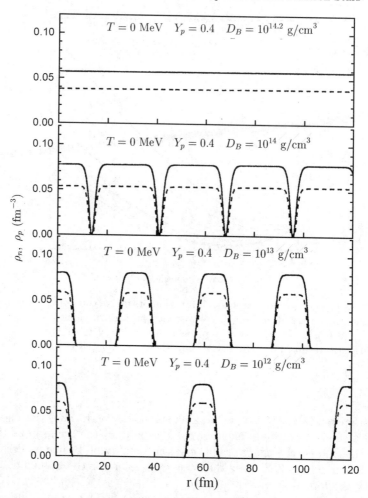

Fig. 11.3. The neutron distribution (*solid curves*) and the proton distribution (*dashed curves*) along a straight line joining the centers of neighboring nuclei in the BCC lattice for $T = 0$ MeV, $Y_p = 0.4$, and the cases $D_B = 10^{12}$, 10^{13}, 10^{14}, and $10^{14.2}$ g/cm^3 [491]

neutron matter) was also calculated for various densities and temperatures in the above range, assuming a uniform distribution under all conditions.

We discuss first the calculated properties of uniform matter. In Fig. 11.2 we show the energy per baryon of homogeneous nuclear matter at zero temperature, as a function of the baryon density. The results of the RMF theory (solid curves) are compared with results obtained using a parameterized form

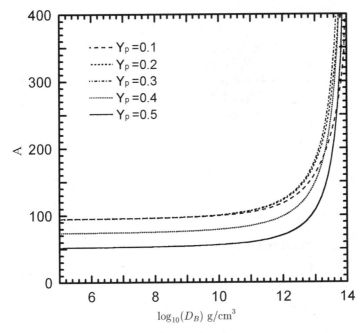

Fig. 11.4. The nuclear mass number A as a function of the baryon mass density D_B for proton fractions $Y_p = 0.1$, 0.2, 0.3, 0.4, and 0.5 at zero temperature [491]

based on the nonrelativistic framework in [426] (dashed curves). The figure shows that the RMF theory provides a stronger density-dependent repulsion than do the nonrelativistic results. A calculation of the free energy of uniform matter at finite temperature has also been done in the same manner. Note that the free energy at zero temperature is equal to the internal energy. The calculated properties of homogeneous nuclear matter at finite temperature have been discussed in [94, 506]. The free energy per baryon of homogeneous nuclear matter was used as an input in the Thomas–Fermi calculation.

By performing the Thomas–Fermi calculations and comparing the free energies of the uniform and nonuniform configurations, the most favorable state of nuclear matter was determined for each condition of density, proton fraction, and temperature. In the zero-temperature case, the thermodynamically favorable state for $D_B \leq 10^{14}$ g/cm^3 with $Y_p \geq 0.3$ is to have a BCC lattice of heavy nuclei without any outside neutron gas. When the proton fraction Y_p decreases to around 0.3, neutrons drip out from the nucleus. In Fig. 11.3 we show the neutron and proton distributions along a straight line joining the centers of neighboring nuclei in the BCC lattice for $T = 0$ MeV, $Y_p = 0.4$, and the cases $D_B = 10^{12}$, 10^{13}, 10^{14}, and $10^{14.2}$ g/cm^3. The phase transition from nonuniform matter takes place around $D_B \sim 10^{14}$ g/cm^3.

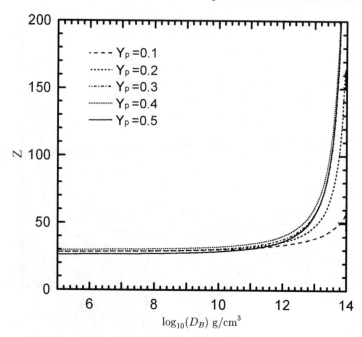

Fig. 11.5. The nuclear proton number Z as a function of the baryon mass density D_B for proton fractions $Y_p = 0.1$, 0.2, 0.3, 0.4, and 0.5 at zero temperature [491]

In [491] the authors assume only a spherical nuclear shape and neglect nonspherical nuclear shapes, which may further reduce the free energy just before the transition to uniform matter. However, these shapes would not cause a significant change in the EOS of matter but would only make the transition smoother. Nonspherical shapes can slightly reduce the surface and Coulomb energy compared with the spherical shape, when the size of the nucleus is comparable to the cell size, as in the case of $D_B = 10^{14}$ g/cm^3 in Fig. 11.3. However, in such cases, the surface and Coulomb energies are much smaller than the bulk energy so that the energy gain due to the shape change is relatively small. The nuclear mass number, $A = Z + N$, was calculated from the higher-density part of the inhomogeneous nuclear matter by integrating the proton number Z and the neutron number N up to the radius $R_{\max}$, which is the maximum between R_p and R_n. Here $R_{\max}$ is considered as the boundary of the nucleus. In Figs. 11.4 and 11.5 we display the nuclear mass number A and the proton number Z as a function of the baryon mass density D_B for various proton fractions Y_p. Both A and Z increase sharply just before the phase transition occurs, reflecting the fact that the nucleus becomes larger and the system turns into the uniform matter phase.

11.2 Equation of State of Nuclear Matter   227

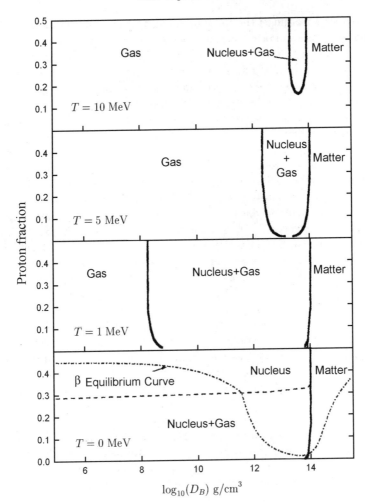

Fig. 11.6. The phase diagrams of nuclear matter at $T = 0$, 1, 5, and 10 MeV in the D_B–Y_p plane. The regions labeled "Matter" correspond to the uniform matter described by the RMF theory. The regions labeled "Nucleus+Gas" and "Nucleus" correspond to nonuniform matter with and without a free-nucleon gas; the boundary between them is shown by the *dashed curve* for $T = 0$ MeV. The regions labeled by "Gas" correspond to uniform matter close to the classical ideal-gas approximation. The β-equilibrium curve at $T = 0$ MeV is also shown by the *dot-dashed* curve [491]

In the finite-temperature case, the second term in (11.4) provides an appreciable contribution to the free-energy density as the temperature increases. In Fig. 11.6, the phase diagram of nonuniform matter and uniform matter at $T = 0$, 1, 5, and 10 MeV is shown. For an extremely low density and in a finite temperature range, the thermodynamically favorable state is a uniform nucleon gas. The uniform nucleon gas can be approximated as a classical noninteracting gas for $D_B < 10^{10}$ g/cm^3. The phase transition from a uniform nucleon gas to nonuniform matter takes place at the density where the system can lower the free energy by making heavy nuclei.

From Fig. 11.6 it can be seen that the lower phase transition density (from a uniform nucleon gas to nonuniform matter) depends on the temperature very strongly, while the higher phase transition density (from nonuniform matter to uniform matter) is nearly independent of the temperature; this latter density is almost constant at $D_B \sim 10^{14}$ g/cm^3. As the temperature increases, the area of the nonuniform matter phase becomes smaller. The nonuniform matter phase disappears when the temperature is higher than $\sim$15 MeV. The neutron and proton distributions for $D_B = 10^{13.5}$ g/cm^3, $Y_p = 0.4$, and the cases $T = 0, 5, 10$, and 15 MeV are shown in Fig. 11.7. From this figure, one can find how the properties of the Wigner–Seitz cell change as the temperature increases. The neutron and proton densities outside the nucleus (ρ_n^{out} and ρ_p^{out}) become large at high temperature, while $\rho_p^{out} = 0$ at $T = 0$ MeV for all D_B and Y_p. There is a tendency for the ratio between the nucleon numbers inside and outside the heavy nuclei decrease as the temperature increases. Furthermore, the heavy nuclei are dissociated when the temperature is high enough.

Figure 11.8 shows the free energy per baryon f/ρ_B as a function of the baryon mass density D_B for various Y_p at $T = 0, 1, 5$, and 10 MeV. In the regime of the uniform nucleon gas phase at low density and finite temperature, the values of f/ρ_B are not significantly altered by variations in Y_p, because the interaction between nucleons is very weak at low density. When the system changes into the nonuniform matter phase, the formation of heavy nuclei can reduce f/ρ_B, and this brings about an appreciable Y_p dependence of f/ρ_B. All curves display a rapid rise in the high-density range, which is due to the strong density-dependent repulsion between nucleons in the RMF theory.

In Fig. 11.9, the results are shown for the entropy per baryon as a function of D_B for various Y_p at $T = 1, 5$, and 10 MeV. The effect of the nonuniform matter phase on the entropy clearly appears in this figure. The drop in the entropy in the nonuniform matter region is due to the formation of heavy nuclei. The Y_p dependence of the entropy in the nonuniform matter region is opposite to the dependence in the uniform matter regions when the temperature is low. This is because more heavy nuclei are formed in the nonuniform matter phase at higher Y_p, and they can then bring about a more rapid drop in the entropy-per-baryon curves compared with the case at lower Y_p. On the other hand, the free-nucleon gas outside is a dominant component of the matter at high temperature, and the effect of the formation of heavy nuclei is weakened as the temperature increases.

11.2 Equation of State of Nuclear Matter   229

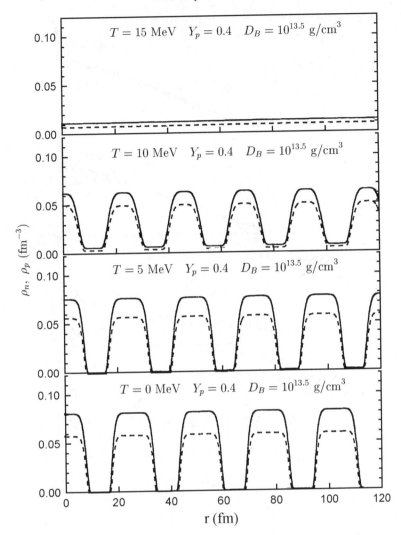

Fig. 11.7. The neutron distribution (*solid curves*) and the proton distribution (*dashed curves*) along the straight lines joining the centers of neighboring nuclei in the BCC lattice for $D_B = 10^{13.5}$ g/cm3, $Y_p = 0.4$ and the cases $T = 0, 5, 10, 15$ MeV [491]

The pressure can be calculated from f/ρ_B by means of the following thermodynamic equilibrium condition:

$$P = \rho_B^2 \left.\frac{\partial (f/\rho_B)}{\partial \rho_B}\right|_{T,Y_p}. \tag{11.9}$$

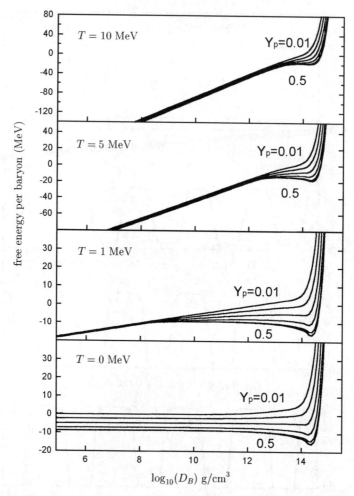

Fig. 11.8. The free energy per baryon as a function of the baryon mass density D_B for the proton fractions $Y_p = 0.01, 0.1, 0.2, 0.3, 0.4$, and 0.5 at $T = 0, 1, 5$, and 10 MeV. (Taken from [491])

We show in Fig. 11.10 the pressure as a function of D_B for various Y_p at $T = 0, 1, 5$, and 10 MeV. In the uniform nucleon gas phase at low density and at finite temperature, the pressure is provided mainly by the thermal pressure, which is given by $P = \rho_B T$. In this case, the pressure is independent of Y_p. When the system goes into the nonuniform matter phase, the reduction of f/ρ_B due to the formation of heavy nuclei brings about a sharp drop in pressure. For high Y_p, the pressure drops to negative values. The drop of the pressure in the nonuniform matter phase has a strong Y_p dependence. In the high-density region, all curves come back to positive values,

11.2 Equation of State of Nuclear Matter

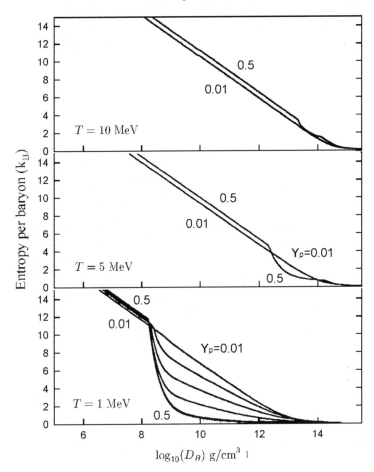

Fig. 11.9. The entropy per baryon as a function of the baryon mass density D_B for the proton fractions $Y_p = 0.01, 0.1, 0.2, 0.3, 0.4$, and 0.5 at $T = 1$ MeV, and for $Y_p = 0.01$ and 0.5 at $T = 5$ and 10 MeV [491]

and have a rapid rise as the density increases. The behavior of the pressure at high density is determined predominantly by the contribution of the vector meson ω and the isovector–vector meson ρ in the RMF theory [506]. Because of the nonlinear term corresponding to the ω meson, the repulsive ω meson field is suppressed at high density, so this repulsive field does not increase linearly with density as it would in an RMF theory without this nonlinear ω meson term. As a result, the pressure is greatly reduced at high density compared with the case in an RMF theory without this nonlinear term. It is necessary to stress that this behavior is in accord with that in

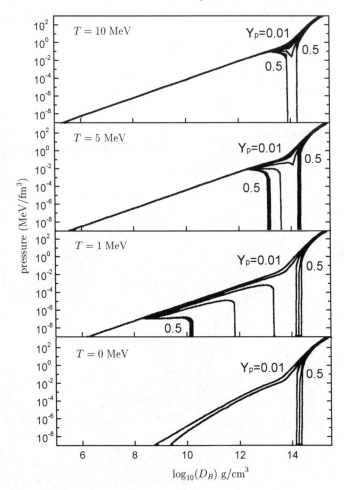

Fig. 11.10. The pressure as a function of the baryon mass density D_B for the proton fractions $Y_p = 0.01, 0.1, 0.2, 0.3, 0.4$, and 0.5 at $T = 0, 1, 5$, and 10 MeV [491]

the relativistic Brueckner–Hartree–Fock theory on which the RMF theory is based.

11.3 Neutron Star Matter and Neutron Star Profiles

In [491], an EOS of nuclear matter with various proton fractions was constructed so that it could be used in the simulation of supernova explosions where β-equilibrium may not be achieved within the relevant timescale. In the study of cold neutron stars, the β-equilibrium condition can be assumed

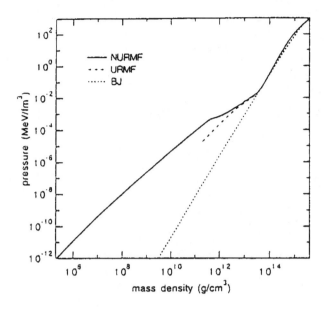

Fig. 11.11. The pressure of neutron star matter at zero temperature as a function of the mass density. The *solid curve* shows the NURMF EOS, which takes account of the nonuniform matter phase in the low-density range, while the *long-dashed curve* shows the URMF EOS, obtained by assuming a uniform matter phase in the same RMF theory. The BJ EOS is also shown by the *short-dashed curve* for comparison [491]

when one is considering static properties. In order to calculate the properties of neutron star matter at zero temperature, one has to incorporate the contribution of leptons into the EOS. The proton fraction of neutron star matter, which is determined by the β-equilibrium condition, is shown by the dot–dashed curve in Fig. 11.6. The large proton fractions at high densities beyond 10^{14} g/cm^3 are remarkable in the RMF theory [94, 116, 506]. In Fig. 11.11, the pressure of the neutron star matter is displayed as a function of the mass density. Here the mass density is equal to the total energy density divided by c^2. Since we have taken account of the nonuniform matter phase, the EOS considered here [491], labeled NURMF (nonuniform relativistic mean field) is very different from the EOS labeled URMF (uniform relativistic mean field), which was obtained by assuming uniform matter in the same RMF theory for the low-density range [506]. The EOS obtained by Bethe and Johnson [507], labeled BJ, is also shown in Fig. 11.11 for comparison.

The profile of a neutron star was calculated in [491] by applying the EOS of neutron star matter and solving the Oppenheimer–Volkoff equation. The gravitational mass of a neutron star as a function of the central mass density is displayed in Fig. 11.12. The mass of the neutron star is not significantly al-

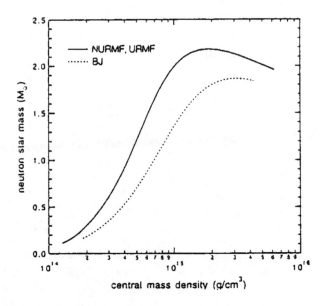

Fig. 11.12. The gravitational mass of a neutron star obtained with the NURMF and URMF EOS is shown by the *solid curve* as a function of the central mass density. There is no obvious difference between the results obtained using the NURMF and URMF EOSs. The mass of a neutron star calculated with the BJ EOS is shown by the *short-dashed curve* for comparison [491]

tered by using the NURMF EOS instead of the URMF EOS. This is because the neutron star mass is determined predominantly by the behavior of the EOS at high density. On the other hand, the properties of the EOS at low density are important in the description of the profile of a neutron star in the surface region, where the matter is inhomogeneous. In order to illustrate the proflie of a neutron star, the baryon mass density distribution of a neutron star with a gravitational mass of $1.4M_\odot$ is shown in Fig. 11.13. The boundaries between uniform matter and nonuniform matter, with and without free dripping neutrons, are displayed by short–dashed and dot–dashed lines. It is difficult to arrive at the low-density region in a calculation of the profile of a neutron star by using the URMF EOS and it is also not appropriate to describe neutron star matter by the uniform RMF approach when $D_B \leq D_0/3$, where heavy nuclei are formed. In Fig. 11.14, the surface region of Fig. 11.13 is expanded in order to make visible the difference in the neutron star profiles between the NURMF and URMF results in the large-radius region. It is interesting that the radius of the neutron star increases because the matter is composed of nuclei in the outermost part of the neutron star. For a neutron star with a gravitational mass of $1.4M_\odot$, the central baryon density is around

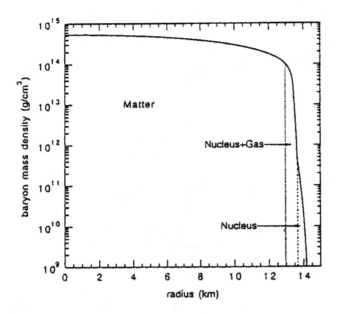

Fig. 11.13. The baryon mass density profiles of a neutron star with a gravitational mass of $1.4 M_\odot$ obtained using the NURMF and URMF EOSs are shown by the *solid curve* and *dashed curve*, respectively. The boundaries between uniform matter and nonuniform matter, with and without free dripping neutrons, are shown by *short-dashed* and *dot-dashed lines*. There is no obvious difference between the results obtained using the NURMF and URMF EOSs in the central core of the neutron star. A difference exists in the surface region, as can be seen in Fig. 11.14 [491]

$1.8 D_0$. In this case, the proton fraction in the central core of the neutron star is around 0.19, as can be seen from the β-equilibrium curve in Fig. 11.6.

The relativistic EOS of nuclear matter considered here [491] is designed for use in supernova simulations and neutron star calculations over a wide density and temperature range with various proton fractions. In [491], an RMF theory with nonlinear σ and ω terms was adopted, which was demonstrated to be successful in describing the properties of both stable and unstable nuclei as well as reproducing the self-energies obtained in the RBHF theory. Relativity plays an essential role in describing nuclear saturation and nuclear structure [53]; it also brings some distinctive properties into the EOS compared with the case in the nonrelativistic framework. Therefore, it will be very interesting and important to study astrophysical phenomena such as supernova explosions, neutron star cooling, and neutron star merging using this relativistic EOS through extensive comparison with the previous studies performed with the nonrelativistic EOSs.

The authors of [491] performed a Thomas–Fermi calculation to describe the nonuniform matter which appears below $D_B \sim D_0/3$. The same RMF

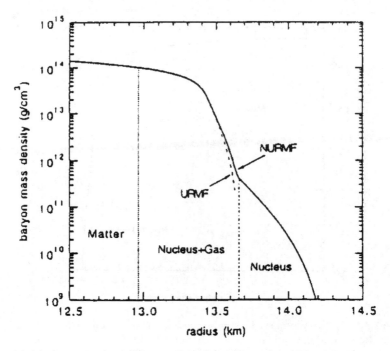

Fig. 11.14. An expansion of Fig. 11.13 in the surface region. The difference between the neutron star profiles obtained with the NURMF and URMF EOSs can be seen in the large-radius region [491]

theory as that used at high density was adopted as the input of the Thomas–Fermi numerical code. Hence, all the thermodynamic quantities, such as the free energy, entropy, and pressure, are found to be smooth in the resulting EOS. A nonuniform matter phase occurs in a wide D_B–Y_p region in the phase diagram at low temperature, while it completely disappears when temperature increases beyond 15 MeV. The formation of heavy nuclei in nonuniform matter leads to remarkable behavior in the EOS.

It is also important to pay attention to the experimental data on nuclei which are being obtained using radioactive nuclear beams, so that we can obtain more information about the properties of asymmetric nuclear matter from experiment [508]. It is definitely important and meaningful to study unstable nuclei and astrophysical phenomena simultaneously in the same framework.

Note that relativistic aspects of nuclear astrophysics have also been investigated in [346, 509] (see also [85] and references therein).

Applications of relativistic many-body physics to the final phases of stellar evolution, supernova collapse, and neutron star cooling, where neutrino

11.3 Neutron Star Matter and Neutron Star Profiles

transport properties may play a crucial role, are considered in [510]. Neutral-current interactions are responsible for the scattering of neutrinos in this case. The authors of [510] analyze the screening of scattering cross sections in a system composed of electrons, protons, and neutrons at $T = 0$, the baryonic ground state being treated in the relativistic Hartree approximation (RHA) with the σ–ω model. The cross section is written down in terms of the imaginary part of the polarization propagators. The effect of correlations is considered in the relativistic RPA framework. A neutrino probe produces excitations of particle–hole and particle–antiparticle pairs. In a dense system, these excitations propagate via the interactions of the constituents, modifying the response to the probe. In the RRPA this effect is taken into account by replacing the Hartree polarizations by RPA polarizations, obtained as the solution of a Dyson equation. Extension of this relativistic many-body calculation to hot matter has also been performed in [510], using real-time finite-temperature field theory.

12 Alternative Relativistic Models

12.1 Quark–Meson Coupling Models

So far, we have been considering current theories of nucleon–nucleon forces and relativistic models of nuclear structure (nuclear matter and finite nuclei) based on the philosophy, common to both cases, of using the concept of meson exchange (PVS models) as a tool for generating the NN interaction directly. In this chapter we consider alternative relativistic approaches. An important question in nuclear physics is whether quarks play any essential role in nuclear structure. The discovery of the EMC effect [166, 167] (the structure function of the nucleon is changed by the nuclear medium) is an indication of the fact that the quark degrees of freedom are necessary to understand deep inelastic scattering at momentum transfers of several GeV. For this reason, it is necessary to develop a theory of nuclear structure incorporating quark–gluon degrees of freedom, but this will be a challenging program. This program has been started in [511–513], the approach developed in these papers being referred to as the quark–meson coupling (QMC) model. It is a relativistic mean-field model, initially proposed for nuclear matter, the nucleons being described by the nonoverlapping MIT bag model, and the interaction between nucleons being produced by coupling of meson fields to the quarks. In this model a mean-field Dirac equation, together with the MIT bag boundary conditions [514], are used to describe quarks; the nucleons are also assumed to be described by a Dirac equation in the effective fields that arise from the coupling of meson fields to the quarks in the nucleons. Similar approaches have been considered by various other groups [515].

In [511–513] it was shown that within the QMC model, one can obtain a reasonable description of the bulk properties of nuclear matter. In the large-quark-mass limit, this model reduces to the Walecka approach. However, the quark degrees of freedom, which are related to the internal structure of the nucleons, generate mechanisms which are not involved in the Walecka model. In particular, at the Hartree level, within the QMC model, one obtains a lower value of the incompressibility of nuclear matter (if the parameters of the model are chosen to reproduce the binding energy and equilibrium density of saturated nuclear matter) than in models with point-like nucleons (the Walecka model, for example) at the same level of sophistication.

Below we discuss briefly the QMC model for nuclear matter at zero temperature [516] (in [517] this model was extended to the description of hot nuclear matter). In this case the nucleons are assumed to be described by a static MIT bag in which quarks interact with the scalar and vector fields φ and ω, which are treated as classical in a mean-field approximation.

The quark field $\psi_q(\boldsymbol{r},t)$ inside the bag then satisfies the equation

$$\left[i\gamma^\mu \partial_\mu - (m_q - g_\sigma^q \varphi) - g_\omega^q \omega \gamma^0\right]\psi_q(\boldsymbol{r},t) = 0, \qquad (12.1)$$

where m_q is the current quark mass and g_σ^q and g_ω^q are the quark couplings with the σ and ω mesons.

The normalized ground state for a quark (in an s state) in the bag is given by

$$\psi_q(\boldsymbol{r},t) = N \exp\left(-i\frac{\epsilon_q t}{R}\right)\begin{pmatrix} j_0(xr/R) \\ i\beta_q \boldsymbol{\sigma}\hat{r} j_1(xr/R) \end{pmatrix}\frac{\chi_q}{\sqrt{4\pi}}. \qquad (12.2)$$

The single-particle quark energy in units of R^{-1} is then

$$\epsilon_q = \Omega_q + g_\omega^q \omega R, \quad \text{and} \quad \beta_q = \sqrt{\frac{\Omega_q - Rm_q^*}{\Omega_q + Rm_q^*}}, \qquad (12.3)$$

where $\Omega_q = (x^2 + R^2 m_q^{*2})^{1/2}$, $m_q^* = m_q - g_\sigma^q \varphi$ is the effective quark mass, R is the bag radius, χ_q is the quark spinor, and N is the normalization factor.

The boundary condition at the bag surface is given by

$$i\gamma n \psi_q = \psi_q. \qquad (12.4)$$

For the ground state, this reduces to

$$j_0(x) = \beta_q j_1(x), \qquad (12.5)$$

which determines the dimensionless quark momentum x. The energy of the nucleon bag is

$$E_\text{bag} = 3\frac{\Omega_q}{R} - \frac{Z}{R} + \frac{4}{3}\pi R^3 B, \qquad (12.6)$$

where Z is a parameter accounting for the zero-point energy and B is the bag constant. After subtracting the spurious center-of-mass motion inside the bag, the effective mass of the nucleon bag at rest is given by [516]

$$M^* = \sqrt{E_\text{bag}^2 - \langle p_\text{c.m.}^2 \rangle}, \qquad (12.7)$$

where $\langle p_\text{c.m.}^2 \rangle = \sum_q \langle p_q^2 \rangle \equiv 3(x/R)^2$. The bag radius R is then obtained from

$$\left.\frac{\partial M^*}{\partial R}\right| = 0. \qquad (12.8)$$

12.1 Quark–Meson Coupling Models

Following the Born–Oppenheimer approximation, the bag is assumed to respond instantaneously to changes in the nuclear environment. Hence the equilibrium condition (12.8) also applies in the medium.

The total energy density of nuclear matter at a baryon density ρ_V is given, in the usual form, by

$$\mathcal{E} = \frac{\gamma}{(2\pi)^3} \int_0^{p_F} d^3k \sqrt{k^2 + M^{*2}} + \frac{g_\omega^2 \rho_V^2}{2m_\omega^2} + \frac{1}{2} m_\sigma^2 \varphi^2 , \qquad (12.9)$$

where $\gamma = 4$ is the spin–isospin degeneracy factor for nuclear matter. The vector mean field ω is determined from

$$\omega = \frac{g_\omega \rho_V}{m_\omega^2} , \qquad (12.10)$$

where $g_\omega = 3 g_\omega^q$. Finally, the scalar mean field is fixed by

$$\frac{\partial \mathcal{E}}{\partial \varphi} = 0 . \qquad (12.11)$$

The scalar and vector couplings g_σ^q and g_ω^q are fitted to the saturation density and binding energy for nuclear matter. For a given baryon density, x, R, and φ are calculated from (12.5), (12.8), and (12.11), respectively.

In the original version of the QMC model [511–513], the bag constant B is taken as B_0, corresponding to the bag parameter for a free nucleon. A possible medium dependence of B and Z is considered in [516]. According to the scaling ansatz advocated in [430], the bag constant should scale like $B/B_0 \simeq \Phi^4$ [431, 518]. Here Φ denotes the universal scaling, and $\Phi \sim m_\rho^*/m_\rho \simeq f_\pi^*/f_\pi \simeq (M^*/M)^{2/3}$ has been suggested in [518]. The scaling ansatz

$$\frac{B}{B_0} = \left(\frac{M^*}{M} \right)^{k/3} \qquad (12.12)$$

has been utilized in [516] for the in-medium bag constant, the value $k = 0$ corresponding to the initial version of this model. The parameter Z was taken to be independent of the density: $Z = Z_0$. For m_q, $m_q = 0$ was chosen in [516]. The coupling constants g_σ and g_ω were chosen to fit the binding energy and the saturation density of nuclear matter.

It is worth noting that this model gives a semiquantitative description of the Okamoto–Nolen–Schiffer anomaly, if the quark mass differences are included [519]. In [520] the QMC model has been utilized in a problem where the quark degrees of freedom are apparently involved, namely the nuclear EMC effect, and was shown to quantitatively reproduce the existing experimental data. In [521–523], the model has been extended and applied to describe the bulk properties of finite nuclei, the ρ and γ quark couplings being added in this case to obtain a realistic treatment of nuclear structure. As shown in

Table 12.1. Coupling constants and calculated properties for symmetric nuclear matter at normal nuclear density. The effective nucleon mass M^*, and the nuclear incompressibility K are in MeV. The bottom row is for QHD. (From [523])

m_q (MeV)	R (fm)	$g_\sigma^2/4\pi$	$g_\omega^2/4\pi$	M^*	K
0	0.6	5.84	6.29	730	293
	0.8	5.38	5.26	756	278
	1.0	5.04	4.50	774	266
5	0.6	5.86	6.34	729	295
	0.8	5.40	5.31	754	280
	1.0	5.07	4.56	773	267
10	0.6	5.87	6.37	728	295
	0.8	5.42	5.36	753	281
	1.0	5.09	4.62	772	269
QHD		7.29	10.8	522	540

[521], the main result in the QMC model is that, in the scalar (φ) and vector (ω) fields, a nucleon behaves essentially as a point-like particle with an effective mass M^* which depends on the position through only the φ field, moving in a vector potential generated by the ω meson. Because of their vector character, the vector interactions have no effect on the nucleon structure; they give only a shift in the nucleon energy [523].

The effective nucleon mass M^* in this approach is given by a model describing the nucleon structure (and depends on the position via only the φ field, as mentioned above). The MIT bag model is utilized in [523]. A relativistic oscillator model (ROM) is used in [522] as an alternative description, with a (scalar) confining potential $V_{\mathrm{con}} = Cr^2$, where $C = 830\,\mathrm{MeV/fm}^2$; further details related to this model can be found in [524]. The nuclear-structure results obtained in the ROM are quite similar to those of the bag model [522].

So the ROM suggests an alternative to the bag models, and has one fewer free parameter than have bag models with a medium-dependent bag constant B (the oscillator constant C can also be considered as medium-dependent). Notice also that the center-of-mass effects can be handled more easily in oscillator models than in bag models.

The coupling constants and some calculated properties of nuclear matter (for $m_\sigma = 550\,\mathrm{MeV}$ and $m_\omega = 783\,\mathrm{MeV}$) at the saturation density are listed in Table 12.1 for the QMC model based on the MIT bag model [523].

In the case of self-consistent calculations for finite nuclei, in the QMC model one has to solve a coupled system of nonlinear differential equations of a Walecka-type model with a coupling constant g_σ depending on the value of the φ field. In [523] this dependence is given by

$$g_\sigma(\varphi(r)) = g_\sigma \left[1 - \frac{a}{2} g_\sigma \varphi(r)\right], \qquad (12.13)$$

12.1 Quark–Meson Coupling Models 243

Table 12.2. Model parameters for finite nuclei [523]

m_q (MeV)	R_B (fm)	$g_\sigma^2/4\pi$	$g_\omega^2/4\pi$	$G_\rho^2/4\pi$	m_σ (MeV)
0	0.6	3.55	6.29	6.79	429
	0.8	2.94	5.26	6.93	407
	1.0	2.51	4.50	7.03	388
5	0.6	3.68	6.34	6.78	436
	0.8	3.12	5.31	6.93	418
	1.0	2.69	4.56	7.02	401
10	0.6	3.81	6.37	6.78	443
	0.8	3.28	5.36	6.92	428
	1.0	2.91	4.62	7.02	416

where $a = (6.6, 8.8, 11) \times 10^{-4}$ for values of the free bag radius equal to (0.6, 0.8, 1.0) fm, respectively.

The physical origin of this field (density) dependence, which produces a new saturation mechanism for nuclear matter, is the relatively rapid increase of the lower component of the Dirac wave function of a confined, light quark.

The self-consistent calculations carried out in [523] contain six fitting parameters, m_q, m_σ, R, g_σ, g_ω, $G_\rho/2 = g_\rho$ (g_ρ being the ρN coupling constant utilized in the previous chapters), while the values of m_ω and m_ρ were chosen to be equal to 783 MeV and 770 MeV, respectively. Three possible values of R (0.6, 0.8, 1.0 fm) were utilized in these calculations for each value of m_q considered (0, 5, 10 MeV). The value of G_ρ was chosen so as to reproduce properly the symmetry energy of nuclear matter; however, the respective value of g_ρ appears to be too large compared with that utilized in the NN scattering problem. The ω–nucleon and σ–nucleon coupling constants used in the QMC models appear to be smaller than in the QHD models, this point causing problems in the description of the observed values of the spin–orbit splittings in finite nuclei [516]. Let us emphasize that obtaining the correct, parameter-free value of the spin–orbit force (see Chap. 8) would provide strong support in favor of any relativistic framework.

The values of the parameters utilized in [523] for calculating finite nuclei are summarized in Table 12.2, and the results are given in Table 12.3.

It should be mentioned that there are some discrepancies in the energy spectra of nuclei, in particular, in the spin–orbit splittings obtained in [523].

In [516, 522] an alternative approach has been considered which allows variations of the bag constant B and of the parameter Z in nuclear matter, which have been suggested by the fact that quarks are presumably deconfined at high enough densities. The authors of [516, 522] suggested that B might decrease with increasing density. Using this idea, the properties of nuclear matter and finite nuclei were studied in [522]. In that paper, B in the medium,

Table 12.3. Binding energy per nucleon E/A (in MeV), r.m.s charge radius r_c (in fm), and difference between r_n and r_p (in fm); $m_q = 5\,\text{MeV}$ and $R_B = 0.8\,\text{fm}$ [523]

	$-E/A$			r_c			$r_n - r_p$		
	QMC	QHD	Exp.	QMC	QHD	Exp.	QMC	QHD	Exp.
^{16}O	5.84	4.89	7.98	2.79	2.75	2.73	−0.03	−0.03	0.0
^{40}Ca	7.36	6.31	8.45	3.48[a]	3.48[a]	3.48	−0.05	−0.06	0.05
^{48}Ca	7.26	6.72	8.57	3.52	3.47	3.47	0.23	0.21	0.2
^{90}Zr	7.79	7.02	8.66	4.27	4.26	4.27	0.11	0.10	0.05
^{208}Pb	7.25	6.57	7.86	5.49	5.46	5.50	0.26	0.27	0.16

[a] Fit

B^*, is given by

$$B^* = B\left[1 - \alpha_B \frac{S(r)}{M}\right], \tag{12.14}$$

where $S(r)$ is an average scalar potential and α_B is an arbitrary parameter. For finite nuclei, the results in this case move towards those discussed in previous chapters, and there is a certain improvement in the spin–orbit splittings.

The essential point of the QMC model is that in this case an attempt is made to include quark degrees of freedom in the context of nuclear structure; this attempt appears to be successful in reproducing many properties of nuclear matter and finite nuclei.

We should mention that in [525] the QMC model was applied to a study of Λ hypernuclei while in [526] model predictions were made concerning the variation of hadron masses in dense matter, and the QMC model was generalized to allow the self-consistent determination of the vector meson masses as well as the nucleon mass (bearing in mind that the ω and ρ are simple $q\bar{q}$ states in QCD). At low density, the behavior of these masses is well approximated by a term linear in the density:

$$\frac{M^*}{M} \simeq 1 - 0.21\frac{\rho_B}{\rho_0} \tag{12.15}$$

and

$$\frac{m_V^*}{m_V} \simeq 1 - 0.17\frac{\rho_B}{\rho_0}. \tag{12.16}$$

Using the QMC model, the variation of the masses of the Λ, Σ, and Ξ has been also calculated. From these calculations, a new, simple scaling relation between the hadron masses has been obtained:

$$\frac{\delta m_V^*}{\delta M^*} \simeq \frac{\delta M_\Lambda^*}{\delta M^*} \simeq \frac{\delta M_\Sigma^*}{\delta M^*} \simeq \frac{2}{3} \quad \text{and} \quad \frac{\delta M_\Xi^*}{\delta M^*} \simeq \frac{1}{3}, \tag{12.17}$$

where $\delta M_j^* \equiv M_j - M_j^*$. Here M is the bare nucleon mass, while M^* is its effective mass. The factors, 2/3 and 1/3 in (12.17) come from the ratio of the number of nonstrange quarks in j to that in the nucleon. This means that the mass of a hadron is primarily determined by the number of nonstrange quarks. These experience the common scalar field generated by the surrounding nucleons in the medium, and the strength of the scalar field. In [526] a shift of the mass of the ρ meson, predicted by the QMC model at the appropriate densities, has also been obtained, at ρ_0 the authors of [526] find $m_\rho - m_\rho^* \approx 140\,\text{MeV}$, while at $2\rho_0$, $m_\rho^* \sim 580\,\text{MeV}$.

The authors of the QMC model consider that the model should be improved in future along the following lines [522, 523]:

1. Aspects of chiral symmetry should be taken into consideration, for example by using the cloudy-bag model of the nucleon, in which the three-quark bag is surrounded by a cloud of pions (see [527], where the authors of those papers performed nuclear-matter studies with density-dependent meson–nucleon coupling constants using a chiral confining model).
2. Effects of correlations between the nucleons should be taken into account, i.e. the model should be extended beyond the mean-field approximation.
3. Modifications of the meson masses in the medium should be incorporated into the theory.

The idea of the QMC model is very attractive. The model relies strongly on the choice of the nucleon model, the MIT bag model being used in the initial version of the QMC model. Another popular model for the nucleon is the constituent quark model [22]. In this model, quarks acquire masses owing to the spontaneous chiral symmetry breaking. It is then very natural to have nearly zero-mass pions, and their coupling to the constituent quarks. This consideration allows a simple interpretation of the direct coupling of other mesons such as σ and ω mesons. The constituent quark model has been used extensively also for nucleon–nucleon interaction and was later extended to baryon–baryon interactions with great success [528–531]. Consequently, it is very interesting to construct a quark–meson coupling model where the nucleon is described in terms of the constituent quarks and their coupling with mesons and gluons. Such a model (referred to as the quark mean-field (QMF) model) has been developed in [532, 533]. It has a direct connection to the OBE formalism, and so it is reasonable to expect quantitative results similar to those of the RBHF theory (see Chap. 9) to be obtained in the framework of the QMF model.

The authors of [532] obtained very good nuclear-matter properties with the use of nonlinear self-energy terms in the meson Lagrangian, and the spin–orbit splitting in finite nuclei is expected to be large owing to the large reduction of the nucleon mass in the QMF model. The nucleon size increases by about 7% at the normal density of nuclear matter.

The relationship between QMC and quantum hydrodynamics is discussed in detail in [534].

Finally in this section, we mention one more possible way to construct effective nuclear models, by taking into account the symmetries of QCD, which determine largely how the hadrons should interact with each other. This possibility has already been discussed in Chap. 7 in the framework of $SU(2) \times SU(2) \times U_{em}(1) \times U_\omega(1)$ symmetry. Another promising approach to this problem is to proceed to a wider symmetry group, such as $SU(3)$, for example. This way has been chosen in [535], where the general scheme of such an approach is suggested, the authors of [535] apply this formalism to the description of nuclear matter and finite nuclei.

12.2 The Relativistic Point-Coupling Model

The relativistic point-coupling model (RPCM) was introduced in [536, 537]; later, it was extended and applied to describe ground-state properties of finite nuclei and nuclear matter in [66, 538, 539]. Essential contributions to this approach have been made in [540, 541]; in these papers the role of QCD scales and chiral symmetry in finite nuclei is examined, the Dirac–Hartree mean-field coupling constants of [66, 538] are scaled in accordance with the QCD-based prescription [542], and the results [540, 541] provide good evidence that QCD and chiral symmetry apply to finite nuclei. In the present section we discuss briefly the basic ideas of the RPCM. The choice of the RPCM Lagrangian was suggested by empirically based improvements[1] to a Walecka-type (σ–ω) model, but using contact (zero-range) interactions to make possible a simpler consideration of the exchange (Fock) terms. The model is determined by a relativistic Lagrangian density in the fermion fields and does not contain any meson fields. The Lagrangian is given by a series of zero-range interactions [66, 538, 539]:

$$\mathcal{L} = \; :\{\mathcal{L}_{\text{free}} + \mathcal{L}_{4\text{f}} + \mathcal{L}_{\text{hot}} + \mathcal{L}_{\text{der}} + \mathcal{L}_{\text{em}}\}: \, , \qquad (12.18)$$

where

---

[1] In the language of perturbative field theory, the Walecka-type models correspond to tree approximations or to special subclasses of one-loop insertions. However, an effort to take the Walecka Lagrangian exactly as a basis for a regular loop-type expansion of a renormalizable theory led to serious problems even at the level of the one-loop approximation [536]. Calculating meson propagators at the one-loop level for the self-energy generates unphysical ("tachyon") poles and vacuum instabilities [89, 470, 543]. To avoid these difficulties, the authors of [536, 537] proposed a more practical approximation of using a relativistic Hartree–Fock model with apparently nonrenormalizable four-fermion-field couplings.

12.2 The Relativistic Point-Coupling Model

$$\mathcal{L}_{\text{free}} = \bar{\psi}\left(i\gamma_\mu \cdot \partial^\mu - M\right)\psi, \tag{12.19}$$

$$\mathcal{L}_{4f} = -\frac{1}{2}\alpha_S(\bar{\psi}\psi)(\bar{\psi}\psi) - \frac{1}{2}\alpha_V(\bar{\psi}\gamma_\mu\psi)(\bar{\psi}\gamma^\mu\psi)$$
$$- \frac{1}{2}\alpha_{TS}(\bar{\psi}\boldsymbol{\tau}\psi)(\bar{\psi}\boldsymbol{\tau}\psi) - \frac{1}{2}\alpha_{TV}(\bar{\psi}\boldsymbol{\tau}\gamma_\mu\psi)(\bar{\psi}\boldsymbol{\tau}\gamma^\mu\psi), \tag{12.20}$$

$$\mathcal{L}_{\text{hot}} = -\frac{1}{3}\beta_S(\bar{\psi}\psi)^3 - \frac{1}{4}\gamma_S(\bar{\psi}\psi)^4 - \frac{1}{4}\gamma_V\left[(\bar{\psi}\gamma_\mu\psi)(\bar{\psi}\gamma^\mu\psi)\right]^2, \tag{12.21}$$

$$\mathcal{L}_{\text{der}} = -\frac{1}{2}\delta_S(\partial_\nu\bar{\psi}\psi)(\partial^\nu\bar{\psi}\psi) - \frac{1}{2}\delta_V(\partial_\nu\bar{\psi}\gamma_\mu\psi)(\partial^\nu\bar{\psi}\gamma^\mu\psi). \tag{12.22}$$

$\mathcal{L}_{\text{em}}$ describes the interaction of the nucleons with the electromagnetic field; it has a conventional form and for this reason is not written down here. In the above equations, ψ is the nucleon field, the subscripts S and V are assotiated with the scalar and vector nucleon fields, respectively, and the subscript T is associated with the isovector fields. In (12.18) $\mathcal{L}_{\text{free}}$ is the kinetic term of the nucleons, and $\mathcal{L}_{4f}$ contains four two-nucleon-force terms (the subscript 4f means that fermion operators enter each term of the corresponding Lagrangian density). $\mathcal{L}_{\text{hot}}$ and $\mathcal{L}_{\text{der}}$ contain higher-order terms and derivatives, respectively, in the nucleon densities. The derivatives act on densities rather than on one of the fields. The first term of $\mathcal{L}_{\text{hot}}$ is a three-nucleon-force term, whereas the remaining two terms are four-nucleon-force terms. $\mathcal{L}_{\text{der}}$ contains two nonlocal two-nucleon-force terms. The colons in (12.18) denote a normal ordering with respect to the vacuum state $|0\rangle$. That is, in all expressions between the colons, creation operators are written on the left and annihilation operators are written on the right (including a minus sign for each transposition of fermion operators).

Note that the Lagrangian (12.19)–(12.22) is treated in a no-sea approximation (only positive-energy states are considered).

Minimizing the expectation value of the Hamiltonian corresponding to (12.19)–(12.22) in the space of Slater determinants, one obtains [538, 539] the following equation for the fermion field operator $\psi(x)$:

$$\{i\gamma_\mu\partial^\mu - M - U_{\text{SC}}\}\psi(x) = 0, \tag{12.23}$$

where U_{SC} is the static self-consistent potential (the Coulomb potential is not included, for brevity of notation), given by

$$U_{\text{SC}} = V_S + V_V\gamma^0 + V_{TS}\tau_3 + V_{TV}\tau_3\gamma^0. \tag{12.24}$$

Here, the isoscalar–scalar and isoscalar–vector potentials are

$$V_S = \alpha_S\rho_S + \beta_S\rho_S^2 + \gamma_S\rho_S^3 + \delta_S\Delta\rho_S, \tag{12.25}$$

$$V_V = \alpha_V\rho_V + \gamma_V\rho_V^3 + \delta_V\Delta\rho_V, \tag{12.26}$$

and the isovector–scalar and isovector–vector potentials are

$$V_{\text{TS}} = \alpha_{\text{TS}} \rho_{\text{TS}} , \qquad (12.27)$$

$$V_{\text{TV}} = \alpha_{\text{TV}} \rho_{\text{TV}}, \qquad (12.28)$$

where

$$\rho_{\text{S}} = \langle \bar{\psi} \psi \rangle , \qquad \rho_{\text{V}} = \langle \bar{\psi} \gamma^0 \psi \rangle , \qquad (12.29)$$

$$\rho_{\text{TS}} = \langle \bar{\psi} \tau_3 \psi \rangle , \qquad \rho_3 \equiv \rho_{\text{TV}} = \langle \bar{\psi} \gamma^0 \tau_3 \psi \rangle . \qquad (12.30)$$

We should mention that U_{SC} depends on the second derivatives $\Delta \rho_{\text{S}}$, $\Delta \rho_{\text{V}}$ of the densities. By use of the Lagrangians (12.18)–(12.22), an effective model can be determined,[2] and mesonic degrees of freedom are not involved. Instead one has an expansion of the nucleon scalar and vector potentials V_{S} and V_{V} in powers and derivatives of the nucleon scalar and vector densities ρ_{S} and ρ_{V}. As is discussed below, the RPCM effective Lagrangian appears to be consistent with chiral symmetry.

The Hartree equations given above are nonlinear equations because the potentials involved depend on the nucleon wave functions. They can be solved by an iteration procedure: starting with some fixed potentials, one solves (12.23). Using the wave functions obtained, one obtains new potentials. The iterations are performed repeatedly until convergence is established [66, 538]. Note that from the computational point of view, the RPCM is a very convenient model for practical use; the calculations are much easier to carry out in this case than in meson–nucleon coupling models.

The calculation procedure for the RPCM involves "damping" to stabilize the convergence, i.e. the average field utilized in the next iteration step is not the field obtained from the eigenfunctions, but the average (maybe with different empirically chosen weights) of the latter field and the average field utilized in the previous iteration. This method of stabilizing the self-consistent procedure is used in meson–nucleon coupling models also, as well as in many other self-consistent calculations.

To obtain the optimal set of coupling constants, a Dirac–Hartree equation solver was used in [538] as a function call in a generalized nonlinear least-squares minimization program based on an adjustment algorithm of Levenberg–Marquardt type (which acts on several nuclei simultaneously) with respect to well-measured nuclear ground-state observables. The experimental ground-state observables appearing in χ^2 were:

(a) the ground-state masses (binding energies),
(b) the r.m.s. charge radii,
(c) the spin–orbit energy splitting of the least-bound neutron and proton spin–orbit pairs,

for each of the chosen nuclei.

---

[2] The model considered here is very similar to the nonrelativistic Hartree–Fock approximation for nuclei with Skyrme forces. Four-fermion coupling may be considered as a zero-range approximation to a finite-range effective interaction; the derivative terms appear (in the higher-order correction terms) to simulate the finite ranges of the (mesonic) interactions.

12.2 The Relativistic Point-Coupling Model

Table 12.4. Optimized coupling constants for the NHM Lagrangian, and corresponding dimensional power-counting coefficients and chiral expansion order [66, 540]

Constant	Magnitude	Dimension	c_{lmn}	Order
α_S	-4.508×10^{-4}	MeV^{-2}	-1.93	Λ^0
α_{TS}	7.403×10^{-7}	MeV^{-2}	0.013	Λ^0
α_V	3.427×10^{-4}	MeV^{-2}	1.47	Λ^0
α_{TV}	3.257×10^{-5}	MeV^{-2}	0.56	Λ^0
β_S	1.110×10^{-11}	MeV^{-5}	0.27	Λ^{-1}
γ_S	5.735×10^{-17}	MeV^{-8}	8.98	Λ^{-2}
γ_V	-4.389×10^{-17}	MeV^{-8}	-6.87	Λ^{-2}
δ_S	-4.239×10^{-10}	MeV^{-4}	-1.81	Λ^{-2}
δ_V	-1.144×10^{-10}	MeV^{-4}	-0.49	Λ^{-2}

The model contains nine coupling constants, which were determined self-consistently by reproducing the experimental values of the ground-state properties for three closed-shell spherical nuclei. The three nuclei chosen were doubly magic ^{16}O and ^{208}Pb and singly magic ^{88}Sr (which has a closed proton subshell). The optimal set of nine coupling constants of the RPCM obtained by Nikolaus, Hoch, and Madland (NHM) [66, 540] is given in Table 12.4. The authors of these papers believe that the nine constants can possibly be reduced to eight, by setting α_{TS} equal to zero (since α_{TS} is only $\sim 2\%$ of the magnitude of α_{TV}). The constants α_{TS} and α_{TV} correspond to the meson fields of δ and ρ mesons, respectively; assuming $\alpha_{TS} = 0$ corresponds to neglecting δ meson exchange.

The coupling constants obtained in this way were then used to calculate [66] observables of the same type as those listed above for a large number of other closed-shell nuclei and a number of properties of saturated nuclear matter. The calculations for nuclear matter were carried out without any parameter fitting. The calculated properties of nuclear matter are given in Table 12.5 (for nuclear matter, there are no Coulomb forces and $\delta_S = \delta_V = 0.0$).

Note that a reasonable value of the incompressibility K of nuclear matter is obtained in [66] owing to the presence of the component $\mathcal{L}_{\text{hot}}$ in the Lagrangian density.

In the simplest version of the RPCM corresponding to the σ–ω model, only the isoscalar–scalar and isoscalar–vector coupling constants α_S and α_V are different from zero. In the Hartree approximation, one obtains for infinite nuclear matter in the RPCM the same results as in the σ–ω mean-field theory if the parameter values are taken as

Table 12.5. Properties of saturated nuclear matter [66]

Quantity	Magnitude
Fermi wave number p_F	1.299 fm^{-1}
Density ρ_{NM}	0.148 fm^{-3}
Binding energy per nucleon	16.126 MeV
Incompressibility K	264.032 MeV
Symmetry energy, a_4	37.194 MeV
Effective nucleon mass M^*/M	0.575

Table 12.6. Average absolute deviations of the predicted observables from the measured observables [539]

Observable	Average absolute deviation
E_B	2.783 MeV
$\langle r^2 \rangle_{\text{charge}}^{1/2}$	0.021 fm
$\Delta E_{\text{s.o.}}^{(n)}$	1.409 MeV
$\Delta E_{\text{s.o.}}^{(p)}$	0.844 MeV

$$\alpha_V = \left(\frac{g_\omega}{m_\omega}\right)^2 \quad \text{and} \quad \alpha_S = -\left(\frac{g_\sigma}{m_\sigma}\right)^2 ; \qquad (12.31)$$

α_{TV} can be related easily to $(g_\rho/m_\rho)^2$ for the ρ meson: $\alpha_{TV} = (g_\rho/m_\rho)^2$.

For finite nuclei, the RPCM predicts the ground-state binding energies and r.m.s. charge radii very well; it also obtains the correct values of the spin–orbit splittings. The main point of the RPCM is that it involves strong potentials V_S and V_V. The central magnitude of $V_S - V_V$ is approximately equal to -725 MeV ($V_S \approx -400$ MeV, $V_V \approx +325$ MeV), and the magnitude of the gradient of this quantity, $|\nabla(V_S - V_V)|$, appears to be essential for explaining the experimentally observed values of the spin–orbit splittings [66]. The results of studying the predictive ability of the RPCM model for 32 nuclei are summarized in Table 12.6 [539].

Note also that in [66] the RPCM was applied also to nuclei far outside the valley of β-stability. The exotic doubly magic nuclei ^{28}O, ^{70}Ca, ^{48}Ni, ^{100}Sn, ^{132}Sn, and ^{164}Pb appear to be bound and to have reasonable ground-state properties.

Along the lines of [536, 537, 539], in Appendix A12 it is shown that one can easily develop a Hartree–Fock procedure based on the generalized form of the four-fermion coupling Lagrangian.

The effects of light degrees of freedom (pions) cannot be included in the RPCM by a redefinition of the four-fermion couplings. However, in [539] extensions of the model to include chiral symmetry and chiral-symmetry breaking were considered. In this way, the light degrees of freedom were recovered using a nonlinear realization of the chiral symmetry [311].

It is difficult to find an appropriate description of nuclear properties in terms of the *natural* degrees of freedom (quarks and gluons) on the basis of QCD. The problem is that at low energy, the *physical* degrees of freedom of nuclei are nucleons and (intranuclear) pions [540]. For this reason, effective-field theories are used in practical calculations at low energies. One expects to pay a high price for this replacement of *natural* degrees of freedom by *effective* ones: one needs to consider an infinite number of interaction terms with unknown coefficients (to be determined from experiment rather than computed from the underlying theory). In this case, an appropriate expansion parameter must be chosen and a "naturalness" assumption adopted [541, 542, 544, 545] (all the couplings of the effective theory presented in dimensionless form should be of order 1). However, QCD symmetries may be used to derive crucial constraints on the effective Lagrangian, providing a dynamical framework for nuclear calculations. These symmetries may impose restrictions on the form of the effective interactions. The symmetries may also be helpful in making the choice of the proper degrees of freedom and for establishing relations between the parameters used in the effective theory.

In accordance with [542], one can scale a generic Lagrangian component as

$$\mathcal{L} \sim -c_{lmn} \left[\frac{\bar{\psi}\Gamma\psi}{f_\pi^2 \Lambda} \right]^l \left[\frac{\pi}{f_\pi} \right]^m \left[\frac{\partial^\mu, m_\pi}{\Lambda} \right]^n f_\pi^2 \Lambda^2 , \qquad (12.32)$$

where ψ and π are the nucleon and pionic fields, respectively; Γ is any Dirac matrix; f_π and m_π are the weak pion decay constant and the pion mass, respectively; Λ is the generic QCD large-mass scale ~ 1 GeV, which also determines the transition region between two sets of degrees of freedom, namely the natural (quark–gluon) and physical (pion–nucleon) degrees; (∂^μ, m_π) stands, for either a derivative or the pion mass (the possibility of chiral-symmetry-breaking terms containing the small parameter m_π/Λ is allowed for). Isospin operators are ignored in (12.32).

One of the basic principles of low-energy QCD is chiral symmetry. This priinciple requires that [546]

$$\Delta = l + n - 2 \geq 0 . \qquad (12.33)$$

This means that the series used includes only positive powers of $1/\Lambda$. If the theory considered is *natural* [542, 544, 545], the dimensionless parameters c_{lmn} should be of order unity. The coefficients c_{lmn} are the origin and the basis of our knowledge about scales. If the coefficients are *natural*, scaling operates.

In [540] a comparison is made of the generic chiral Lagrangian (see (12.32), (12.33)) and the Lagrangian of the RPCM given by (12.18)–(12.22).

Note that in (12.18)–(12.22), $\mathcal{L}_{4f}$ corresponds to $\Delta = 0$, the first term of $\mathcal{L}_{hot}$ corresponds to $\Delta = 1$, and the remaining two terms correspond to $\Delta = 2$;

$\mathcal{L}_{\mathrm{der}}$ also corresponds to $\Delta = 2$. As was mentioned earlier, the light degrees of freedom (pions) are not involved[3] in (12.18)–(12.22), and the Lagrangian does not contain all components: it has four operators in order $(1/\Lambda)^0$, one operator in order $(1/\Lambda)^1$ and four operators in order $(1/\Lambda)^2$, presenting an incomplete combination of three different orders in $(1/\Lambda)$.

However, a comparison of two Lagrangians is still meaningful, since for testing the naturalness assumption it is not essential whether a certain c_{lmn} parameter is equal to 0.5 or 2.0 or any other value of order unity. If the model is modified by introducing additional terms, all of the parameters c_{lmn} will be changed. However, the assumption of naturalness can still be applied.

The nine parameters of the RPCM Lagrangian are given in Table 12.4 [66, 540] in dimensional (α_i) and dimensionless (c_{lmn}) forms. The dimensionless values are obtained by making (12.19)–(12.22) equal to (12.32) with $\Lambda = 1\,\mathrm{GeV}$, introducing isospin operators into (12.32), and expressing c_{lmn} in terms of α, β, γ, and δ. As can be seen from this table, in the dimensional form they span 13 orders of magnitude, whereas in the dimensionless form six of the nine parameters can be considered as natural. Only the very small α_{TS} and large γ_{S} and γ_{V} are unnatural (however, the sum of γ_{S} and γ_{V} appears to be natural). The authors of [540] guess that the unnaturally small value of α_{TS}, if it is correct, requires a symmetry to explain this small value.

However, a more recent calculation[4] [370] (utilizing 40 observables to determine nine parameters) carried out within the RPCM demonstrated that actually seven of the nine and the sum of the remaining two parameters were natural (when scaled in accordance with QCD mass scales and taking into account the constraint of chiral symmetry); α_{TS} appeared also to be natural in this calculation.

This fact demonstrates an essential point, that the various components of the Lagrangian feel different nuclear qualities. For example, slight modifications of the larger parameters do not change strongly the values of the small parameters. The components α, β, γ, and δ of the Lagrangian mix only within each type; different types of terms do not mix. As is shown in [539] (see also Appendix A12), the exchange (Fock) mixing within the α components does not eliminate naturalness.

The values of α_{S}, α_{V} and α_{TV} given in Table 12.4 were obtained in the framework of relativistic Hartree calculations within the RPCM. They may also be derived from calculations of the same type that utilize meson–nucleon coupling (see (12.31)). In [81], a compilation is given of the results of 12 calculations of this type. From these, one obtains the average values

---

[3] The way to treat pions is discussed in [539], in an extended version of the RPCM including chiral symmetry.
[4] Note also that in [370], pseudospin symmetry (see Chap. 7) was investigated in the RPCM.

$$\alpha_S = -3.93 \times 10^{-4} , \qquad (12.34)$$

$$\alpha_V = 2.78 \times 10^{-4} , \qquad (12.35)$$

$$\alpha_{TV} = 3.65 \times 10^{-5} . \qquad (12.36)$$

These values agree well with those given in Table 12.4, providing confirmation of naturalness in finite nuclei (see, however, [213]). The examples considered above indicate the role of chiral symmetry and QCD in finite nuclei.

Concluding this section, let us mention also the results obtained in [547] in a refined the RPCM. One of the advantages of RPCM is that it avoids the use of mesonic degrees of freedom, in particular the σ meson. Also the RPCM enables one to easily treat the Fock (exchange) components in terms of the Hartree (direct) potentials via the Fierz rearragement technique (see Appendix A12).

12.3 Scalar Derivative Coupling Models

As was mentioned earlier, the Walecka model in its original form gives a value of the incompressibility of nuclear matter ($K \simeq 550$ MeV) about twice as large as the empirical values ($K \simeq 210$–300 MeV), and it is necessary to introduce nonlinear self-couplings of the meson fields to obtain a reasonable value of K.

With regard to these problems the relativistic (nonlinear) derivative scalar coupling model (RDSCM) suggested by Zimányi and Moszkowski [548] in the framework of the RMF theory has recently also attracted much attention [494, 549–561]. The nonlinearity of the RDSCM is related to the derivative coupling between the scalar and fermion fields (this nonlinearity does not introduce any extra free parameters). In the original form of the RDSCM, the coupling of the scalar meson σ to the nucleons ψ in the Walecka model is replaced by [548, 557]

$$g_{NN\sigma}\bar{\psi}\psi\varphi \to g_{NN\sigma}\bar{\psi}(\slashed{p}/M)\psi\varphi . \qquad (12.37)$$

In [548] this derivative scalar coupling is expressed in terms of an equivalent field-dependent constant by rescaling the nucleon eigenfunction and Lagrangian. The same aim can be achieved by the following procedure:

$$\slashed{p} \to M^* . \qquad (12.38)$$

If we put (12.38) into the right-hand side of (12.38), we can see that the replacement (12.37) is equivalent to considering an effective coupling constant

$$g_{NN\sigma} \to g^*_{NN\sigma} \equiv (M^*/M)g_{NN\sigma} . \qquad (12.39)$$

This replacement is not a unique procedure. For example, the authors of [551] considered another version of a field-dependent coupling which also solves the problem of obtaining the proper value of K. Also, as was mentioned above, the RBHF theory [152, 415, 416] and the relativistic QMC model [511–516] yield density-dependent coupling constants starting from the $\bar{\psi}\psi\varphi$ coupling.

In what follows, we discuss briefly the basic points of the RDSCM [548] and some of its applications; we follow also the comparative analysis of the Walecka model and RDSCMs given in [558]. So we consider in this section the Walecka model with the Lagrangian $\mathcal{L}_W$, given by

$$\mathcal{L}_W = \bar{\psi}i\gamma_\mu\partial^\mu\psi - \bar{\psi}M\psi + \frac{1}{2}(\partial_\mu\varphi\partial^\mu\varphi - m_\sigma^2\varphi^2)$$
$$+ g_\sigma\bar{\psi}\psi\varphi - \frac{1}{4}\omega^{\mu\nu}\omega_{\mu\nu} + \frac{1}{2}m_\omega^2\omega_\mu\omega^\mu - g_\omega\bar{\psi}\gamma_\mu\psi\omega^\mu \,, \quad (12.40)$$

the original version of the RDSCM (to be referred to as the ZM model),

$$\mathcal{L}_{ZM} = -\bar{\psi}M\psi + m^{*-1}\left[\bar{\psi}i\gamma_\mu\partial^\mu\psi - g_\omega\bar{\psi}\gamma_\mu\psi\omega^\mu\right]$$
$$- \frac{1}{4}\omega^{\mu\nu}\omega_{\mu\nu} + \frac{1}{2}m_\omega^2\omega_\mu\omega^\mu + \frac{1}{2}(\partial_\mu\varphi\cdot\partial^\mu\varphi - m_\sigma^2\varphi^2) \,, \quad (12.41)$$

and two modifications of the RDSCM (to be referred to as the ZM2 and ZM3 models),

$$\mathcal{L}_{ZM2} = -\bar{\psi}M\psi + m^{*-1}\left[\bar{\psi}i\gamma_\mu\partial^\mu\psi - g_\omega\bar{\psi}\gamma_\mu\psi\omega^\mu\right.$$
$$\left. - \frac{1}{4}\omega^{\mu\nu}\omega_{\mu\nu} + \frac{1}{2}m_\omega^2\omega_\mu\omega^\mu\right] + \frac{1}{2}(\partial_\mu\varphi\cdot\partial^\mu\varphi - m_\sigma^2\varphi^2) \quad (12.42)$$

and

$$\mathcal{L}_{ZM3} = -\bar{\psi}M\psi + m^{*-1}\bar{\psi}i\gamma_\mu\partial^\mu\psi - g_\omega\bar{\psi}\gamma_\mu\psi\omega^\mu$$
$$- \frac{1}{4}\omega^{\mu\nu}\omega_{\mu\nu} + \frac{1}{2}m_\omega^2\omega_\mu\omega^\mu + \frac{1}{2}(\partial_\mu\varphi\cdot\partial^\mu\varphi - m_\sigma^2\varphi^2) \,. \quad (12.43)$$

Further, we scale the fields ψ and ω_μ in accordance with the following prescriptions [548, 557]: $\psi \to m^{*1/2}\psi$ for all ZM models and $\omega_\mu \to m^*\omega_\mu$ for the ZM2 and ZM3 models. If we consider the case of nuclear matter, we obtain the following rescaled unified Lagrangian densities:

$$\mathcal{L}_R = \bar{\psi}i\gamma_\mu\partial^\mu\psi - \bar{\psi}(M - m^{*\beta}g_\sigma\varphi)\psi$$
$$+ m^{*\alpha}\left[-g_\omega\bar{\psi}\gamma_\mu\psi\omega^\mu - \frac{1}{4}\omega_{\mu\nu}\omega^{\mu\nu} + \frac{1}{2}m_\omega^2\omega_\mu\omega^\mu\right] + \frac{1}{2}(\partial_\mu\varphi\cdot\partial^\mu\varphi - m_\sigma^2\varphi^2) \,,$$
$$(12.44)$$

with the following values for the parameters α and β for different models:

$$\text{W}: \quad \alpha = 0\,, \quad \beta = 0\,, \quad (12.45)$$
$$\text{ZM}: \quad \alpha = 0\,, \quad \beta = 1\,, \quad (12.46)$$
$$\text{ZM2}: \quad \alpha = 1\,, \quad \beta = 1\,, \quad (12.47)$$
$$\text{ZM3}: \quad \alpha = 2\,, \quad \beta = 1\,. \quad (12.48)$$

12.3 Scalar Derivative Coupling Models

From (12.44), we obtain the following system of Euler–Lagrange equations

$$\{i\gamma_\mu \partial^\mu - (M - m^{*\beta}g_\sigma\varphi) - m^{*\alpha}g_\omega\gamma_\mu\omega^\mu\}\psi = 0 , \quad (12.49)$$

$$\partial_\nu \omega^{\nu\mu} + m_\omega^2 \omega^\mu = g_\omega \bar\psi \gamma^\mu \psi , \quad (12.50)$$

$$(\partial_\mu \partial^\mu + m_\sigma^2)\varphi = g_\sigma m^{*\beta}\{1 - \beta(1 - m^*)\}\bar\psi\psi$$
$$- \frac{\alpha}{M} g_\sigma m^{*\alpha+1}\left\{-g_\omega\bar\psi\gamma_\mu\psi\omega^\mu - \frac{1}{4}\omega^{\mu\nu}\omega_{\mu\nu} + \frac{1}{2}m_\omega^2\omega_\mu\omega^\mu\right\} . \quad (12.51)$$

In the mean-field approximation, we have for nuclear matter

$$\omega_0 = \frac{g_\omega}{m_\omega^2}\rho_B , \quad (12.52)$$

$$\varphi = \frac{g_\sigma}{m_\sigma^2}m^{*2\beta}\rho_S + \frac{\alpha}{2}\frac{C_\sigma^2 C_\omega^2}{g_\sigma M^5}m^{*\alpha+1}\rho_B^2 , \quad (12.53)$$

where the values C_i^2 ($i = \sigma, \omega$) are defined in a conventional way,

$$C_i^2 = \frac{g_i^2 M^2}{m_i^2} , \quad (12.54)$$

and use has been made of the equation [558]

$$[1 - \beta(1 - m^*)] = m^{*\beta} . \quad (12.55)$$

From (12.42), it can be seen that in the RDSCM the baryon vector density ($\rho_B = B/V$) is the only source of the time component of the vector field ω^μ. The field φ in the general case (when $\alpha \neq 0$, i.e. for ZM2 and ZM3) has two sources related to ρ_S and ρ_B. Note that the versions of the RDSCM considered here (i.e. all ZM models) require two parameters (being nonlinear) to describe the properties of nuclear matter.

We introduce the effective nucleon mass by the following equation:

$$M^* = M - g_\sigma\varphi\left(1 + \frac{g_\sigma\varphi}{M}\right)^{-\beta} = M - m^{*\beta}g_\sigma\varphi . \quad (12.56)$$

If we define the effective nucleon mass by $M^* = M + S$, then we obtain

$$S = -m^{*\beta}g_\sigma\varphi = -m^{*\beta}S_0 = -\frac{S_0}{(1 + S_0/M)^\beta} , \qquad S_0 = g_\sigma\varphi . \quad (12.57)$$

We should mention that in the original paper [548], the definition of S is slightly different from that given by (12.57). In fact, Zimányi and Moszkowski used the following definition:

$$S = S_0 = g_\sigma\varphi , \qquad m^* = \frac{1}{(1 + S_0/M)} . \quad (12.58)$$

The definitions (12.57) and (12.58) coincide in the first order in S_0. In this approximation, the Walecka and ZM models are identical. If we define the vector field V as a quantity which shifts the single-nucleon energy,

$$V = m^{*\alpha} g_\omega \omega_0 = \left(\frac{1}{1 + g_\sigma \varphi/M}\right)^\alpha g_\omega \omega_0 , \quad (12.59)$$

(12.59) reproduces the results of the Walecka model and the ZM model ($\alpha = 0$). In the models with $\alpha \neq 0$, V appears to be coupled to the scalar field.

Using a conventional procedure, one can obtain the following expressions for the energy density and pressure at a given temperature:

$$\mathcal{E} = \frac{C_\omega^2}{2M^2} m^{*\alpha} \rho_B^2 + \frac{M^4}{2C_\sigma^2} \left(\frac{1 - m^*}{m^{*\beta}}\right)^2$$
$$+ \frac{\gamma}{(2\pi)^3} \int d^3k \, \widetilde{E}(k)(f_k + f_{\bar{k}}) , \quad (12.60)$$

$$P = \frac{C_\omega^2}{2M^2} m^{*\alpha} \rho_B^2 - \frac{M^4}{2C_\sigma^2} \left(\frac{1 - m^*}{m^{*\beta}}\right)^2$$
$$+ \frac{1}{3} \frac{\gamma}{(2\pi)^3} \int d^3k \, \frac{k^2}{\widetilde{E}(k)} (f_k + f_{\bar{k}}) , \quad (12.61)$$

where

$$\rho_B = \frac{\gamma}{(2\pi)^3} \int d^3k \, (f_k - f_{\bar{k}}) , \quad (12.62)$$

$$\rho_S = \frac{\gamma}{(2\pi)^3} \int d^3k \, \frac{M^*}{\widetilde{E}(k)} . \quad (12.63)$$

The Fermi–Dirac distribution functions for baryons and antibaryons, f_k and $f_{\bar{k}}$, are defined by (4.38) and (4.39); $\widetilde{E}(k)$ is defined by (8.38); and the kinetic part of the chemical potential ν is defined via the thermodynamic chemical potential μ by

$$\nu = \mu - V . \quad (12.64)$$

Note that the ρ meson is not involved in our considerations in the present section. The entropy density s can be calculated using the thermodynamic potential Ω and the following equation:

$$\frac{\Omega}{V} = -P = \mathcal{E} - Ts - \mu \rho_B . \quad (12.65)$$

At zero temperature, all the nucleon states in nuclear matter are occupied up to the Fermi energy E_F (or Fermi momentum p_F) and the distribution functions f_k are reduced to Θ-functions: $f_{is} = \Theta(E_{Fi} - E_{is})$, and $f_{\bar{i}s} = 0$. At $T = 0$ the incompressibility K is defined by

12.3 Scalar Derivative Coupling Models 257

Table 12.7. Coupling constants C_σ^2 and C_ω^2, binding energy E_B (MeV) at the density ρ_0 (fm^{-3}), ρ_S (fm^{-3}), and m^* for the models indicated [558]

Model	C_σ^2	C_ω^2	E/B	ρ_0	ρ_S	m^*
Walecka	357.4	273.8	−15.75	0.148	0.138	0.54
ZM	169.2	59.1	−15.9	0.160	0.155	0.85
ZM2	219.3	100.5	−15.77	0.152	0.147	0.82
ZM3	443.3	305.5	−15.76	0.149	0.143	0.72

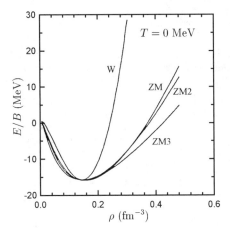

Fig. 12.1. Energy per baryon as a function of baryon density for the Walecka model (W) and Zimányi–Moszkowski model (ZM, ZM2, ZM3) [558]

$$K = 9\rho_0^2 \frac{\partial^2}{\partial \rho^2}\left(\frac{\mathcal{E}}{\rho}\right)\bigg|_{\rho=\rho_0} = 9\rho_0 \frac{\partial^2 \mathcal{E}}{\partial \rho^2}\bigg|_{\rho=\rho_0} , \qquad (12.66)$$

where ρ stands for the baryon density ρ_B.

Suitable closed forms for the values of the incompressibility K can be found in [558]. In Table 12.7, some results of calculations, including the saturation properties of nuclear matter, are presented for the values of the parameters utilized in [558].

In Fig. 12.1, the density dependence of the binding energy E/B, as a function of ρ, is shown. To provide an idea of the softness of the EOS for different models, the pressure as a function of density is given in Fig. 12.2. From this figure it can be seen that the ZM3 model gives the softest equation of state (the value of K being directly related to $dp/d\rho$). In Table 12.8, the fields and incompressibility are given for the values of ρ_0 listed in Table 12.7.

As can be seen from the calculations of [558], the RDSCMs considered here yield a reasonable incompressibility (at any rate, in the case of ZM and ZM2) for nuclear matter and consequently produce good binding energies

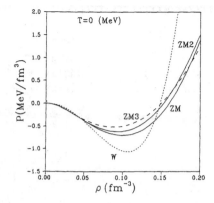

Fig. 12.2. Pressure as a function of baryon density for the various models [558]

Table 12.8. Values (in MeV) for S, V, $V + S$, $V - S$, and K at the values of ρ_0 given in Table 12.7 for the models considered [558]

Model	S	V	$V+S$	$V-S$	K
Walecka	−431.02	354.15	−76.87	785.18	550.82
ZM	−140.64	82.50	−58.13	223.13	224.71
ZM2	−167.83	109.73	−58.09	277.56	198.32
ZM3	−267.00	203.71	−63.28	470.71	155.74

for finite nuclei. All of the RDSCMs generate a soft equation of state in comparison with the Walecka model; the ZM3 version provides the softest one and generates a value of $V - S$ which comes close to the result of the Walecka model. One should bear in mind that the spin–orbit splitting in finite nuclei is directly related to the value of $(d/dr)(V - S)$. In the case of the Walecka-type models one obtains reasonable spin–orbit splittings and it is known how to treat the problem of the incompressibility (the problem was discussed earlier; see also [562, 563]), without spoiling the spin–orbit force. However, in the case of RDSCMs one encounters a problem of reproduction of the proper values of the spin–orbit splittings in finite nuclei, at any rate in the mean-field approximation [494, 559]. RDSCMs have also been applied to study the properties of multilambda matter [549], neutron stars [553], Δ-excited nuclear matter [555], and some thermodynamic properties of nuclear matter [554, 564]. The author of [557] derived an effective NNφ interaction for Dirac nucleons which includes a derivative coupling of the ZM type, the method being based on the relativistic $SU(6)$ model of the meson–baryon couplings. Further discussion of the properties of RDSCMs can be found in [565].

12.3 Scalar Derivative Coupling Models

Concluding this chapter, we can say that several types of relativistic models for nuclear matter and finite nuclei have been introduced and discussed in the preceding sections:

1. Models of the PVS type (with pseudoscalar, vector, and scalar mesons), having all the general features of the OBE potentials and containing also self-interactions of mesonic fields. These models include meson and nucleon degrees of freedom.
2. Models including
   (a) quark degrees of freedom (QMC models);
   (b) only fermion degrees of freedom (RPC models);
   (c) mesons and nucleons but with a more complicated form of the σ-meson–nucleon interaction (DSC models).

The models of the first type are well developed; these models reproduce many nuclear phenomena and may be considered at the present stage as standard models for the description of various properties of nuclear matter and finite nuclei. They have also been utilized successfully to describe interactions of nuclei with various types of particles (nucleons, $\bar{p}$, Λ, Σ, etc.). The basic feature of all these models is that they involve two strong mesonic fields S and V with the transformation properties of a world scalar and of a four-vector (the time component), which make the relativistic description visible and which are responsible, in particular, for reproducing the shell-model spin–orbit potential.

The relativistic point-coupling model does not contain mesonic degrees of freedom directly; only fermion fields are taken into consideration explicitly. However, this model has many features in common with PVS models (more specifically, nonlinear σ–ω models) that utilize meson and nucleon degrees of freedom. In particular, the RPCM includes two strong relativistic potentials with the transformation properties of a relativistic scalar ($-400\,\text{MeV}$) and a relativistic vector ($+325\,\text{MeV}$), which, when they are used in the relativistic self-consistent procedure, also generate the proper value of the spin–orbit splitting. This makes the RPCM suitable for further applications in finite nuclei. Note that the nature of these two strong relativistic fields appearing in the RPCM is not related to the mesonic potentials.

In the case of the QMC model, an attempt has been made to include quark degrees of freedom in the context of nuclear structure. The QMC and DSC models both reproduce reasonable values of the nuclear binding energies and the incompressibility of nuclear matter at the mean-field level. For this reason, these two models can be treated as phenomenological effective models, which are useful for many calculations but present a problem in recovering the correct spin–orbit potential for finite nuclei.

13 Some Recent Applications of Relativistic Nuclear Theory

The Dirac approach to nuclear theory was introduced as an alternative to the conventional nonrelativistic framework which has been used in nuclear physics for almost 50 years. The relativistic approach has its opponents [566–571] and has its support. This support comes, as discussed in [67, 432], from QCD, which is now considered as the underlying theory of strong interactions, and also from a large amount of experimental data that can be described in the relativistic framework on solid theoretical grounds. Some of these phenomena (e.g. saturation in the BHF scheme) cannot be reproduced at all at the level of the nonrelativistic theory, and others (e.g. the spin–orbit force) need extra fitting parameters. The most compelling support for the relativistic treatment of finite nuclei is obtained from calculations of spin observables (the analyzing power and the spin rotation).

The relativistic approach to nuclear-structure theory started about 30 years ago and is rapidly developing at present; the results obtained in this field are impressive. In the framework of a single textbook, it is impossible not merely to discuss but even to mention all contributions to the relativistic theory of nuclear structure. In what follows, we only make reference to some of the basic results obtained quite recently.

13.1 Mean Fields in Colliding Nuclear Matter

The studies of heavy-ion collisions at energies of $\sim 1\,\text{GeV}$ now being carried out are to a large extent induced by interest in the EOS of nuclear matter, i.e. the properties of nuclear matter at high density and nonzero temperature. Nuclear matter under extreme conditions may occur in astrophysical objects. One can try to infer information about the EOS from heavy-ion collisions. Heavy-ion collisions may be studied in the framework of transport theories. Relativistic transport models are usually obtained on the basis of the effective meson–nucleon theories considered above, the self-energy of a nucleon in a medium being an essential ingredient of this approach. The relativistic description appears to be preferable because the Lorentz structure of the nucleon self-energy generates naturally a momentum dependence of the forces.

A large amount of interest has been shown recently in relativistic formulations of quantum kinetic equations in investigations of nuclear dynamics in heavy-ion collisions. Covariant transport equations for nucleons have been developed [572–583] in the mean-field approximation (Vlasov equation) and taking into account collision terms (Vlasov–Uhlenbeck equation). The self-consistent field for the Vlasov equation (i.e. the drift term of the Vlasov–Uehling–Uhlenbeck equation) has been obtained in the Hartree approximation in [572–579]. In [580, 581] a more general theory has been developed. From the equations given in those references, it is possible to obtain kinetic equations corresponding to different levels of approximations.

In [582], a relativistic kinetic equation with a self-consistent mean field was obtained for a field theory with the Walecka Lagrangian. The authors of [582] studied the covariant one-particle Wigner function, this function being the quantum analogue of the classical one-particle distribution function in phase space. Using the Hartree–Fock method, these authors obtained an equation for the Wigner function in a closed form, a covariant kinetic equation with self-consistent mean field (without collision terms). This equation was applied to investigations of collective modes and of stability conditions for the equilibrium state of nuclear matter. Owing to exchange terms, the kinetic equation has solutions corresponding to oscillations of the spin density.

In [584], an approach was suggested in which the nucleon self-energy in colliding nuclear matter was constructed by an approximate method starting from the relativistic Brueckner results for equilibrated nuclear matter (without solving the Bethe–Salpeter equation for an anisotropic configuration). The main tool in this case is the momentum- and density-dependent real part of the nucleon self-energy (rather than the T-matrix) for spherical configurations.

In [585, 586], in a similar way, the imaginary part of the self-energy was used to construct the in-medium cross section for such configurations. The authors of [585, 586] utilize the fact that the self-energy components in nuclear matter are almost linear in the respective densities, but with momentum-dependent and weakly density-dependent factors instead of proportionality constants. These *coupling functions* give the mean value of T-matrix averaged over one Fermi sphere. These functions are used then to obtain the average self-energy for two Fermi ellipsoids. Using this method, the authors of [585, 586] respect covariance and the self-consistency of the mean fields. However, the T-matrix in colliding nuclear matter is replaced by that in a single body of nuclear matter. The self-energy includes effectively the correlation and exchange effects for each nuclear-matter stream. These effects produce a softening of the Schrödinger equivalent optical potential for a nucleon in colliding nuclear matter.

The self-energies for the colliding-nuclear-matter configuration are presented again in the form of dynamical coupling functions and total densities (as was done for one Fermi sphere). These have been used in a relativistic transport theory of the Landau–Vlasov type in a *local phase-space configuration* approximation [583]. This method gives an improved description of the mean-field dynamics.

In [585], the mean field for use in relativistic transport calculations was obtained from microscopic Dirac–Brueckner self-energies, the method going beyond the local density approximation (LDA). It was shown that this method is superior to the LDA, as well as to a standard nonlinear version of the σ–ω model, in describing various in-plane and out-of-plane flow observables for Au-on-Au reactions at incident energies ranging from 250 to 800 A MeV.

13.2 Hartree–Fock–Bogoliubov Approximation

As mentioned above, correlations are essential in a description of the nuclear structure. There are two ways to improve the independent-particle approach: (a) to introduce long-range correlations related to collective phenomena; and (b) to take into account short-range correlations, which are responsible, for example, for pairing of nucleons in the nuclear medium. As shown in Chap. 10, the long-range correlations have been treated successfully in the relativistic version of the random-phase approximation. However, the relativistic description of pairing correlations in nuclear structures[1] still remains a challenging problem. To get an idea of the current status of the problem, the reader should consult [84, 587–593]. Pairing has an essential role in describing all nuclei with open shells. One of the possible ways to include pairing correlations in finite nuclei is to use a constant-gap approximation. There are no problems in using this approximation, which has been utilized in the most of the practical calculations for stable nuclei [84]. Some essential progress in developing a relativistic theory of pairing correlations (the Hartree–Fock–Bogoliubov approximation) in nuclear matter has been achieved recently in [589–592] (for $^1S_0$ pairing). The approach developed in those papers enables one to obtain a simultaneous description of the energy gap in the single-particle spectrum and of the saturation point of nuclear matter. The common starting point in [589–591] is a relativistic generalization of the Gorkov method, developed for the description of superconductivity in metals [593]. In the framework of this method, nuclear matter is considered as a system of quasiparticles, moving in the Hartree–Fock field plus a self-consistent pairing field. In the relativistic treatment, the pairing field has two different components: a Lorentz scalar Δ_s and the time component of a four-vector Δ_0. The different space–time transformation properties of these two components must be considered properly in the description of a relativistic nuclear system in the superfluid phase. The energy gap appears to be the result of a strong destructive interference between the scalar and vector components of the pairing field. In [591] it is

---

[1] For nuclear structures, there is no natural cutoff of the momentum in the gap equation. For this reason, a relativistic treatment is appropriate in this case.

shown that a proper relativistic treatment of the problem introduces a new, specific feature: the pair amplitude vanishes beyond a certain value of the momentum of the paired nucleons $|\boldsymbol{p}|$. This determines the occurrence of a cutoff in the relativistic gap equations. This cutoff removes the contributions of high-momentum components of the interaction to the gap equation. The value of this cutoff gives an energy gap in agreement with the estimates of nonrelativistic theory.

In [592], relativistic effects on $^3S_1$–$^3D_1$ pairing in nuclear matter are considered. The authors of [592] incorporate minimal relativity in the gap equation, i.e. they use DBHF single-particle energies in the energy denominators and modify the free NN interaction by a factor $M^{*2}/\widetilde{E}_k\widetilde{E}_{k'}$. As the main result, they find that relativistic effects decrease the value of the gap at the saturation density (1.36 fm^{-1}) greatly, in accordance with the lack of evidence for strong neutron–proton pairing in finite nuclei.

As for the description of finite nuclei, the following points should be emphasized. It was mentioned above that for nuclei close to the β-stability line, pairing is usually incorporated into RMF theory in the form of the BCS method (see Chap. 4). However, the BCS approximation is a poor method for nuclei far from stability. In drip-line nuclei, the Fermi energy is close to the particle continuum. The lowest particle–hole or particle–particle modes are often embedded in the continuum, and one should consider explicitly the coupling between bound and continuum states. The BCS approach does not give a proper description of the scattering of nucleonic pairs from bound states to the (positive-energy) continuum, which leads to an unbound system (the levels in the continuum being partially occupied). If the system is put into a box of finite size, one obtains unreliable results for nuclear radii that depend on the box size.

A unified theory of mean-field and pairing correlations in the nonrelativistic case is given by the Hartree–Fock–Bogoliubov method in coordinate space (see [594]). A relativistic extension of the Hartree–Fock–Bogoliubov method was obtained in [588]. In practical applications for finite nuclei the relativistic Hartree–Bogoliubov (RHB) theory is mostly used at present. An illustration of the continuum Hartree–Bogoliubov problem is given in Fig. 13.1. In fact, the authors of [588] start from a relativistic Lagrangian containing σ, ω, and ρ mesons and develop an RHB procedure based on the relativistic Hartree approximation for the ground state. However, if the coupling constants obtained within standard relativistic Hartree parameterizations are used for the one-meson-exchange pairing interaction in the RHB procedure, the resulting pairing correlations appear to be too strong. For this reason, in practical RHB calculations pairing correlations are often described by a two-body force of finite range of Gogny type [595]. This force has been fitted reasonably to the pairing properties of finite nuclei for a wide range of A; the finite radius of the Gogny force automatically ensures a proper cutoff in momentum space.

13.2 Hartree–Fock–Bogoliubov Approximation

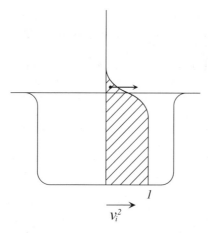

Fig. 13.1. Illustration of the continuum Hartree–Bogoliubov problem [596]

In [597], the RHB method is applied to describe the properties of light nuclei with a large neutron excess, pairing correlations and the coupling to particle-continuum states being described by finite-range two-body forces. In [598], ground-state properties of deformed proton-rich odd-Z nuclei in the region $59 \leq Z \leq 69$ are described in the framework of the RHB theory. The RHB theory in coordinate space is used in [599] to describe the Zr isotopic chain from ^{116}Zr to the drip line nucleus ^{140}Zr. This theory provides a self-consistent treatment of pairing correlations in the presence of the continuum. The authors of [599] predict neutron halos in the Zr nuclei close to the neutron drip line, which contain not one or two neutrons, as is the case in the halos investigated so far in light p-shell nuclei, but up to six neutrons. This phenomenon has not been observed experimentally so far. The prediction made in [599] is based on an RHB theory of the continuum which allows, in the canonical basis, the scattering of Cooper pairs to low-lying resonances in the continuum.

In [600, 601], the RHB theory in coordinate space was used to describe the chain of even–even nickel isotopes reaching from the proton drip line to the neutron drip line; in [378] the RHB theory was used to study tin isotopes. These investigations will be discussed in more detail in Sect. 13.11.

We should mention also [602], where shell effects in nuclei around the stability line were investigated within the framework of the RHB theory with self-consistent finite-range pairing. Using the two-neutron separation energies of Ni and Sn isotopes, the role of σ and ω-meson self-couplings (see (4.2)) in the shell effects in nuclei was investigated. It was observed that the existing successful theories of nuclear forces (Lagrangian parameter sets) based upon the nonlinear scalar self-coupling of the σ meson exhibit shell effects which are stronger than those suggested by the experimental data. The authors of [602] introduced nonlinear vector self-coupling of the ω meson into the RHB theory. It was shown that the inclusion of the vector self-coupling of the ω meson in addition to the nonlinear scalar coupling of the σ meson provides good agreement with the experimental data on shell effects in nuclei around the stability line. The Dirac–Hartree–Bogoliubov approximation for finite nuclei has received further development in [603].

In [604], the RHB approach has been extended to be able to include nonlinear coupling terms of mesons. In that paper, the theory is applied to nuclear matter to observe the effect of nonlinear terms on the nuclear gap.

Further investigations and applications of the relativistic theory of pairing and of the RHB method can be found in [605–612].

13.3 Spin–Orbit Splitting for Single-Particle and Single-Hole Energies

Here we would like to draw attention to [613–615]. These investigations were motivated by the following problem in particular. Experimental data show that the $p_{3/2}^{-1}$–$p_{1/2}^{-1}$ single-hole splitting in ^{15}O is larger than the $d_{3/2}$–$d_{5/2}$ single-particle splitting in ^{17}O, the respective numbers being 6.176 and 5.086 MeV [614]. If one uses a single-particle spin–orbit potential of the form $-\zeta \boldsymbol{l}\cdot\boldsymbol{s}$ (ζ being a constant), the value of the $d_{3/2}$–$d_{5/2}$ splitting produced by this force should be 5/3 times larger than the $p_{1/2}$–$p_{3/2}$ splitting, since the spin–orbit splitting of the two partners $j = l \mp 1/2$ is proportional to $(2l+1)$.

In [614], the following mechanisms are suggested to provide the reduction of the spin–orbit splittings of the particle states in comparison with the hole states:

1. As is discussed above, the significant point of the Dirac framework, and of the DBHF method in particular, is that the relativistic spinors in the nuclear medium differ considerably from the free relativistic spinors, since the effective nucleon mass in the nucleus is much smaller than that of a bare nucleon (for this reason, the lower component of the relativistic spinor in the nucleus is strongly enhanced). The Dirac effective mass in the interior of the nucleus is also much smaller than that at the nuclear surface. The spin–orbit splitting is, to first order, proportional to the inverse of the Dirac effective nucleon mass [63, 132, 616]. This fact enhances the spin–orbit splittings of the single-hole states in comparison with the spin–orbit splittings of the single-particle states.
2. The authors of [614] argue also that for states close to the Fermi level, it is necessary to consider the coupling of single-particle states to two-particle–one-hole (2p–1h) and three-particle–two-hole (3p–2h) configurations at low energies. The aim of [614] was to study the interplay of the relativistic

effects and core polarization terms in determining single-particle energies and spin–orbit splittings.
3. The self-consistent wave functions for the particle states are less localized than those of hole states, yet outside the limits of an oscillator model. The single-particle wave functions in [614] were approximated by harmonic-oscillator waves with realistic radii, and the authors of that paper obtained results in fair agreement with experiment, both for the spin–orbit splittings and for the energies of single-particle and single-hole states. The Hartree–Fock insertions play an essential role in these calculations.

13.4 The Anomalous Kink in the Isotope Shifts of Pb Nuclei

Standard nonrelativistic models provide reasonable results for many properties of stable nuclei. However, there are cases where the nonrelativistic treatment appears to be not appropriate. Some such cases have been considered above. Another example is the anomalous kink found in the isotope shifts of Pb nuclei[2] [617, 618], which cannot be recovered in the framework or of the standard Skyrme approach or of the Gogny method. In contrast, the relativistic approach provides a very good account of the experimental data [617].

If the spin–orbit force is greatly modified [619], the Skyrme approach generates isotope shifts similar to those derived by the relativistic calculation. In fact, in [620] a modification of the Skyrme energy functional has been suggested, with a spin–orbit force containing an extra parameter, this parameter being fitted to reproduce the kink. In this case one obtains an isobaric-spin dependence of the spin–orbit force very different from that obtained in the standard Skyrme approach [84].

Some essential progress in understanding the origin of the kink effect in the behavior of the r.m.s. charge radii in the Pb isotopic chain has been made in [621]. As is shown in that paper, the kink effect appears to be an inherent feature of the standard relativistic mean-field theory. It is found that relativistic self-consistent effects, as well as contributions from the ρ meson, determine the actual structure of the kink effect [621].

The kink effect is displayed in both linear and nonlinear relativistic mean-field models, and is more pronounced in the models with smaller values of the compressibility modulus K. However, the spin–orbit force generated by the ρ meson has no significant influence on the kink effect [621]. We should mention that the kink effect can also be reproduced by point-coupling models (see Sect. 12.2).

---

[2] The kink in the charge radii has been observed also in the isotope chains of Kr, Rb, Sr and Zr.

As for the isobaric-spin dependence of the spin–orbit force (ISDSOF) in general, the investigation of its role in the relativistic framework started in [60]. Its role can be greatly modified in exotic nuclei. We should mention [622, 623], where ISDSOF has been studied in light neutron-rich nuclei using the RMFA. In those papers it was shown that in drip-line nuclei the strength of the spin–orbit interaction is greatly reduced, generating small values of the spin–orbit splittings. In [623], spin–orbit interactions in the proton drip-line nucleus ^{72}Kr were discussed in terms of Skyrme–Hartree–Fock and relativistic mean-field theories using various parameter sets. It was found that the two models gave different results for the spin–orbit splittings and also for the orbital-angular-momentum dependence of the single-particle energies.

It should be emphasized, however, that the isobaric-spin dependence of the spin–orbit potential may be strongly influenced by including the gradient coupling of the ρ meson in the Hartree approximation [51, 59] or by considering the exchange (Fock) terms [51, 60], for pions in particular [62, 353].

13.5 Electroweak Interactions in Nuclei in the Relativistic Framework

This subject is discussed explicitly in [76, 77, 83, 85, 624] along the following lines:

1. *Electromagnetic currents.* It is possible to define an electromagnetic current that takes into account some effects of the internal structure of a nucleon, incorporating, for example, the proper value of the nucleon magnetic moment. This current is local and it has the proper space–time transformation properties, its four-divergence being equal to zero. The dimensions of nucleons are taken into account phenomenologically in RMF calculations by introducing a single-nucleon form factor $f_{\text{s.n.}}(q^2)$, where q^μ is the four-momentum transfer.[3] Describing the composite structure of the nucleons in more detail requires consideration of the derivative expansion for $Q^2 \geq M^2$ (where $Q^2 = -q^2$), which needs serious justification. The effective weak currents are worked out similarly. It is necessary to take care over the partial conservation of the axial-vector current in this case.

2. *Weak interactions* [625–628]. The authors of [625] the authors consider weak–axial transitions in $A = 16$ nuclei. The considerations are based on the Dirac equation, with scalar and vector potentials. The relativistic effects appear to be important, and produce new qualitative features which improve on the standard nonrelativistic approach. In particular, the authors of [625] show that the relativistic description strongly enhances the

---

[3] See, however, Chap. 5, where the electromagnetic structure of the nucleon is described via the mechanism of vector meson dominance.

13.5 Electroweak Interactions in Nuclei in the Relativistic Framework

anomalously reduced nonrelativistic β-decay rate, bringing it closer to the experimental rate. Using the same approach, the inverse process, muon capture, has also been studied. It is shown that the μ-capture rate may be reproduced simultaneously with the β-decay rate.

3. *Electroweak exchange currents.* References [76, 85] contain explicit expressions for the long-range pion and pair currents. Meson exchange currents are important in reproducing the high-momentum-transfer dependence of electromagnetic observables in light nuclei. Exchange currents are important also in describing pion production in the p $\to$ pπ^0 reaction near threshold (see Sect. 13.9, item 3 in the list).

The problem of the axial vector current in nuclear many-body physics is discussed in [629].

4. *Recent developments in electromagnetic and electroweak interactions* [83, 85, 265, 624, 630]. The interaction of electrons with nuclei is an important and reliable source of data on nuclear many-body structures. In this case, the elementary interaction is weak but has been intensively investigated. High-quality measurements have beene performed in this field for a large number of nuclei over a wide range of energy and momentum transfer; the data are available from both inclusive and exclusive measurements. The continuous-wave electron beams at NIKHEF, Mainz, MIT-BATES, Saskatoon, and CEBAF are very suitable for investigating new kinematical regions. The relativistic models discussed above are perfect candidates for performing investigations in electronuclear physics. There are two excellent reviews [265, 630] summarizing the theoretical results obtained in this field.

In [265] a review is given of electromagnetic response functions evaluated by means of relativistic models of nuclear structure; the theoretical results obtained are compared with existing data, the constraints of current conservation and self-consistency being emphasized. RPA correlations are shown to be important and to play different roles in different kinematical regimes.

In [630], nuclear phenomena related to inclusive and semi-inclusive electron scattering at intermediate energy are reviewed; much emphasis is put on the derivation of the general formulae and the validity of the approximations involved.

Photoproduction of pions on nuclei has also been considered in the framework of the relativistic theory [85, 253]. In the case of coherent photo-production of pions, the initial and final states of the target nucleus coincide. This case is of special interest because, for its theoretical description in the local approximation (i.e. in the high-energy limit for the nucleon and the Δ-isobar Green function), one needs only the density distribution in the nucleus, i.e. the influence of the uncertainty related to the nuclear structure is minimized in this case.

5. *Parity violation in* (e, e'). Pioneering experiments in this field have been done at MIT-BATES and Mainz. Parity violation effects in electron scattering will be one of the most important directions of future investigations at CEBAF. Theoretical contributions in this field are discussed in [83, 85] (see also references therein).

For recent publications on parity-violating effects see, for example, [631–633].

13.6 Hypernuclei in the Relativistic Framework

Normal nuclei are composed of neutrons and protons. It is possible to envisage also bound many-body systems containing other baryons, such as Λ, Σ, and Ξ hyperons. The relativistic mean-field description of these systems involves the central potentials U_B and the spin–orbit potentials U_B^{SO} ($B = \Lambda, \Sigma, \Xi$) [634]. It may involve also an isobaric-spin component, attributed usually to coupling of the baryon (N, Σ, or Ξ) to the ρ meson; the Λ does not couple to the isovector field. For Λ-hypernuclei, the relativistic central potential is given by

$$U_\Lambda \approx 27\text{–}30 \text{ MeV}, \tag{13.1}$$

while the Λ–nucleus spin–orbit force is at least one order of magnitude smaller than that in the nucleon–nucleus case. Studies of the Σ^- atomic levels suggest

$$U_\Sigma \approx 20\text{–}30 \text{ MeV}. \tag{13.2}$$

Atomic data demonstrate also the presence of an imaginary part $W_\Sigma \approx 10\text{–}15$ MeV. More recent investigations of Σ^- atoms show that the isoscalar potential for Σ is repulsive; however, the isovector potential can be attractive.

In emulsion experiments with K^- beams, there are some events related to the formation of Ξ^- hypernuclei, these data can be treated in terms of a relativistic single-particle potential with a depth

$$U_\Xi \approx 28 \text{ MeV}. \tag{13.3}$$

Note that the authors of [635] investigated an extreme assumption $U_\Xi \approx 300$ MeV; in this case Ξ's would be favored in nuclear matter rather than Λ's.

RMF studies of hypernuclei started in [636]. The authors of that paper suggested that one should obtain the weak spin–orbit force for Λ-hypernuclei on a phenomenological basis. In [637, 638] this problem was treated in the Hartree approximation, and in [639] the weak spin–orbit force for the Λ-hyperon has been confirmed by experiment. In [640] an explanation of the small value of the spin–orbit potential for Λ-hypernuclei was given on the basis of quark and gluon dynamics. In [641] the possibility of introducing an $\omega \Lambda \Lambda$ tensor interaction was considered; this problem has been investigated further in [642–644], where it was shown that the Λ–nucleus spin–orbit force could be made arbitrarily small by adding a tensor coupling.

The relativistic mean-field description of hypernuclei starts with the following Lagrangian [645]:

$$\mathcal{L} = \mathcal{L}_{\text{baryon}} + \mathcal{L}_{\text{meson}} + \mathcal{L}_{\text{coupling}} + \mathcal{L}_{\text{Coulomb}}, \quad (13.4)$$

where $\mathcal{L}_{\text{baryon}}$ is given by

$$\mathcal{L}_{\text{baryon}} = \sum_B \bar{\psi}_B (i\gamma^\mu \partial_\mu - M_B)\psi, \quad (13.5)$$

the sum running over all baryons of the baryon octet (p, n, Λ, Σ^+, Σ^0, Σ^-, Ξ^0, Ξ^-). $\mathcal{L}_{\text{meson}}$ is given in a conventional form, including, in the general case, φ, ω_μ, and $\boldsymbol{\rho}_\mu$ fields, as well as a self-interaction term for the φ field. The interaction of baryons with meson fields is determined by minimal coupling requirements:

$$\mathcal{L}_{\text{coupling}} = -\sum_B g_{\sigma B} \bar{\psi}_B \psi_B \varphi - \sum_B g_{\omega B} \bar{\psi}_B \gamma^\mu \psi_B \omega_\mu - \sum_B g_{\rho B} \bar{\psi}_B \gamma^\mu \boldsymbol{\tau}_B \psi_B \boldsymbol{\rho}_\mu. \quad (13.6)$$

For finite nuclei, $\mathcal{L}_{\text{Coulomb}}$ is also taken into consideration; it contains a part corresponding to the free electromagnetic field and a term

$$-\sum_B q_B e \bar{\psi}_B \gamma_\mu \psi_B A^\mu$$

(where q_B is the charge number of the baryon) describing the interaction of the baryon with the electromagnetic field. The coupling constants g entering (13.6) are determined either empirically or using $SU(6)$ symmetry. A lot of calculations of the properties of hypernuclei with $S = -1$ have been performed in the framework of the RMF theory; we shall mention only some of the references [646–656] (see also the review paper [645]). The current status of multiply strange nuclear systems is reviewed in [634], while in [657] the QMC model described above is applied to describe the properties of Λ, Σ, and Ξ hypernuclei.

13.7 Theoretical Analysis of $A(e, e'p)B$ Reactions

Much work has been done in this field recently [658–667]. Calculations based on the standard nonrelativistic distorted-wave impulse approximation (DWIA) contain two major difficulties (see [658] and references therein). The spectroscopic factors derived from DWIA analyses of low-p_m ($p_m \leq 300\,\text{MeV}$) data are too small in comparison with theoretical predictions, p_m being the proton momentum. The second point is that the high-p_m data ($300\,\text{MeV} \leq p_m < 600\,\text{MeV}$) on $\tilde{\rho}(p_m)$ for the same levels are much larger than the results of those DWIA calculations which fit the low-p_m data [663], where $\widetilde{\rho}(p_\mathrm{m})$ is an effective proton momentum distribution (reduced cross section). In the relativistic distorted-wave impulse approximation (RDWIA), the nucleons (both bound and scattered) are represented by relativistic spinors obtained as solutions of the Dirac equation (with scalar and vector potentials).

The RDWIA successfully solves the problems mentioned above. The reasons why the relativistic treatment generates smaller cross sections in the low-p_m region were discussed in [662, 665]. In [665], the success of the relativistic theory is related to the proper description of distortion effects in the electron and outgoing proton waves. For the high-p_m region, other reasons become important. The presence of higher-momentum components in the relativistic bound-nucleon eigenfunction and in the nucleon current operator is believed to play an important role in this case. These points are discussed in more detail in [658].

The authors of [667] investigate dynamical relativistic effects in quasielastic 1p-shell proton knockout from ^{16}O; the data are well described by RDWIA calculations. At large missing momentum, the structure observed in the left–right asymmetry indicates the existence of dynamical relativistic effects.

13.8 Exclusive Pion Production in Nucleon–Nucleus Scattering

It is expected that pion production in nucleon–nucleus scattering at high energy will supply us with information about the nuclear wave function at short distances and about the in-medium effects of intermediate processes, since it includes a high-momentum-transfer process and isobar states. In [668] the pion production cross section in the p–^{12}C reaction is derived and calculated on the basis of the relativistic impulse approximation. The elementary scattering amplitude for nucleon–nucleon collisions is given in a totally covariant form, starting from totally covariant interaction Lagrangians. Coupling constants and cutoff parameters are derived by adjusting elastic nucleon–nucleon scattering at 0.8 GeV.

In [668] the contributions of exchanged mesons to the total cross sections for pion production in nucleon–nucleon collision at 0.8 GeV were calculated. The calculations were carried out in the center-of-mass frame and included both direct and exchange diagrams. These calculations utilized coupling constants and parameters for meson–NN and meson–nΔ vertices obtained by fitting the elastic NN scattering amplitudes at 0.8 GeV. It was found that the inclusion of ρ meson exchange significantly reduces the cross section from the value calculated with pion exchange only in the Δ excitation mechanism. The same effect was established using cutoff parameters for πNN and πNΔ vertices smaller than 1.005 GeV. The diagram involving the exchanged pion and the induced Δ-isobar gives the total cross section as 16.58, 13.20, and

13.8 Exclusive Pion Production in Nucleon–Nucleus Scattering

9.88 mb for values of the cutoff parameter $\Lambda = 0.90$, 0.80, and 0.70, respectively. However, σ, ω, and a_1 (axial-vector) mesons increase the cross section and destroy the ρ meson effect. For this reason alone, the ρ meson and others were excluded from consideration in [668]. To obtain the pion production cross section in nucleon–nucleus scattering, the diagram is calculated as being composed of an exchanged pion and a Δ-isobar in [668].

The coincidence cross sections for the zero-degree spectra of the reaction ^{12}C $(p,n\pi^+)$ ^{12}C$_{\text{g.s.}}$ at 0.82 GeV incident energy were obtained in [668], two types of results being given. In the first case the calculations were performed by integrating triple differential cross sections over the pion angle over the acceptance range of $12° \leq \theta \leq 141°$ of the detector, while in the second case the results were obtained by integrating differential cross sections over the whole range of the pion angle. In both cases the results were compared with the experimental data translated into the spectrum of excitation energy ω_L.

The excitation energy ω_L is defined by $\omega_L = \omega_\pi + E_R$, where ω_π is the pion energy, and E_R is the target recoil energy, which depends on the angle of the produced pion. For this reason E_R was obtained by averaging over the pion angle, its value being about several MeV around the peak. The cross sections were calculated with distorted waves for the produced pion and free states of the incoming and outgoing nucleons, and also taking account of distortion of the states of both the produced pion and the nucleons. The constants and parameters were originally obtained by fitting the elastic NN scattering amplitudes at 0.8 GeV.

In the calculations of both types, the coincidence cross sections have a peak at about 0.22 GeV, while the experimental data have a corresponding peak at around 0.3 GeV. In [668] it was shown that the calculations may provide an 8.5% larger cross section at most, because of the approximate attenuation factor, which was evaluated from the distortion effects of the incoming and outgoing nucleons. If one takes this point into account, the calculations of the first type show that the contribution of coherent pion production is about 50% of the data, measured as π^+ events, or even less than half of the data depending on the coupling constants. This characteristic of the spectra agrees with the data described in papers concerning coherent pion production. If the pion distorted wave is replaced by a plane wave, the peak manifests itself at around 0.35 GeV. In [668] it was shown that the main contribution to the optical potential of pion–nucleus elastic scattering, which is the main reason for the distortion of the pion wave, is Δ–hole excitation in the target nucleus. For this reason, the peak position of the coherent pion component is shifted towards lower excitation energies by the Δ–hole correlations.

It is known that the experimental cross section contains false contributions, for example ^{12}C$(p,n\pi^+n)$ and ^{12}C$(p,n\pi^+p)$ reactions. The excitation of the projectile, when the latter is excited to a Δ^+ (decaying into $n + \pi^+$), also makes a contribution to the pion cross section. The difference between the calculated result and the experiment is believed to arise from these contributions.

The angular distribution of the cross section has a large value at forward angles of the produced pion (in the laboratory frame). These values increase with increasing energy of the produced pion. For this reason, the cross sections obtained by integrating the differential cross section over the whole range of the pion angle are much larger at large pion energies than those obtained by integrating over the acceptance range. This forward-peaked structure has been manifested also in the angular distribution of the spin-longitudinal component of the coherent pion production (see [668] and references therein).

The medium effect of the intermediate delta state is not taken into consideration in [668]. The self-energy of the Δ-isobar state is expected to shift the resonance peak position by several MeV in a nonrelativistic treatment.

13.9 The Role of Relativity in Few-Body Systems

Relativistic effects can be manifested also in few-body systems at relatively low energies. A lot of investigations have been done in this field. The results are discussed in [34, 669–685] and are related to the following subjects:

1. *The low-energy theorem for scalar and vector interactions* [168, 169, 669–671]. Scalar and vector interactions are essential components of the nucleon–nucleon force, probably the most important after pion-exchange interactions. In particular, they determine, as discussed above (see Chap. 6), the elastic scattering of protons by nuclei, because this process demonstrates isoscalar interactions that are scalar and vector by their nature. Important relativistic effects, called Z diagram contributions (see Fig. 9.7 and the discussion in Chap. 9), arise from the use of the scalar interaction in the Dirac equation [230]. The contribution of Z diagrams generates a potential

$$V^{(2)} = \frac{\boldsymbol{p}^2 S^2}{2(\boldsymbol{p}^2 + M^2)^{3/2}}, \tag{13.7}$$

of the second order in the strength of the scalar interaction (the coupling constant is involved in S), $\boldsymbol{p}$ being the proton momentum. The potential $V^{(2)}$ is a repulsive one and is equal to ~ 25 MeV for protons with energies from 200 to 800 MeV; it does not arise in a nonrelativistic approach.

It is known that the Z diagram contribution related to use of the Dirac equation can be misleading [669]. This effect is manifested in πN scattering if the γ_5 interaction is considered: in this case the low-energy scattering is not consistent with the soft-pion theorems. The use of pseudovector coupling is necessary for the results to be consistent with soft-pion theorems based on chiral symmetry. Another concern about Z diagram effects related to the Dirac equation is that they could be misleading for composite particles such as nucleons [566].

However, recent low-energy theorems (see [669] and references therein) for scalar and vector interactions demonstrate that the Z diagram effects appearing when the Dirac equation is considered must be present owing only to Lorentz invariance, and they are not influenced by the complicated nucleon structure (in terms of quarks and gluons). In a more general form, the low-energy theorem, based upon Lorentz invariance, demonstrates that utilizing the Dirac propagator for a nucleon is an efficient tool for treating relativistic effects.

2. *Relativistic effects in a bound nucleon's electromagnetic current.* Relativistic effects in the electromagnetic current of a bound nucleon have been found in a number of recent electron scattering experiments. These experiments observe protons knocked out of a deuterium target, and measure the longitudinal–transverse asymmetry

$$A_\phi = \frac{\sigma(\phi=0) - \sigma(\phi=\pi)}{\sigma(\phi=0) + \sigma(\phi=\pi)}, \tag{13.8}$$

where $\sigma(\phi=0)$ and $\sigma(\phi=\pi)$ are differential cross sections for protons ejected in the scattering plane of the electron with their momenta to the left and right, respectively, of the momentum transfer $\boldsymbol{Q}$ of the electron. This asymmetry involves the interference of a charge-current amplitude and a transverse current amplitude.

The recent results on the longitudinal–transverse asymmetry in electron scattering from deuterium show distinctly better agreement with relativistic calculations that include Dirac spinors in the electromagnetic current (see [669, 672] and references therein).

3. *Threshold production of neutral pions.* Very interesting experimental results on neutral-pion production in proton–proton collisions have been obtained very close to the threshold energy for this reaction, utilizing a cooled proton beam and a hydrogen target. These experiments were performed at the Indiana University Cyclotron Facility and the results have been confirmed at CELSIUS (see [669]). The main result of these experiments is that the measured cross sections for $\mathrm{p}+\mathrm{p} \to \mathrm{p}+\mathrm{p}+\pi^0$ are about five times as large as the theoretical predictions. This fact has generated a large number of theoretical investigations of this process (see [669, 673–677], for example).

It appears that the $\mathrm{p}+\mathrm{p} \to \mathrm{p}+\mathrm{p}+\pi^0$ reaction involving soft pions is not determined or, at any rate, not dominated by contributions which are controlled by soft-pion theorems. The recent theoretical calculations conclusively demonstrate that pionic mechanisms are not sufficient and a short-range mechanism is also needed. Relativistic corrections of the type discussed in item 1 of the present list have been shown to be one of the important short-range contributions to threshold production of π^0 mesons in proton–proton collisions.

4. *Relativistic effects in bremsstrahlung* [669, 678–680]. The recent theoretical results are consistent with existing bremsstrahlung data from TRI-

UMF (see [669] and references therein), but the errors in the existing data are too large to draw conclusions about relativistic and off-shell effects. New experimental efforts are being made at present to study relativistic and off-shell effects in the NN interaction by means of precise bremsstrahlung experiments.

5. *Elastic electron-deuteron scattering* [669]. The large momentum transfers involved in recent experiments on elastic electron–deuteron scattering require a relativistic description of the deuteron and its electromagnetic interactions. As mentioned above, measurements of the asymmetry A_ϕ, in (e, e'p) reactions show that a nucleon in a deuteron should be described in the framework of a relativistic theory that includes at least positive-energy Dirac spinors (or equivalent relativistic effects). However, the low-energy theorem (item 1 of the present list) suggests that negative-energy components of the Dirac spinors should also be taken into consideration. Three recent theoretical analyses of electron–deuteron elastic scattering are based on relativistic equations derived from quantum field theory and applied in the framework of meson exchange dynamics [681–684]. They involve Dirac-spinor wave functions for the interacting nucleons, the negative-energy components being taken into account. These relativistic calculations of e + D scattering provide predictions of the tensor alignment parameter t_{20}. Predictions of t_{20} are of great significance in view of the Jefferson Laboratory experiment that is in progress.

6. *Relativistic effects in three-nucleon binding* [669] (see also [121]). There is a strong interest in relativistic effects in three-nucleon binding also. The experimental value of the binding energy of the triton is 8.48 MeV, while nonrelativistic calculations based on local NN potentials that provide optimal ($\chi^2 \approx 1$) fits to NN scattering data up to 350 MeV generate a value of about 7.7 MeV, so that about 0.8 MeV remains unexplained.

The missing binding energy is attributed to some combination of nonlocalities in the NN potential, three-body forces, and relativistic effects. However, it is worth mentioning that negative-energy components of the nucleon wave function generate the same effect as a certain subclass of three-body forces. Also, a relativistic approach motivates one to consider specific nonlocalities. Typically, the contribution of relativistic effects to the triton binding energy has been found to be about +0.5 MeV; some calculations generating larger effects are discussed in [670].

13.10 Systematic Study of Even–Even Nuclei up to the Drip Lines

The nuclei far from the stability valley, which have extreme N/Z ratios (extreme isospin values) and are typically very short-lived (β-unstable), are referred to as exotic or radioactive nuclei. Recent experiments [155–157] have

13.10 Systematic Study of Even–Even Nuclei up to the Drip Lines

demonstrated the existence of new phenomena in these nuclei, such as neutron halos in exotic light nuclei near the neutron drip line.[4]

In [687] a calculation was performed of the ground-state properties of about 2000 even–even nuclei ranging from $Z = 8$ to $Z = 120$ up to the proton and neutron drip lines in the relativistic mean-field approximation. This contribution is the first microscopic and self-consistent many-body calculation covering all even–even nuclides up to the proton and neutron drip lines. Axially symmetric deformation was taken into account in [687] using a constrained method with a deformation parameter. All possible ground-state configurations were extracted from the energy curves obtained by the constrained calculations. The parameter set TMA was adopted.

Table 13.1. The parameter set TMA: the meson masses and the coupling constants

M (MeV)	938.900
m_σ (MeV)	519.151
m_ω (MeV)	781.950
m_ρ (MeV)	768.100
g_σ	$10.055 + 3.050/A^{0.4}$
g_ω	$12.842 + 3.191/A^{0.4}$
g_ρ	$3.800 + 4.644/A^{0.4}$
g_3	$-0.328 - 27.879/A^{0.4}$
g_4	$38.862 - 184.191/A^{0.4}$
c_3	$151.590 - 378.004/A^{0.4}$

This parameter set was determined by fitting the experimental data on masses and charge radii over a wide mass range [688]; the parameters of this set are listed in Table 13.1. A novel feature of this parameter set is its mass dependence, which is introduced so as to reproduce nuclear properties quantitatively from the light-mass region to the superheavy region. Overall reproduction of nuclear properties over a wide mass range cannot be achieved with a unique parameter set unless one introduces a mass dependence [688]. The bulk properties of nuclear matter at saturation, calculated for uniform matter with the parameter set TMA in the limit of infinite mass number, are shown in Table 13.2.

In order to take axially symmetric deformation into account, the fields were expanded in [687] in terms of the eigenfunctions of a deformed axially symmetric harmonic-oscillator potential [92]; constrained calculations of the quadrupole moment of the nucleon distribution, Q_m, were performed to sur-

---

[4] By definition, the neutron and proton drip lines are localized where the separation energy of a neutron or proton, respectively, drops to zero and nuclei become unstable to neutron or proton emission (see, for example, [686])

Table 13.2. Bulk properties of nuclear matter calculated with the parameter set TMA

ρ_0	(fm^{-3})	0.147
E/A	(MeV)	-16.0
a_{sym}	(MeV)	30.68
K	(MeV)	318
M^*/M		0.635

vey the coexistence of multiple shapes [689]. In [687] the calculated nuclear properties for all equilibrium deformations of all even–even nuclei are tabulated, the general trend of masses, radii and deformations over the whole region of the nuclear chart being explored. The agreement with experimental data and predictions such as magicness beyond the experimental frontier are also discussed. The possible appearance of triaxial deformations was studied in [687] in the RMF model with a triaxial deformation being found in cases of the coexistence of prolate and oblate deformations (see also [690]). It is also to be noted that the calculated mass table [690] covering the nuclear chart has profound implications for astrophysical problems such as the determination of the nuclides in the outer crust of neutron stars and the path of r-process nucleosynthesis. These astrophysical applications are currently being pursued [492].

Note that in [690] it was shown that nuclear radii and neutron skins directly reflect the saturation density of asymmetric nuclear matter. The proton distributions in nuclei have been found to be remarkably independent of the equation of state of asymmetric matter. It is the neutron distributions that are dependent on the EOS. Macroscopic model calculations have been performed over the entire range of the nuclear chart on the basis of two popular phenomenological, but distinctively different EOSs, derived from the SIII parameter set for the nonrelativistic Skyrme–Hartree–Fock theory and the TM1 parameter set for the relativistic mean-field theory. The saturation density for a small proton fraction remains almost the same as the normal nuclear matter density for the SIII EOS, but it becomes significantly smaller for the TM1 EOS. The key EOS parameters used to describe the saturation density are the density derivative of the symmetry energy and the incompressibility of symmetric nuclear matter, and the saturation energy can be written using the symmetry energy alone as a good approximation. It is concluded that a systematic experimental study of heavy unstable nuclei would enable us to determine the EOS of asymmetric nuclear matter at about the normal nuclear-matter density with a fixed proton fraction down to about 0.3.

13.11 Exotic Nuclei and Superheavy Nuclei

For radioactive nuclei with a large neutron excess, the exotic phenomena involve (a) weak binding of the outermost neutrons, (b) pronounced effects of the coupling between bound states and the particle continuum, and (c) regions of neutron halos with very diffuse neutron densities and major modifications in the shell structures. As mentioned above (see Sect. 13.2), the basic problem in the theoretical investigation of drip-line nuclei is related to the closeness of the Fermi level to the particle continuum, a fully self-consistent relativistic Hartree–Bogoliubov theory in coordinate space [597] does describe correctly the coupling between bound and continuum states.

Theoretical investigations of exotic nuclei are also of interest because extrapolations of different (in particular, nonrelativistic and relativistic) nuclear models to this region give very different results. Although conventional nonrelativistic models and RMF theory generate many similar features for stable nuclei, the two approaches appear to be significantly different for drip line nuclei, in particular in the way they describe the spin–orbit term and its isospin dependence.

The RHB theory has been applied in [597] to the description of the neutron drip line in light nuclei; this theory predicts the location of the neutron drip line, the reduction of the spin–orbit interaction, the r.m.s. radii, the changes in surface properties, and the formation of neutron skins and of neutron halos (without introducing new adjustable parameters).

The existence of neutron halos in light neutron-rich nuclei has been widely known since the pioneering work by Tanihata et al. [155, 156]. The question has been asked also of whether proton halos exist in light proton-rich nuclei. An experiment in this field has been done by Morlock et al. [691] which definitely has shown the existence of proton halos. In [692], the newly observed proton halo in the first excited state $1/2^+$ of ^{17}F was studied and is shown to be reproduced without readjustment of any parameters. The authors of [692] also predict the existence of proton halos in ^{18}Ne.

A detailed investigation of the ground-state properties of proton-rich nuclei in the framework of the RMF + BCS theory was carried out in [693, 694]. For proton-rich nuclei, the Coulomb barrier confines the protons in the interior of the nucleus. For this reason, the effect of the coupling to the continuum is weaker than for neutron-rich nuclei and for nuclei close to the proton drip line, the RMF + BCS approach is a reasonable approximation.

In [598] the ground-state properties of deformed proton-rich odd-Z nuclei in the region $59 \leq Z \leq 69$ were described in the framework of RHB theory. The model predicts the location of the proton drip line, and the properties of proton emitters beyond the drip line and provides information about the deformed single-particle orbitals occupied by the odd valence proton.

A detailed, systematic investigation of the RHB theory is described in [378, 601]. Here, the pairing correlations are taken into account by both a density-dependent force of zero range and a finite-range Gogny force. This theory was applied to describe the chains of nickel and tin isotopes extending from the proton drip line to the neutron drip line. The following conclusions were made [378, 601]:

1. The RHB theory is one of the most successful models for exotic nuclei.
2. The success of the RHB theory arises from the proper treatment of the particle shell structure and the coupling between the bound states and the continuum (the contribution of the continuum appears to be important for nuclei near the drip line).
3. The development of the halo is essentially related to the particle shell structure (the orbitals with small orbital angular momentum near the threshold play a considerable role).
4. Within the RHB model, the halo should exist not only in light nuclei but also in heavy systems.
5. The diffuseness of the neutron potential $(S+V)$ changes greatly from the proton to the neutron drip line, in contrast to that of the proton potential. (The diffuseness in this latter case is not changed strongly, owing to the contribution of the Coulomb potential).
6. As was shown earlier (see Chap. 8, for example), the spin–orbit potential in the RMF is given by $(S-V)'$ in particular. For this reason, the spin–orbit splitting in exotic nuclei is weaker than in stable nuclei (however, it remains comparatively strong even near the drip line).

In relation to the subject of the present section, let us mention also [342, 416]. The authors of [342] describe the properties of exotic nuclei by using the RMF Hartree theory with density-dependent meson–nucleon coupling constants together with the BCS theory, while in [416] the density-dependent RH theory is utilized (with and without pairing correlations). The reader should also refer to [695–697] for information on this subject.

Superheavy nuclei [698] have large mass and charge numbers. The strong Coulomb potential produces large modifications in the proton shell structure: single-particle states with large angular momentum and small overlap with the nuclear center are lowered in comparison with small-j states. The Coulomb interaction also pushes protons to large radii, modifying in a complicated way the distribution of density and the single-particle proton and neutron potentials. The large value of A of superheavy nuclei produces a high average density of single-particle levels. For this reason, investigations of shell effects in superheavy nuclei depend, with very high sensitivity, on the detailed relations between the single-particle states.

A detailed, systematic investigation of superheavy nuclei with various relativistic and nonrelativistic models has been carried out in [699–701] in the meanfield + BCS approximation. The authors of those papers have studied the roles of the isospin dependence of the spin–orbit potential and of the nucleon effective mass in the predictions for spherical closures in superheavy nuclei.

One of the basic problems in studying superheavy elements is the prediction (and production) of the nexy doubly magic nucleus after $Z = 82$, $N = 126$. In the 1960s predictions were made pointing an island of long-lived superheavy elements around $Z = 114$, $N = 184$. More recent calculations of superheavy elements have predicted new magic numbers for both protons and neutrons. In a spherical relativistic mean-field calculation [702, 703], a wide range of nuclei in the superheavy region has been investigated using various parameter sets, and $Z = 120$ and $N = 172$ have been suggested as the next spherical magic shells.

On the other hand, on the basis of a deformed relativistic mean-field calculation using the NL1 parameter set, the authors of [704, 705] predicted $Z = 120$ and $N = 184$ as the next possible magic numbers in the superheavy region.

The $N = 172$ gap predicted by the RMF calculations originates from the large energy splitting between the $2g_{7/2}$ and $3d_{5/2}$ shells. In nonrelativistic models, these two orbitals are very close in energy, and this degeneracy is related to pseudospin symmetry (see Sect. 8.7). It is interesting [706] that in SHF calculations, the pseudospin degeneracy holds in most cases. Considering that the idea of PSS has a relativistic origin, it is surprising to see that this symmetry is so dramatically violated in RMF theory. As a matter of fact, the presence in superheavy elements of pronounced magic gaps, for example at $Z = 120$ and $N = 172$, in RMF models is a direct manifestation of the breaking of PSS [706].

13.12 Dilepton Production by Bremsstrahlung of Meson Fields in Nuclear Collisions

The relativistic mean-field approach suggests that strong time-dependent meson fields are formed in the process of a relativistic heavy-ion collision. In [707, 708] this approach has been developed for pion and photon bremsstrahlung. Using this approach, the authors of [709] suggested a mechanism of particle production by the collective bremsstrahlung and decay of classical meson fields in relativistic heavy-ion collisions. This mechanism may be important [710] in central collisions of the heaviest nuclei at the SPS bombarding energy of $160A$ GeV.

In [711], it is shown that the collective bremsstrahlung of the vector meson field can provide an important source of dilepton production in high-energy heavy-ion collisions.

13.13 (p, n) Spin Experiments and Relativity in Nuclear Physics

The recent (p, n) spin experiments on various nuclear targets at forward angles performed at RCNP at Osaka University have attracted strong theoretical interest [712]. The multipole decomposition of the angular distributions up to high excitation energy provides a Gamow–Teller strength S_{β^-} close to the sum-rule value, $3(N - Z)$, where N and Z are the neutron and proton numbers of the target nucleus [713].[5] Together with a small strength observed up to ~ 10 MeV in (n, p) reactions on ^{90}Zr [715], which should extend to high excitation energies as in the (p, n) case, this fact suggests that the Landau–Migdal parameter for delta–nucleon coupling is small; the result $g'_\Delta \leq 0.2$ can be obtained [712, 716]. This upper bound can be extracted from the relation that the Gamow–Teller operators, $O(\beta_\pm)$, are reduced by the delta–nucleon coupling [717]:

$$\widetilde{O}(\beta_\pm) = \frac{O(\beta_\pm)}{1 + g'_\Delta U_\Delta} \ . \tag{13.9}$$

Here, U_Δ is the delta–hole polarization function and

$$U_\Delta = (8/9) \left[f_\pi^* \right)^2 / m_\pi^2 \right] (\rho/\omega_\Delta) \sim 0.8$$

at the normal matter density [712]. To quantify this, if S_β is $(90 \pm 5)\%$ of the Ikeda sum-rule value [712], then $g'_\Delta \leq 0.1$ at the normal matter density, even if we use the delta–hole polarization function at half the normal matter density as an effect of finite nuclear size [247, 717]. This value of g'_Δ is much smaller than the estimate obtained using the G-matrix and induced interactions [718, 719].

This interpretation leads to a contradictory consequence for pion condensation and its precritical phenomena, which are expected at large momenta in the pion channel. The small g'_Δ makes the attractive contribution of the delta–nucleon correlations very large and increases the pionic collectivity at high momenta. This role of g'_Δ can be seen in the dimesic function due to a delta isobar at finite momenta [720],

$$\epsilon_\Delta = 1 + \left(g'_\Delta - \frac{q^2}{m_\pi^2 + q^2} \right) U_\Delta \ . \tag{13.10}$$

When the repulsion g'_Δ is reduced, the attractive contribution due to pion exchange, $q^2/(m_\pi^2+q^2)$, dominates the delta dimesic function, which enhances the pionic collectivity. In fact, a study of the effect of g'_Δ on pion condensation and its precritical phenomena by Shiino et al. [721] shows that for the values $g' = g'_{\Delta\Delta} = 0.6$, pion condensation takes place below $g'_\Delta \leq 0.26$ at the normal

---
[5] Let us mention that the authors of [714] perform a self-consistent relativistic RPA calculation for the isobaric analogue and Gamow–Teller resonances based on relativistic mean-field theory results for the ground states of finite nuclei.

matter density. The present value, $g'_\Delta \leq 0.2$, is definitely inconsistent with the experimental observations of no pion condensation in a finite nucleus [247, 722].

Since there is a further ambiguity in the Landau–Migdal parameter $g'_{\Delta\Delta}$ for delta–delta correlations and also in the nucleon effective mass, the condition for pion condensation may not be fulfilled [723]. An estimate of g', g'_Δ, and $g'_{\Delta\Delta}$ taking account of the relativistic effects and the induced interaction was performed by Krewald et al. [724]. Even in this case, if the value for g'_Δ is as small as that indicated by the (p, n) experiments, nuclei should be very close to the critical condition, and hence a precritical phenomenon has to take place [720]. The enhancement of the differential inelastic cross sections leading to unnatural (abnormal) parity states at high momenta should be greatly enhanced, as the cross sections are modified as $\tilde{\sigma} = \sigma/\epsilon^2$, where the entire dimesic function due to the pionic correlations, including the nucleon particle–hole excitations, becomes close to zero. However, such a precritical enhancement was not observed in various experiments performed around 1980 using (p,p') reactions leading to unnatural (abnormal) parity states [247].

Hence the recent (p, n) experiments suggest a small g'_Δ, while the absence of pion condensation and its precritical phenomenon indicate a large g'_Δ. The solution to this puzzle is considered in [725, 726]. It is related to precisely the relativistic theory of nuclear structure described in the preceding chapters.

The relativistic description of nuclei provides an additional interesting consequence in the pion channel. The relativistic pion Lindhard function Π^{rel} can be written as [39] (see also (7.42))

$$\Pi^{\text{rel}} = \left(\frac{M^*}{M}\right)^2 \Pi , \qquad (13.11)$$

when the widely accepted pseudovector pion–nucleon coupling is adopted ($x = 0$ in (7.1)). Hence, the lowest pion self-energy is reduced by about 50% from the nonrelativistic value. With the use of Π^{rel}, the collectivity that would otherwise be necessary for pion condensation (without isobars) provides no pion condensation, even without the Landau–Migdal short-range correlations, i.e. if $g' = 0$ [39].

If we use the relativistic description, another puzzle about the spin response functions may also be solved. The spin experiments performed at Los Alamos and the recent RCNP experiments using (p, n) reactions provide $R_L/R_T \leq 1$ [727–729]. This ratio R_L/R_T, where R_L and R_T are the longitudinal and transverse spin response functions, respectively, is expected to be larger than 1 in the nonrelativistic framework. This is because the interaction in the pion channel is [247]

$$V_\pi = \frac{f_\pi^2}{m_\pi^2}\left(g' - \frac{q^2}{m_\pi^2 + q^2 - \omega^2}\right) \boldsymbol{\sigma}_1 \cdot \hat{q} \boldsymbol{\sigma}_2 \cdot \hat{q} \boldsymbol{\tau}_1 \cdot \boldsymbol{\tau}_2 , \qquad (13.12)$$

while

$$V_\rho = \frac{f_\pi^2}{m_\pi^2} \left(g' - C_\rho \frac{q^2}{m_\rho^2 + q^2 - \omega^2} \right) \sigma_1 \times \hat{q} \cdot \sigma_2 \times \hat{q} \tau_1 \cdot \tau_2 \qquad (13.13)$$

in the rho meson channel, where $C_\rho \sim 2$. At the large momenta, $q \sim 1.7 \, \text{fm}^{-1}$, at which the experiments were performed, V_π is negative (attractive), but V_ρ is positive (repulsive) if $g' \sim 0.6$. Hence the pion (longitudinal) response is enhanced and the rho meson (transverse) response is quenched. This fact provides $R_L/R_T \geq 1$. However, the relativistic Lindhard function in the pion channel is reduced by about 50% compared with the nonrelativistic value and the expected enhancement in the pion channel is not large enough to overcome this large reduction. The rho meson response is not reduced so much as the pion response, and the ratio could be less than 1 in the relativistic case [730].

In summary, we have discussed the implications of the recent (p, n) spin experiments performed at RCNP at Osaka University [712]. The RCNP data were used to deduce that the Landau–Migdal parameter g'_Δ for the delta–nucleon is less than or equal to 0.2. In the nonrelativistic framework, this small g'_Δ causes pionic collective phenomena in nuclei. This contradicts the results of high-momentum-transfer experiments leading to isovector unnatural (abnormal) parity states, which show no anomaly. We have shown that the relativistic description of nuclei can accommodate these two seemingly contradictory results (large $g'_\Delta = 0.6$ from (p,p') experiments at large momentum transfer, and small $g'_\Delta \leq 0.2$ from (p, n) experiments at small momentum transfer]. The small nucleon effective mass $M*$ due to the deep scalar potential in the relativistic model accommodates the small g'_Δ and the nonexistence of collective pionic effects in nuclei. It has been shown also that the small effective nucleon mass explains the observed abnormal behavior of the ratio of the longitudinal to the transverse spin response, R_L/R_T, which is less than one [726]. We should mention that the possible occurrence of surface pion condensation in finite nuclei has been studied with RMF theory in [731].

13.14 Role of Currents (ω and ρ Fields)

In the RMF theory, the physical observables are mostly sensitive to the time-even fields (S, ω_0, ρ_0). However, the properties of the time-odd fields $(\boldsymbol{\omega}, \boldsymbol{\rho})$, which occur only in nuclear systems with broken time-reversal symmetry, have been investigated to a lesser extent. Consider an odd-A system. In this case the nuclear wave function does not possess time-reversal symmetry, owing to the presence of the last (odd) nucleon. Because of this fact, a nonvanishing nuclear current appears,

$$\boldsymbol{j} = \sum_{i=1}^{A} \bar{\psi} \boldsymbol{\gamma} \psi, \qquad (13.14)$$

yielding spatial components of the vector fields $\boldsymbol{\omega}$ and $\boldsymbol{\rho}$ which would otherwise not be present. We neglect the ρ meson contribution since it is small, and obtain an additional equation for $\boldsymbol{\omega}$ [732]:

$$(-\Delta + m_\omega^2)\boldsymbol{\omega}(\boldsymbol{r}) = g_\omega \boldsymbol{j}_\omega(\boldsymbol{r}) \,. \tag{13.15}$$

The Dirac equation is also modified in this case owing to a nonzero "magnetic" field $g_\omega \boldsymbol{\omega}(\boldsymbol{r})$:

$$\{\boldsymbol{\alpha}(-\mathrm{i}\nabla - g_\omega\boldsymbol{\omega}(\boldsymbol{r})) + \beta M^*(r) + V(r)\}\psi_i(\boldsymbol{r}) = E_i\psi_i(\boldsymbol{r}) \,. \tag{13.16}$$

For an odd-A nucleus this modified Dirac equation should be solved self-consistently together with the Klein–Gordon equations for all the boson fields (the $\boldsymbol{\omega}$ field, in particular). To do this, one needs to work with a deformed basis [423]; see also [727]. From such calculations, it follows that the time-reversal breaking does not have an important effect on properties such as the total binding energy and the radii (it should be noted also that calculations in a deformed basis for heavy A-odd systems are rather tedious).

On the other hand, it is known at present that the time-odd fields are very important in the RMF approach for a proper description of rotating nuclei [478], magnetic moments [191], and pairing correlations [733], and in the relativistic description of the dynamics of giant resonances [473, 475]. In [478] it is shown that nuclear magnetism modifies the expectation values of the single-particle spin, orbital, and total angular momenta along the rotational axis, effectively creating additional angular momentum.

13.15 Fission Barriers

It has always been a challenging task to develop microscopic methods for nuclear collective motion that can be applied to the problem of nuclear fission. Since the 1970s, fully self-consistent mean-field models have become available for nuclei namely nonrelativistic Hartree–Fock models using the Skyrme force or the Gogny force, and the relativistic mean-field theories. In particular, it has been established that the earlier versions of the Skyrme forces need to be modified to obtain agreement with experimental data, the fission barriers being strongly sensitive to the surface, i.e. density-dependent, properties of the effective force. In [734], collective potential energy surfaces for fission have been calculated within the relativistic mean-field approach. These calculations have demonstrated that well-fitted parameterizations within the RMF approximation can provide reasonable fission barriers, which are comparable to those obtained by nonrelativistic methods. Investigations of fission barriers with parameter sets that generate different effective nucleon masses reveal that all relativistic models with an effective mass bigger than $0.7M$ should be excluded from consideration and that the standard best-fit forces with a low effective nucleon mass are preferable. In [735], the investigations of [734] were extended to asymmetric fission and to a larger variety of actinides (^{240}Pu, ^{232}Th, and ^{226}Ra). Standard parameterizations which are well fitted to nuclear ground-state properties were found to provide a reasonable drscription similar to that obtained from nonrelativistic calculations. Furthermore, stable octupole deformations in the ground states of Ra isotopes were considered. These deformations were obtained for a series of isotopes, in qualitative agreement with the results of nonrelativistic approaches. However, the quantitative features are different for the relativistic and nonrelativistic schemes and for different relativistic models.

13.16 Chiral Dynamics and Saturation of Nuclear Structure

Recently, a new method to describe nuclear forces has been developed. It is based an effective-field theory, in particular, chiral perturbation theory (see, for example, [32] and references therein). It has been emphasized in [736–738] that explicit consideration of pion dynamics is also very important for obtaining saturation in nuclear structures. We present here the main results obtained in this field [738–740]. We should mention that in nuclear matter, the Fermi momentum k_F is the relevant momentum scale. At the saturation point one has $k_{F0} \simeq 2m_\pi$, so that both values are of comparable magnitude. The authors of [738] have calculated the equation of state of symmetric nuclear matter in the three-loop approximation of chiral perturbation theory (this approach is closely related to the results obtained in [737]). The contributions to the energy per particle from one- and two-pion exchange are ordered in powers of k_F. It is shown that even at order $O(k_F^4)$, two-pion exchange ensures nuclear binding, the saturation mechanism being very simple (in the chiral limit). This mechanism is revealed to arise via the combination of an attractive k_F^3 term and a repulsive k_F^4 term. The saturation point of nuclear matter and the incompressibility of nuclear matter $K \simeq 250$ MeV are obtained at order $O(k_F^5)$. One fitting parameter is utilized, namely a momentum cutoff $\Lambda \simeq 0.65$ GeV, which parameterizes all necessary short-range dynamics. The prediction of the asymmetry energy 33.8 MeV is in good agreement with empirical values. Pure neutron matter is predicted to be unbound [738, 739]. This approach has been developed further in [739]. In particular, the authors of [739] find that at the equilibrium nuclear-matter density, chiral one- and two-pion exchanges produce an attractive average nuclear field with a depth -53.2 MeV. The momentum dependence of the real part of the single-particle nuclear potential is nonmonotonic, generating an effective nucleon mass $\simeq 0.8M$.

The approach suggested in [737, 738] has been applied in [740] to finite nuclei. In practice, the authors of [740] start with a relativistic point-coupling Lagrangian of the form given by (12.18)–(12.22) (the values α_{TS} and α_{TV} are not taken into consideration, calculations being restricted to the Hartree approximation for $N = Z$ nuclei). $\mathcal{L}_{hot}$ is not taken into account; however, an extra density dependence of the Lagrangian utilized in [740] is generated

13.16 Chiral Dynamics and Saturation of Nuclear Structure

by the density dependence of the coupling constants $\alpha_S(\rho)$, $\alpha_V(\rho)$, $\delta_S(\rho)$, $\delta_V(\rho)$; the vertex functions $\alpha_S(\rho)$ and $\alpha_V(\rho)$ were chosen to be in a form composed of two parts each. The components $\alpha_{S,V}^0$, which do not depending on density, are governed by the QCD condensates, while the components $\Delta\alpha_{S,V}(\rho)$ refer to the pionic fluctuations; these are reexpressed as density-dependent corrections to the mean fields. The choice of $\delta_S(\rho)$ and $\delta_V(\rho)$ was made in [740] on the basis of dimensional considerations. The calculations in [740] were performed for only two nuclei, ^{16}O and ^{40}Ca. The accuracy of the calculations of binding energies is about 8% and 5%, respectively, while for the charge radii the accuracy is about 2% and 4%, respectively. The authors of [740] are planning to improve the results for the spin–orbit splittings by considering the exchange (Fock) contributions and to extend their calculations to $N \neq Z$ nuclei.

We should mention also that in [741], problems related to the nuclear spin–orbit force are considered using the two-loop approximation of chiral perturbation theory. In [742], shell-model calculations were performed in which low-momentum vertices derived from a chiral NN potential (Idaho B [743]) were used as input instead of G-matrix vertices. The authors of [742] calculated spectra and binding energies for the three nuclei ^{18}O, ^{134}Te, and ^{210}Po, the results being in good agreement with experiment.

14 Summary and Outlook

We have discussed the relativistic framework for describing nuclear structures and reactions, a new approach in nuclear physics at low and intermediate energies, born about 30 years ago. The basic feature of this framework is the existence in the nucleus of two strong fields with the transformation properties of a world scalar (an attractive field, with a depth $\approx -420\,\mathrm{MeV}$) and of the time-component of a four-vector (a repulsive field, with a value $\approx +330\,\mathrm{MeV}$). These values were obtained by an empirical method, making no assumptions concerning the nature of the fields (they can be derived from the known values of the depth of the shell-model potential and the strength of the spin–orbit force in finite nuclei). The RBHF theory also provides similar values for the scalar and vector fields.

Under these conditions the Schrödinger equation for single-particle nucleon states becomes inapplicable (the values of the fields being comparable to the bare nucleon mass). This leads us to use the Dirac equation rather than the Schrödinger equation, and we encounter the necessity to treat the nucleus as a relativistic system.

The relativistic description provides a new saturation mechanism in nuclear structure even in the mean-field approximation. The repulsive vector field has a value proportional to the density of the system. The scalar field generates attraction and its depth, above a certain density, becomes constant, i.e. the scalar field becomes "frozen". This property of the scalar field is provided by the presence of the lower component of the nucleon Dirac spinor, which is ruled out from consideration from the beginning in the nonrelativistic theory of the atomic nucleus.

The lower component of the single-particle wave function, in turn, appears to be greatly enhanced in the relativistic theory owing to the small value of the nucleon effective mass in the nuclear medium.

The relativistic treatment of the stability problem in weak external fields (the problem of pion condensation) brings new, important features into the solution of this problem. In particular, in the case of pseudovector pion–nucleon coupling, for a certain choice of parameters (determining the nuclear-matter ground state) the system is stable at all densities, such behavior being related to the relativistic treatment. The polarization integral appears in this case with a factor (the squared ratio of the effective nucleon mass to the bare mass) that suppressing greatly the contribution of the polarization integral. This effect brings stability to the system. Relativistic effects (of a different character) are also manifested in investigations of the stability problem for pseudoscalar pion–nucleon coupling. New relativistic effects appear in the problem of pion condensation because the transition operator in this case mixes the upper and lower components of the nucleon wave function, the latter component being greatly enhanced.

The dominant principles utilized in constructing an effective relativistic field theory for nuclear structures are related to the basic symmetries (Lorentz covariance, isotopic symmetry, and parity conservation) and to the constraints imposed by causality, unitarity, vector dominance, chiral symmetry, and renormalizability.[1]

Relativistic theory provides in this case a natural basis for studying nuclear ground-state properties. Using a small number of fitting parameters, having a clear physical meaning, the relativistic mean-field approximation reproduces the nuclear bulk properties reasonably well. A version of this approximation extended by including nonlinear meson self-interaction terms is now used as a standard model for the description of finite nuclei. It has been shown that the RMF model reproduces the properties of stable nuclei over a wide mass range of the periodic table. Extensive studies in the relativistic framework have been performed on nuclei away from the stability line; these studies have been motivated by recent advances in the experimental study of unstable nuclei using radioactive-nuclear-beam facilities. It is remarkable that the relativistic models reproduce the properties of both nuclei far away from stability and stable nuclei.

The relativistic models have also been applied to the study of deformed nuclei. It has been demonstrated that these models describe successfully the deformation and other properties of both stable and unstable nuclei. The equilibrium shapes of isotopes have been investigated through comparisons with experimental information on deformations and charge radii, which has become available recently from isotope shift measurements, even for unstable nuclei. The coexistence of more than one shape has been predicted for some nuclei, and the change of shape between prolate and oblate along sequences of isotopes has been explored.

The recent development of radioactive nuclear beams has begun a new era for studying unstable nuclei and for exploring new physical phenomena.

---

[1] The condition of renormalizability may exclude from consideration some classes of interaction Lagrangians; it leads to certain restrictions on the character of self-interactions of the boson fields. The initial belief was that in the case of a renormalizable theory, the parameters fitted to the properties of nuclear structures under normal conditions could be directly extrapolated into regions of extreme values. However, at present, renormalizability is considered as a heuristic rather than an obligatory requirement of the effective-field theory. The current point of view is that a nonrenormalizable effective theory can generate a self-consistent description of nuclear structures [82, 84, 85, 116].

One of the most interesting discoveries made with radioactive beams is the existence of a neutron halo in some nuclei, containing neutrons with a very low separation energy and having a far-extending neutron distribution. The experiments by Tanihata et al. [155, 156] showed for the first time quite large reaction cross sections in collisions of ^{11}Li with high-Z targets, an indication that the matter radius of ^{11}Li is much larger than what is predicted by the usual $r_0 A^{-1/3}$ law. The relativistic approach may be considered as a useful and fruitful tool for theoretical investigations of exotic nuclei.

Although conventional nonrelativistic models and RMF theory generate many similar features for stable nuclei, the two approaches appear to be significantly different for drip-line nuclei, in particular, in the description of the spin–orbit term and its isospin dependence.

The success of the relativistic Hartree and Hartree–Fock approximations has ensured a good basis for the study of nuclear excitations in relativistic models. Relativistic RPA calculations have been extended to include nonlinear self-interaction terms.

So we may say in summary that the relativistic theory is strongly supported; this support comes from a large amount of experimental data, which can be described in the relativistic framework on solid theoretical grounds. Some of these phenomena cannot be reproduced at all (saturation in the BHF scheme, for example) at the level of the nonrelativistic theory, and other phenomena (e.g. the spin–orbit interaction) need extra fitting parameters. The strongest support for the relativistic approximation is obtained from calculations of spin observables (the analyzing power and the spin rotation).

Recently, it has been shown also that pseudospin symmetry can be treated naturally as a relativistic symmetry of the Dirac Hamiltonian generated by the current theories of nuclear structure.

The relativistic approach has many astrophysical applications also. In order to clarify the mechanism of supernova explosions and the associated phenomena, numerical simulations are necessary. For this purpose, we need an equation of state of nuclear matter that covers a wide density and temperature range with various proton fractions.

To calculate an EOS for astrophysical purposes, one has to perform consistent calculations for high-density nuclear matter, inhomogeneous nuclear matter, and finite nuclei for the wide range of density, proton fraction, and temperature, which occurs inside neutron stars and supernovas. For this purpose, one has to study the properties of dense matter with both homogeneous and inhomogeneous distributions in the relativistic framework. Relativity leads also to some distinctive properties in the EOS compared with the case in the nonrelativistic framework. Therefore, it is very interesting and important to study astrophysical phenomena such as supernova explosions, neutron star cooling, and neutron star merging using a relativistic EOS.

Support for the relativistic approach also comes from QCD, which is considered as the underlying theory of strong interactions (the existence in the nucleus of two strong fields, S and V, close in magnitude, follows from QCD sum rules).

Vector dominance and chiral symmetry, two basic principles of low-energy QCD, can be successfully incorporated into the structure of the relativistic theory. The composite character of the nucleons (their electromagnetic structure) is quite naturally taken into account in the framework of the RMF approach via the mechanism of vector dominance.

Models respecting chiral symmetry are of outstanding importance in strong-interaction physics. Many attempts have been made to explain nuclear structure using relativistic chiral models. We have considered here the consequences of the assumption that the Lagrangian for interacting meson and baryon fields is invariant under chiral symmetry in addition to the conventional isotopic symmetry, since in this case the notion of the scalar field can be utilized widely. This fact enables one to make a link between the chiral approach and the nuclear Dirac phenomenology. The possibility of the existence of bubble configurations in finite nuclei has been discussed in this connection.

In view of the great success achieved in describing electroweak interactions on the basis of elevating the global symmetry to a local one and the fundamental role of local symmetry in quantum chromodynamics, it appears to be important to formulate the nuclear chiral model in such a way that the local gauge symmetry related to the group $U(1) \times SU(2) \times SU(2)$ (which consists of phase transformations, rotations in isospin space, and chiral rotations) is taken into account. In this case the vector fields are considered as the gauge fields. We have introduced a gauge model with all vector fields appearing in the nuclear-structure context (ρ, a, ω, A) treated as Yang–Mills fields. Eight gauge fields are obtained in this case, the pion being the only Goldstone particle of the model.

The relativistic approach is a rapidly developing theory. The number of publications on this subject is increasing like an avalanche. We would like to hope the current book will help in the further study and development of this field.

A Appendices

A.1 Four-Dimensional Notation and the Dirac Matrices

Four-dimensional tensorial indices are designated by Greek letters $\mu, \gamma, \delta, \ldots$, which may have the values 0, 1, 2, 3. The space–time metric is determined by the metric tensor $g_{\mu\nu}$ ($g_{00} = 1$, $g_{11} = g_{22} = g_{33} = -1$). The coordinates of a four-vector have the following ordering: $A^\mu = (A^0, A^1, A^2, A^3)$. The coordinates of a three-vector are designated by Latin letters, for example,

$$\boldsymbol{A} = (A^1, A^2, A^3) = \{A^i\}, \qquad i = 1, 2, 3. \tag{A.1}$$

The contravariant coordinates A^μ and covariant coordinates A_μ of a four-vector are related by the following relation:

$$A_\mu = g_{\mu\nu} A^\nu, \tag{A.2}$$

where summation over indices that appear twice is assumed in the left-hand side of this equation. The four-dimensional scalar product is defined as follows:

$$A_\mu B^\mu = A_0 B_0 - \boldsymbol{A} \cdot \boldsymbol{B}, \tag{A.3}$$

$$x^2 = x_\mu x^\mu = t^2 - \boldsymbol{x}^2. \tag{A.4}$$

The operators for calculating derivatives over four-coordinates are given by

$$\partial_\mu \equiv \frac{\partial}{\partial x^\mu}, \tag{A.5}$$

$$\Box \equiv \frac{\partial}{\partial x_\mu} \cdot \frac{\partial}{\partial x^\mu}. \tag{A.6}$$

We use here the following standard representations of the Dirac matrices:

$$\{\gamma^i\} = \boldsymbol{\gamma} = \begin{pmatrix} 0 & \boldsymbol{\sigma} \\ -\boldsymbol{\sigma} & 0 \end{pmatrix}, \qquad i = 1, 2, 3, \tag{A.7}$$

$$\beta = \gamma^0 = \gamma_0 = \begin{pmatrix} I & 0 \\ 0 & -I \end{pmatrix}, \tag{A.8}$$

$$\gamma^r = \begin{pmatrix} 0 & \sigma_r \\ -\sigma_r & 0 \end{pmatrix}, \qquad (A.9)$$

$$\boldsymbol{\alpha} = \gamma^0 \boldsymbol{\gamma} = \begin{pmatrix} 0 & \boldsymbol{\sigma} \\ \boldsymbol{\sigma} & 0 \end{pmatrix}. \qquad (A.10)$$

Since the Dirac Hamiltonian must be Hermitian, we have $\alpha^\dagger = \alpha$, $\beta^\dagger = \beta$. Also,

$$\gamma_5 = \gamma^5 = i\gamma^0 \gamma^1 \gamma^2 \gamma^3 = \begin{pmatrix} 0 & 1 \\ 1 & 0 \end{pmatrix}. \qquad (A.11)$$

It should be emphasized that some authors use the matrix given by (A.11) multiplied by a factor i for matrix γ_5. Further,

$$\sigma^{\mu\nu} = \frac{i}{2}[\gamma^\mu, \gamma^\nu], \quad \sigma_{\mu\nu} = -\sigma_{\nu\mu} \quad \mu, \nu = 0, 1, 2, 3, \qquad (A.12)$$

where I is the 2×2 identity matrix, and $\boldsymbol{\sigma}$ is the 2×2 Pauli spin matrix, such that

$$\sigma_x = \sigma^1 = \begin{pmatrix} 0 & 1 \\ 1 & 0 \end{pmatrix},$$

$$\sigma_y = \sigma^2 = \begin{pmatrix} 0 & -i \\ i & 0 \end{pmatrix},$$

$$\sigma_z = \sigma^3 = \begin{pmatrix} 1 & 0 \\ 0 & -1 \end{pmatrix}. \qquad (A.13)$$

This matrix can be represented in spherical coordinates in the following form:

$$\boldsymbol{\sigma} = \begin{pmatrix} \cos\theta & \sin\theta\, e^{-i\Phi} \\ \sin\theta\, e^{i\Phi} & -\cos\theta \end{pmatrix} \widehat{\boldsymbol{u}}_r + \begin{pmatrix} -\sin\theta & \cos\theta\, e^{-i\Phi} \\ \cos\theta\, e^{i\Phi} & \sin\theta \end{pmatrix} \widehat{\boldsymbol{u}}_\theta$$

$$+ \begin{pmatrix} 0 & -ie^{-i\Phi} \\ ie^{i\Phi} & 0 \end{pmatrix} \widehat{\boldsymbol{u}}_\Phi, \qquad (A.14)$$

where $\widehat{\boldsymbol{u}}_r$, $\widehat{\boldsymbol{u}}_\theta$, $\widehat{\boldsymbol{u}}_\Phi$ are the unit vectors in the r, θ, Φ directions, respectively, of spherical coordinates.

The rules for operating with the Dirac matrices are determined totally by the commutation relations. Some of these relations that are useful in practical calculations are given below. All pairs of different matrices γ^μ anticommute,

$$\gamma^\mu \gamma^\nu + \gamma^\nu \gamma^\mu = 2g^{\mu\nu}, \qquad (A.15)$$

while the squared value of each of them is given by

$$(\gamma^1)^2 = (\gamma^2)^2 = (\gamma^3)^2 = -1, \qquad (\gamma^0)^2 = 1. \qquad (A.16)$$

The matrix γ^0 is Hermitian, while all $\boldsymbol{\gamma}$'s are anti-hermitian matrices:

$$\boldsymbol{\gamma}^\dagger = -\boldsymbol{\gamma}, \qquad \gamma^{0\dagger} = \gamma^0. \qquad (A.17)$$

A.1 Four-Dimensional Notation and the Dirac Matrices

The following property is valid also for γ^μ matrices: $(\gamma^\mu)^\dagger = \gamma^0 \gamma^\mu \gamma^0$. Notice that all matrices $\boldsymbol{\alpha}$, β anticommute:

$$\alpha_i \alpha_k + \alpha_k \alpha_i = 2\delta_{ik} , \tag{A.18}$$

$$\beta\boldsymbol{\alpha} + \boldsymbol{\alpha}\beta = 0 , \qquad \alpha_i^2 = \beta^2 = 1 . \tag{A.19}$$

They are Hermitian matrices.

It is easy to see also that

$$\gamma^5 \gamma^\mu + \gamma^\mu \gamma^5 = 0 , \qquad (\gamma^5)^2 = 1 , \tag{A.20}$$

$$\boldsymbol{\alpha}\gamma^5 - \gamma^5 \boldsymbol{\alpha} = 0 , \qquad \beta\gamma^5 + \gamma^5 \beta = 0 , \tag{A.21}$$

$$(\gamma^5)^\dagger = \gamma^5 , \qquad [\gamma_5, \sigma_{\mu\nu}] = 0 . \tag{A.22}$$

We should mention that there are 16 linearly independent matrices Γ_A given by the following entities:

$$1, \quad \gamma_5, \quad \gamma_\mu, \quad \gamma_5 \gamma_\mu, \quad \sigma_{\mu\nu} \tag{A.23}$$

Any 4×4 matrix X may be written as a combination of these matrices:

$$X = \sum_{A=1}^{16} x_A \Gamma_A , \tag{A.24}$$

where

$$x_A = \frac{1}{4} \text{Tr}(X \Gamma_A) . \tag{A.25}$$

The relativistic parity and total-angular-momentum operators are defined as follows:

$$P = \gamma^0 P_{\text{NR}} \tag{A.26}$$

and

$$\boldsymbol{J} = \boldsymbol{r} \times \boldsymbol{p} \begin{pmatrix} I & 0 \\ 0 & I \end{pmatrix} + \frac{\hbar}{2} \begin{pmatrix} \boldsymbol{\sigma} & 0 \\ 0 & \boldsymbol{\sigma} \end{pmatrix} , \tag{A.27}$$

where the spherical-coordinate form for the orbital-angular-momentum operator is given by

$$\boldsymbol{r} \times \boldsymbol{p} = -\mathrm{i}\hbar \left\{ \widehat{\boldsymbol{u}}_\Phi \frac{\partial}{\partial \theta} - \frac{1}{\sin\theta} \widehat{\boldsymbol{u}}_\theta \frac{\partial}{\partial \Phi} \right\} . \tag{A.28}$$

The time-reversal operator for the Dirac representation has the following form:

$$T = \mathrm{i} \begin{pmatrix} \sigma_y & 0 \\ 0 & \sigma_y \end{pmatrix} K , \tag{A.29}$$

where K stands for the complex conjugation operator.

A bar over a wave function corresponds to the conventional bar notation for Dirac spinors:

$$\bar{u} = (\gamma^0 u)^\dagger . \tag{A.30}$$

A.2 Properties of the Ground State of Nuclear Matter in the Walecka Model

The ground-state properties of relativistic nuclear matter in the Walecka model [74] are obtained from the Lagrangian (4.2), containing only the scalar–isoscalar field φ, the vector–isoscalar field ω_μ, and the nucleon field ψ_τ, where $\tau = \pm 1/2$ is the nucleon isospin projection (the self-interactions of the φ field are not taken into account). Only isoscalar fields are present in the ground state owing to isotopic symmetry.

The ground-state properties are determined by dimensionless parameters $M^2 g_i^2/m_i^2$. According to [74], these parameter are

$$C_\sigma^2 = M^2 \frac{g_\sigma^2}{m_\sigma^2} = 266.9\,, \quad C_\omega^2 = M^2 \frac{g_\omega^2}{m_\omega^2} = 195.7\,. \quad (A.31)$$

For these values, the equilibrium ground-state density is $\rho_0 = 0.19$ fm^{-3} and the binding energy per nucleon is $E/A = -15.75$ MeV (the nucleon mass is taken to be equal to 938 MeV). The quasiparticle energy is

$$E_p = g_\omega \omega_0 + \sqrt{p^2 + M^{*2}}\,, \quad M^* = M + g_\sigma \varphi\,. \quad (A.32)$$

The ground-state meson fields are

$$g_\omega \omega_\mu = \delta_{\mu 0} \frac{g_\omega^2}{m_\omega^2} \rho_V = \delta_{\mu 0} \frac{g_\omega^2}{m_\omega^2} \sum_{\tau,\lambda,|\boldsymbol{p}| \le p_F} \bar\psi_{\tau\boldsymbol{p}\lambda} \gamma^0 \psi_{\tau\boldsymbol{p}\lambda}\,, \quad (A.33)$$

$$g_\sigma \varphi = -\frac{g_\sigma^2}{m_\sigma^2} \rho_S = -\frac{g_\sigma^2}{m_\sigma^2} \sum_{\tau,\lambda,|\boldsymbol{p}| \le p_F} \bar\psi_{\tau\boldsymbol{p}\lambda} \psi_{\tau\boldsymbol{p}\lambda}\,, \quad (A.34)$$

where $\boldsymbol{p}$ and λ stand for the quasiparticle momentum and polarization, and p_F is the Fermi momentum. The sums on the right-hand sides of (A.33) and (A.34) include both protons and neutrons. Such sums are calculated according to the following prescription:

$$\sum_{\tau,\lambda,|\boldsymbol{p}| \le p_F} \bar\psi_{\tau\boldsymbol{p}\lambda} O \psi_{\tau\boldsymbol{p}\lambda} = \int_{|\boldsymbol{p}| \le p_F} \frac{d^3 p}{(2\pi)^3 \sqrt{p^2 + M^{*2}}} \operatorname{Tr} O(\hat{p} + M^*)\,, \quad (A.35)$$

where O is an arbitrary operator, $\hat{p} = \gamma^\mu p_\mu$, $p_\mu(l_p, \boldsymbol{p})$, and $l_p = \sqrt{p^2 + M^{*2}}$, p_μ being the quasiparticle four-momentum. Putting $O = \gamma^0$, we obtain a conventional expression for the density (time component of the nuclear current):

$$\rho_V = \int_{|\boldsymbol{p}| \le p_F} \frac{d^3 p}{(2\pi)^3} \operatorname{Tr} \frac{\gamma^0 (\hat{p} + M^*)}{l_p} = \frac{2 p_F^3}{3 \pi^2}\,. \quad (A.36)$$

Putting $O = 1$, we obtain the following expression for the scalar density:

$$\rho_S = \int_{|\boldsymbol{p}| \leq p_F} \frac{d^3 p}{(2\pi)^3} \, \text{Tr} \, \frac{\hat{p} + M^*}{l_p}$$

$$= \frac{M^*}{\pi^2} \left(p_F (p_F^2 + M^{*2})^{1/2} - M^{*2} \ln \frac{p_F + (p_F^2 + M^{*2})^{1/2}}{M^*} \right). \quad (A.37)$$

This is the equation for ρ_S since, according to (A.32) and (A.34),

$$M^* = M - \frac{g_\sigma^2}{m_\sigma^2} \rho_S. \quad (A.38)$$

By solving (A.37), we obtain the density dependence of ρ_S and M^*/M. It is very important that the effective mass M^* is a decreasing function of the density. In the ultrarelativistic limit $\rho_V \to \infty$, M^* is zero. Indeed, as follows from (A.37), the ultrarelativistic scalar density is

$$\rho_{S\infty} = \frac{m_\sigma^2}{g_\sigma^2} M = 0.4025 \text{ fm}^{-3}. \quad (A.39)$$

If we take the ground-state density as $\rho_0 = 0.19$ fm^{-3}, we obtain

$$\rho_{S0} = 0.176 \text{ fm}^{-3} \qquad M_0^*/M = 0.56. \quad (A.40)$$

Thus the "observed" scalar density is about half of the ultrarelativistic value.

The significance of this fact manifests itself in the response of the system to weak external fields. In such fields, the virtual particle–hole pairs are excited. According to (A.32) the energy of such a pair is

$$E_{ph} = E_{p_1 > p_F} - E_{p_2 < p_F} = (p_1^2 + M^{*2})^{1/2} - (p_2^2 + M^{*2})^{1/2}$$

$$\simeq (|p_1 - p_2| \ll p_F) \simeq \frac{p_F (p_1 - p_2)}{(p_F^2 + M^{*2})^{1/2}} \approx \left(\frac{v_F}{c}\right)_{\text{eff}} (p_1 - p_2). \quad (A.41)$$

As can be seen from (A.41), the vector field does not enter E_{ph} and thus the behavior of the system in an external field is determined by the scalar field only, through the nucleon effective mass. Putting $p_F = 1.42$ fm^{-1} and $M^*/M = 0.56$ in (A.41), we obtain the following value for the Fermi velocity:

$$\left(\frac{v_F}{c}\right)_{\text{eff}} = \frac{p_F}{\sqrt{p_F^2 + M^{*2}}} = 0.47, \quad (A.42)$$

which is not small compared with unity.

A.3 General Form of Local Dirac Equation

Consider the most general form of the local stationary Dirac equation [58]. This contains scalar (S), vector (V), pseudoscalar (PS), axial-vector (A), and tensor (T) components:

$$\{\boldsymbol{\alpha}\cdot\boldsymbol{p}+\beta[M+S(\boldsymbol{r})+\gamma^{\mu}U_{\mathrm{V}\mu}(\boldsymbol{r})+\gamma^{5}U_{\mathrm{PS}}(\boldsymbol{r})$$
$$+\gamma_{\mu}\gamma_{5}U_{\mathrm{A}\mu}(\boldsymbol{r})-\sigma_{\mu\nu}U_{\mathrm{T}\mu\nu}(\boldsymbol{r})]\}\psi(\boldsymbol{r})=E\psi(\boldsymbol{r})\,. \quad (\mathrm{A.43})$$

We make the conventional assumptions of good parity and good total angular momentum for the single-particle states. This actually means the invariance of the Hamiltonian in (A.43) under rotation and reflection in space and time, which in turn means that each component in (A.43) should commute with the relativistic-parity (P) and total-angular-momentum ($\boldsymbol{J}$) operators (see Appendix A1):

$$[U(\boldsymbol{r}),\boldsymbol{J}]=[U(\boldsymbol{r}),P]=0\,. \quad (\mathrm{A.44})$$

Application of the assumptions made above to the scalar component of the (A.43) requires that the potential S is independent of angle, i.e. $S(\boldsymbol{r})=S(r)$. Using the definition of the scalar product, we obtain

$$\gamma^{\mu}U_{\mathrm{V}\mu}(\boldsymbol{r})=\gamma^{0}V(\boldsymbol{r})-\boldsymbol{\gamma}\cdot\boldsymbol{U}_{\mathrm{V}}(\boldsymbol{r})\,. \quad (\mathrm{A.45})$$

Owing to the rotation and reflection invariance, $V(\boldsymbol{r})=V(r)$, i.e. $V(r)$ is also angle-independent. For the same reasons, $U_{\mathrm{PS}}(\boldsymbol{r})=U_{\mathrm{A}}(\boldsymbol{r})=0$. However, the radial components of the two three-vectors $U_{\mathrm{V}}^{j}(\boldsymbol{r})$ and $U_{\mathrm{T}}^{0j}(\boldsymbol{r})$ ($j=1,2,3$) may still contribute (see [58] for more details). These radial components are denoted by $U_{\mathrm{V}}^{r}(r)$ and $U_{\mathrm{T}}^{r}(r)$ for the vector and tensor potentials, respectively:

$$\boldsymbol{U}_{\mathrm{V}}(\boldsymbol{r})=\widehat{\boldsymbol{u}}_{r}U_{\mathrm{V}}^{r}(r)\,, \quad (\mathrm{A.46})$$

$$\sigma^{\mu\nu}U_{\mathrm{T}\mu\nu}(\boldsymbol{r})=+\gamma^{0}\boldsymbol{\gamma}\boldsymbol{U}_{\mathrm{T}}(\boldsymbol{r})\,, \quad (\mathrm{A.47})$$

$$\boldsymbol{U}_{\mathrm{T}}(\boldsymbol{r})=\widehat{\boldsymbol{u}}_{r}U_{\mathrm{T}}^{r}(r)\,. \quad (\mathrm{A.48})$$

So, finally, the single-nucleon relativistic potential $U(r)$ has only four components:

$$U(r)=\beta[S(r)+\beta V(r)-\gamma^{r}U_{\mathrm{V}}^{r}(r)-\beta\gamma^{r}U_{\mathrm{T}}^{r}(r)]\,. \quad (\mathrm{A.49})$$

The expression for the radial Dirac matrix γ^{r} is given in Appendix A1.

Further restrictions on the various terms in (A.49) are obtained by requiring that the single-particle Hamiltonian be Hermitian. From this requirement, it follows that S, V, and U_{V}^{r} are real and that U_{T}^{r} is pure imaginary, $U_{\mathrm{T}}^{r}(r)=\mathrm{i}T(r)$, so that the tensor component in (A.49) is given by $-\mathrm{i}\beta\boldsymbol{\alpha}T(r)\widehat{\boldsymbol{u}}_{r}$.

Now is an appropriate place to note a very specific feature of the term $U_{\mathrm{V}}^{r}(r)$. If the various components in (A.49) are not state-dependent, then requiring that $U(r)$ be Hermitian is equivalent to requiring that $U_{\mathrm{V}}^{r}(r)$ be real, in contrast to the requirements of time-reversal invariance:

$$[U(r),T]=0\,, \quad (\mathrm{A.50})$$

where T is the time-reversal operator for the Dirac representation [58]. To be in accordance with time-reversal invariance, $U_V^r(r)$ must be pure imaginary.[1] We should mention that the Hartree–Fock potentials are state-dependent and in this case hermiticity is restored, $U_V^r(r)$ still being pure imaginary. A more detailed discussion of this problem is given in [57, 58].

A.4 Equivalent Local Dirac Nuclear Models

The Dirac phenomenology is consistent with a Dirac equation of the following form [744]:

$$[\boldsymbol{\alpha} \cdot \boldsymbol{p} + \beta(M + S) - (E - V - V_\mathrm{C}) - \mathrm{i}\beta\boldsymbol{\alpha}\widehat{\boldsymbol{u}}_r T]\psi(\boldsymbol{r}) = 0 , \quad (\mathrm{A.51})$$

where S is a world scalar, V is the time component of a four-vector, V_C is the Coulomb potential, T is the tensor component of the potential (see Appendix A3) ($\mathrm{i}T(r) = U_\mathrm{T}^r(r)$), and $\widehat{\boldsymbol{u}}_r$ is the unit vector in the r-direction of the spherical coordinates.

In most applications of the Dirac phenomenology (bound-state problems and scattering), a scalar–vector (SV) model is used that includes attractive (S) and repulsive (V) components only. The empirical evidence requires these components to be strong enough in magnitude and close in absolute value. The tensor component in this case is related to the vector meson tensor interaction, it is omitted in many applications (one should bear in mind that this type of potential influences the value of the spin–orbit splitting).

In [744] it is shown that ST or VT models (alongside the SV model) can fit the experimental data (the vector and scalar potentials being redefined).

Consider the following transformation of a single-nucleon wave function satisfying (A.51):

$$\psi(\boldsymbol{r}) = \mathrm{e}^{\mathrm{i}\sum_A \Gamma^A F_A(r)}\Phi(\boldsymbol{r}) , \quad (\mathrm{A.52})$$

where Γ^A is any of the 16 linearly independent 4×4 matrices which appear often in applications of the Dirac theory (see Appendix A1), and $F_A(r)$ are complex functions which tend to zero for $r \to \infty$. In particular, let us introduce

$$\psi(\boldsymbol{r}) = \mathrm{e}^{\mathrm{i}\gamma^0 F(r)}\Phi(\boldsymbol{r}) . \quad (\mathrm{A.53})$$

The transformations (A.52) and (A.53) leave the asymptotic behavior of the wave function unaltered. Using (A.53), we obtain

$$\left[(E - V - V_\mathrm{C})\gamma^0 \cos 2F - (M + S)\cos 2F + \mathrm{i}(E - V - V_\mathrm{C})\sin 2F \right.$$
$$\left. - \mathrm{i}(M + S)\gamma^0 \sin 2F + \mathrm{i}\gamma^0\boldsymbol{\gamma}\widehat{\boldsymbol{u}}_r T + \gamma^0\boldsymbol{\gamma}\widehat{\boldsymbol{u}}_r \frac{\partial F}{\partial r} - \boldsymbol{\gamma} \cdot \boldsymbol{p}\right]\Phi(\boldsymbol{r}) = 0 . \quad (\mathrm{A.54})$$

---

[1] For the tensor component $U_\mathrm{T}^r(r)$, both hermiticity and time-reversal invariance require this component to be pure imaginary. Note that in the Schrödinger theory, time-reversal invariance and hermiticity generate identical restrictions for time-independent local interactions.

By making different choice for $F(r)$, we can change from an SV model to a totally equivalent ST, VT, or SVT model. To obtain the ST model, we can remove the V component by imposing the following condition:

$$(E - V - V_C)\cos 2F - i(M + S)\sin 2F = (E - V_C). \tag{A.55}$$

So we obtain a new scalar potential S',

$$M + S' = (M + S)\cos 2F - i(E - V - V_C)\sin 2F, \tag{A.56}$$

and a new tensor potential T',

$$T' = T - i\frac{\partial F}{\partial r}. \tag{A.57}$$

Alternatively, we can eliminate the S-component and obtain a vector–tensor model using the following condition:

$$(M + S)\cos 2F - i(E - V - V_C)\sin 2F = M. \tag{A.58}$$

In this case we obtain a new vector potential V',

$$E - V' - V_C = (E - V - V_C)\cos 2F - i(M + S)\sin 2F, \tag{A.59}$$

and a new tensor potential T',

$$T' = T - i\frac{\partial F}{\partial r}. \tag{A.60}$$

Bearing (A.55 and A.58) in mind, we may conclude that $F(r) = 0$ outside the nucleus.

The scalar–vector combination is the preferable model. As was mentioned earlier, this choice of phenomenology is strongly supported theoretically. A more detailed discussion of this problem can be found in [744].

A.5 Nucleon Effective Mass in the Nuclear Medium

Various definitions of the nucleon effective mass have been introduced in the relativistic theory:

1. The natural relativistic effective nucleon mass is given by

$$M^* = M + S; \tag{A.61}$$

   this definition follows directly from the structure of the Dirac equation.
2. The energy dependence of the optical model potential is characterized by the following value (see [128] and references therein):

$$\frac{\overline{M}}{M} = 1 - \frac{dU_e}{d\varepsilon} = 1 - \frac{V}{M}. \tag{A.62}$$

3. Arnold and Clark [131] introduced yet another effective mass (which follows from the equation relating the lower and upper components of the Dirac spinor):

$$\frac{\mathcal{M}}{M} = 1 - \frac{V-S}{2M} = \frac{M^* + \overline{M}}{2M}. \tag{A.63}$$

Further discussion of the different definitions of the nucleon effective mass can be found in [123]. For the original Walecka model (without nonlinear terms), one has

$$\frac{M^*}{M} = 0.61, \quad \frac{\overline{M}}{M} = 0.70, \quad \frac{\mathcal{M}}{M} = 0.655. \tag{A.64}$$

In the case where the nonlinear terms are included, an accurate description of observables in finite nuclei places the value of M^*/M in the range $0.58 \leq M^*/M \leq 0.64$ [745]. This result is supported by investigations of superheavy nuclei [699]. It should be mentioned here that the value of M^*/M in the relativistic theory is directly related (see Chap. 4, for example) to the strength of the spin–orbit force in nuclei [132] (the spin–orbit splittings in nuclei are independent of the specific form of the nonlinear interactions).

The value of M^*/M mentioned above appears to be rather small (for example, a phenomenological value $M^*/M \approx 0.8$ is obtained from the position of the giant quadrupole resonance in heavy nuclei [699]). Also, the level density near the Fermi surface obtained in the RMF approximation is too small, since M^* is too small.

For this reason, attempts have been made to slightly increase the value of M^*/M obtained at the relativistic mean-field level.[2] This point is discussed in [745] (see also the review paper by Ring and Afanasjev [486]).

For example, the authors of [745] consider the possibility of generating some extra spin–orbit force (in addition to the conventional relativistic spin–orbit force $\sim (1/r)(\mathrm{d}/\mathrm{d}r)(S-V)\boldsymbol{l}\cdot\boldsymbol{\sigma}$) by the tensor coupling of the ω meson. This coupling of the ω meson is usually taken to be zero in OBE models or to have a small value of $f_\omega/g_\omega = -0.12$ in nuclear-structure calculations (this value is determined by the anomalous magnetic moments [170]). To obtain a sensitive modification of the conventional spin–orbit force in finite nuclei, one needs to use a value of f_ω/g_ω that is large enough. However, the authors of [745] argue that, as an effective coupling in nuclei (absorbing higher-order effects at the mean-field level), the ω meson tensor coupling could be much larger than that obtained with $f_\omega/g_\omega = -0.12$ and still be of natural size. However, the effect of the isovector (ρ meson) tensor coupling should be taken into consideration and investigated in this connection also (see Chap. 5).

The decrease of the nucleon effective mass inside the nucleus due to the presence of the strong attractive scalar field is hard to check unambiguously.

---

[2] It should be mentioned, however, that in the point-coupling model the value $M^*/M = 0.74$ is obtained.

The problem is that in many cases the observables are sensitive to the nuclear surface, where the change of the nucleon mass is weaker.

However, in [746] an attempt has been made to investigate this problem through photonuclear sum rules. Theoretical sum rule predictions, based on the ground-state expectation values, may be more sensitive to the nuclear interior, i.e. to the region where the effective nucleon mass could actually manifest itself. The authors of [746] have considered the electric-dipole sum rule. In this case the experimental value for the total integrated photoabsorption cross section is larger by a factor of two than the classical Thomas–Reiche–Kuhn (TRK) sum rule result. These authors used a relativistic self-consistent theory, allowing modifications from the filled Dirac sea in the Hartree wave functions, and have shown that the lower nucleon effective mass results in a large increase over the TRK result.

We should mention also [747], where a relation is established that links the effective nucleon mass, the incompressibility, and the effective meson mass (see also [748]).

Further discussion of the problem of the in-medium nucleon effective mass in the RMF theory can be found in [749]. In [750] a calculational method is suggested that introduces phenomenologically an energy dependence of the effective nucleon mass and aims to improve on the single-nucleon spectra (the energy-level densities) in the vicinity of the Fermi surface obtained in the standard RMF approach. In [750] this aim is achieved in calculations for ^{208}Pb and ^{132}Sn, where the Fermi surface is known reliably from experiment (see also [751]).

In [752] the relativistic Hartree approximation used to describe the bound states of both nucleons and antinucleons (see also Sect. 6.4 and references therein) has been extended to include tensor couplings for both vector–isoscalar and vector–isovector mesons (see Sect. 5.1 and 8.85). After the parameters of the model are adjusted to fit the properties of spherical nuclei, the effect of tensor-coupling terms increases the spin–orbit force by a factor of two, while a large effective nucleon mass of $0.8\,M$ is obtained. The overall nucleon spectra of shell-model states are improved in a way that is evident (the predicted antinucleon spectra are deepened by about 20–30 MeV).

A.6 Radial Equations for the Upper and Lower Components $G(r)$ and $F(r)$

For spherical nuclei, the two first-order coupled differential equations for $G_\kappa(r)$ and $F_\kappa(r)$ are given by

$$G'_\kappa = \left(-\frac{\kappa}{r} - \Sigma_{\mathrm{T}}^{\mathrm{D}}\right)G_\kappa + (M + E + S - V)F_\kappa \,, \tag{A.65}$$

$$F'_\kappa = (M - E + S + V)G_\kappa + \left(\frac{\kappa}{r} + \Sigma_{\mathrm{T}}^{\mathrm{D}}\right)F_\kappa \,, \tag{A.66}$$

A.6 Radial Equations for the Upper and Lower Components $G(r)$ and $F(r)$

where $E = M + \varepsilon$, and Σ_T^D arises from the tensor coupling; an explicit expression for this potential can be found in [335].

We introduce the following notation:

$$\Lambda_G = [M + E + S - V]^{-1}, \qquad \Lambda_F = [M - E + S + V]^{-1}. \qquad (A.67)$$

From (A.65) and (A.66), it is easy to obtain two decoupled equations of the second order for G_κ and F_κ:

$$G_\kappa'' + \frac{\Lambda_G'}{\Lambda_G} G_\kappa' + \left[\frac{\Lambda_G'}{\Lambda_G} \left(\frac{\kappa}{r} + \Sigma_T^D \right) + \left(-\frac{\kappa}{r^2} + \Sigma_T^{D'} \right) \right.$$
$$\left. - \frac{(M - E + S + V)}{\Lambda_G} - \left(\frac{\kappa}{r} + \Sigma_T^D \right)^2 \right] G_\kappa = 0, \qquad (A.68)$$

$$F_\kappa'' + \frac{\Lambda_F'}{\Lambda_F} F_\kappa' + \left[-\frac{\Lambda_F'}{\Lambda_F} \left(\frac{\kappa}{r} + \Sigma_T^D \right) - \left(-\frac{\kappa}{r^2} + \Sigma_T^{D'} \right) \right.$$
$$\left. - \frac{(M + E + S - V)}{\Lambda_F} - \left(\frac{\kappa}{r} + \Sigma_T^D \right)^2 \right] F_\kappa = 0. \qquad (A.69)$$

These equations contain the wave functions and their first and second derivatives. To obtain a Schrödinger-type equation for the upper component, we perform a transformation to the modified functions given by $G_\kappa = \Lambda_G^{-1/2} \widetilde{G}_\kappa$. The functions G and $\widetilde{G}$ are identical asymptotically, while the differential equation for $\widetilde{G}_\kappa$ contains no first derivative.

We have the following equation for $\widetilde{G}_\kappa(r)$:

$$\widetilde{G}'' - \left\{ \frac{l(l+1)}{r^2} + \frac{2M}{\hbar^2} \left[(S+V) \right. \right.$$
$$+ \frac{\varepsilon}{M} V + \frac{1}{2M} (S^2 - V^2) - \frac{1}{2M} \left(\Sigma_T^{D'} - 2\frac{\kappa}{r} \Sigma_T^D - (\Sigma_T^D)^2 \right)$$
$$- \frac{2(S'' - V'')(2M + \varepsilon + S - V) - 3(S' - V')^2}{8M(2M + \varepsilon + S - V)^2}$$
$$\left. \left. + \frac{1}{2M} \frac{(S' - V')}{(2M + \varepsilon + S - V)} \left(\frac{\kappa}{r} + \Sigma_T^D \right) - \varepsilon \left(1 + \frac{\varepsilon}{2M} \right) \right] \right\} \widetilde{G} = 0. \qquad (A.70)$$

The second term in (A.70) gives the centrifugal barrier, the third term represents the central potential, the fourth term reproduces the energy dependence of the central potential, the fifth term corresponds to a relativistic correction to the central potential, the sixth term is related to the contribution of the tensor coupling to the central potential, the seventh term introduces some extra energy dependence of the central potential, and the eighth term generates the spin–orbit force (the potential Σ_T^D in the eighth term of (A.70) determines the contribution of the tensor coupling to the spin–orbit potential).

We should mention that the procedure of introducing $\widetilde{F}_\kappa$ in the same fashion as $\widetilde{G}_\kappa$ encounters some problems, since Λ_F'/Λ_F is singular.

A.7 Boundary Conditions for Wave Functions and Meson Potentials

The behavior of the solution of the Dirac equation at the origin is determined by the following equations:

$$\kappa < 0, \quad \begin{cases} G_\kappa \sim r^{-\kappa} \\ F_\kappa \sim r^{-\kappa+1} \end{cases}, \qquad (A.71)$$

$$\kappa > 0, \quad \begin{cases} G_\kappa \sim r^{\kappa+1} \\ F_\kappa \sim r^{\kappa} \end{cases}. \qquad (A.72)$$

The asymptotic behavior of the functions $G_\kappa(r) = r g_\kappa(r)$ and $F_\kappa(r) = r f_\kappa(r)$ for $r \to \infty$ is determined by the asymptotic behavior of the spherical Hankel function of the first type.

Consider the Klein–Gordon equation, for example, for the vector–isoscalar field V:

$$\frac{d^2 V}{dr^2} + \frac{2}{r}\frac{dV}{dr} - m_V^2 V(r) = -g_V^2 \rho_V . \qquad (A.73)$$

We multiply this equation by $4\pi r^2$ and integrate the equation obtained from $r = 0$ up to $r = r_0$, where r_0 is the radius of a small sphere with its center at the origin. We obtain

$$r_0^2 \frac{dV}{dr}\bigg|_{r=r_0} - m_V^2 V(0) \frac{r_0^3}{3} = -g_V^2 \rho_V(0) \frac{r_0^3}{3} . \qquad (A.74)$$

If we set r_0 to zero, we obtain the boundary condition for the function $V(r)$ at the origin,

$$\frac{dV}{dr}\bigg|_{r=0} = 0 , \qquad (A.75)$$

and similarly

$$\frac{dS}{dr}\bigg|_{r=0} = 0 . \qquad (A.76)$$

These conditions correspond to zero slope of the potentials at the origin. The presence of nonlinear terms (in the meson fields) in the Klein–Gordon equation does not change these conditions.

At large distances (from the nucleus), the sources of the meson fields vanish and we obtain the following solutions:

$$(V, S) = A_{(V,S)} \frac{\exp(-m_{(V,S)} r)}{r} , \qquad (A.77)$$

where $A_{(V,S)}$ are constants which may be either calculated or not considered at all, by introducing the following boundary conditions at infinity:

$$\frac{1}{(V,S)} \cdot \frac{d(V,S)}{dr} = -m_{(V,S)} . \qquad (A.78)$$

A.8 Generalized Weinberg Transformation

We start from the ps coupling Lagrangian. Its nucleonic part, which is of interest to us, is given by [274]

$$L_{\text{ps}} = \bar{\psi}\, \not{p}\psi - g_\sigma \bar{\psi}(\sigma + i\gamma_5 \boldsymbol{\pi}\boldsymbol{\tau})\psi \;. \tag{A.79}$$

Before the Weinberg transformation is done, it is useful to bring L_{ps} into the form

$$L_{\text{ps}} = \bar{\psi}\, \not{p}\psi - g_\sigma \sigma \bar{\psi} e^{i/\sigma \gamma_5 \boldsymbol{\pi}\cdot\boldsymbol{\tau}}\psi + g_\sigma \sigma \bar{\psi}\psi \left(\cos \frac{|\boldsymbol{\pi}|}{\sigma} - 1 \right)$$
$$+ ig_\sigma \bar{\psi}\boldsymbol{\tau}\cdot\boldsymbol{\pi}\gamma_5 \psi \left(\frac{\sigma}{|\boldsymbol{\pi}|} \sin \frac{|\boldsymbol{\pi}|}{\sigma} - 1 \right) \;. \tag{A.80}$$

The following relation is used:

$$\exp\left\{ \frac{i}{\sigma}\gamma_5 \boldsymbol{\pi}\cdot\boldsymbol{\tau} \right\}$$
$$= \cos\gamma_5 \frac{\boldsymbol{\tau}\cdot\boldsymbol{\pi}}{\sigma} + i\sin\gamma_5 \frac{\boldsymbol{\tau}\cdot\boldsymbol{\pi}}{\sigma} = \cos\frac{|\boldsymbol{\pi}|}{\sigma} + i\gamma_5 \frac{\boldsymbol{\tau}\cdot\boldsymbol{\pi}}{|\boldsymbol{\pi}|} \sin\frac{|\boldsymbol{\pi}|}{\sigma} \;. \tag{A.81}$$

The generalized Weinberg transformation (GWT) has the following form:

$$\psi' = \exp\left\{ -\frac{i}{2\sigma}\lambda\gamma_5 \boldsymbol{\tau}\cdot\boldsymbol{\pi} \right\}\psi \;. \tag{A.82}$$

After the GWT has been performed, we obtain

$$\bar{\psi}'\not{p}\psi' - g_\sigma \bar{\psi}'(\sigma + i\gamma_5 \boldsymbol{\tau}\cdot\boldsymbol{\pi})\psi' = \bar{\psi}\not{p}\psi + \frac{\lambda}{2}\bar{\psi}\gamma^\mu \partial_\mu \left(\frac{\boldsymbol{\tau}\cdot\boldsymbol{\pi}}{\sigma} \right)\gamma_5 \psi$$
$$- g_\sigma \sigma \bar{\psi} \exp\left\{ \frac{i}{\sigma}(1-\lambda)\boldsymbol{\tau}\cdot\boldsymbol{\pi}\gamma_5 \right\}\psi$$
$$+ g_\sigma \sigma \bar{\psi}\exp\left\{ -\frac{i}{\sigma}\lambda\boldsymbol{\tau}\cdot\boldsymbol{\pi}\gamma_5 \right\}\psi \left(\cos\frac{|\boldsymbol{\pi}|}{\sigma} - 1 \right)$$
$$+ ig_\sigma \bar{\psi}\exp\left\{ -\frac{i}{\sigma}\lambda\boldsymbol{\tau}\cdot\boldsymbol{\pi}\gamma_5 \right\}\boldsymbol{\tau}\cdot\boldsymbol{\pi}\gamma_5 \psi \left(\frac{\sigma}{|\boldsymbol{\pi}|}\sin\frac{|\boldsymbol{\pi}|}{\sigma} - 1 \right)$$
$$= \bar{\psi}\not{p}\psi + \lambda \frac{g_\sigma}{2M^*} \bar{\psi}\gamma_\mu \boldsymbol{\tau}\gamma_5 \psi \partial_\mu \boldsymbol{\pi} - \frac{\lambda g_\sigma}{2M^{*2}}\partial_\mu(M^*)\bar{\psi}\gamma^\mu \boldsymbol{\tau}\gamma_5 \psi \boldsymbol{\pi}$$
$$- M^*\bar{\psi}\psi - i(1-\lambda)g_\sigma \bar{\psi}\gamma_5 \boldsymbol{\tau}\psi\boldsymbol{\pi} - \frac{g_\sigma^2}{2M^*}\bar{\psi}\psi\lambda(2-\lambda)\boldsymbol{\pi}^2 + O(\boldsymbol{\pi}^3) \;. \tag{A.83}$$

So we have started from a pure pseudoscalar coupling and, after the GWT, arrived at a form containing both types of πN interaction, the pion mass being renormalized as follows:

$$m_{\pi W}^{*2} = m_\pi^2 + \frac{g_\sigma^2}{M^*}\lambda(2-\lambda)\rho_S \ . \tag{A.84}$$

For the mixing parameter (ps/pv), we obtain $x = (1-\lambda)/\lambda$ (see Chap. 7). Let us note an interesting observation connected with the total contribution to m_π^{*2} from the GWT, polarization effects (see Chap. 7), and pion–scalar and pion–vector mesonic interactions [74]. The total value of m_π^{*2}, for symmetric infinite nuclear matter, is given in this case by [274]

$$m_\pi^{*2} \approx \frac{M}{M^*}m_\pi^2 \ , \tag{A.85}$$

i.e. m_π^* is independent of λ, if the polarization is taken into account through the mixing parameter $x = (1-\lambda)/\lambda$.

A.9 Expansions of the Vertex Functions for Various Mesons

The nonrelativistic expansion of the vertex functions corresponding to $\mathcal{L}_{int}$ (see (8.3)) for the various mesons will be obtained in this Appendix (see [335]). The nucleon spinor is taken to be the free Dirac spinor in the limit $p \ll M$:

$$u(\boldsymbol{p}, s) \simeq \begin{pmatrix} 1 \\ \boldsymbol{\sigma} \cdot \boldsymbol{p}/2M \end{pmatrix} \chi_s \ , \tag{A.86}$$

χ being the two-component spin wave function (the isospin indices are omitted). For example, for the σ meson and the time component of the ω field, we have

$$(2E_\sigma^\omega)^{1/2} \left\langle {}^\omega_\sigma N(p_f) \right| \mathcal{L}_{int}^{\overset{\omega}{\sigma}} \left| N(p_i) \right\rangle$$
$$= g_\sigma^\omega \chi_f^\dagger \left[1 - \frac{(\boldsymbol{p}_f \mp \boldsymbol{p}_i)^2}{8M^2} \pm i\boldsymbol{\sigma} \cdot \frac{\boldsymbol{p}_f \times \boldsymbol{p}_i}{4M^2} \right] \chi_i \ , \tag{A.87}$$

whereas for the π and ρ mesons, we obtain

$$\sqrt{2E_\pi} \left\langle \pi N(p_f) \right| \mathcal{L}_{int}^\pi \left| N(p_i) \right\rangle = ig_\pi \chi_f^\dagger \frac{\boldsymbol{\sigma} \cdot (\boldsymbol{p}_f - \boldsymbol{p}_i)}{2M} \chi_i \ , \tag{A.88}$$

$$\sqrt{2E_\rho} \left\langle \rho N(p_f) \right| \mathcal{L}_{int}^\rho \left| N(p_i) \right\rangle = ig_\rho \chi_f^\dagger \boldsymbol{\epsilon} \frac{\boldsymbol{\sigma} \times (\boldsymbol{p}_f - \boldsymbol{p}_i)}{2M} \chi_i \ , \tag{A.89}$$

$\boldsymbol{\epsilon}$ being the polarization of the ρ meson.

The σ and ω contributions are of zeroth order (in a $1/M$ expansion) and, for this reason, they are dominant. The third term in (A.87) generates the spin–orbit potential in finite nuclei [59].

The NN potential produced by these mesons is given in the lowest order by

$$V_{\substack{\sigma\\\omega}}(\boldsymbol{r}_1,\boldsymbol{r}_2) = \mp \frac{\exp(-m_{\substack{\sigma\\\omega}}|\boldsymbol{r}_1-\boldsymbol{r}_2|)}{|\boldsymbol{r}_1-\boldsymbol{r}_2|}\frac{g^2_{\substack{\sigma\\\omega}}}{4\pi}. \tag{A.90}$$

In infinite nuclear matter, these contributions to the nucleon–nucleon potential give rise to scalar and timelike vector self-energies (illustrated in Fig. 8.2) of the order of $-420\,\mathrm{MeV}$ and $+330\,\mathrm{MeV}$ in the nucleon equation of motion, for standard values of g_σ and g_ω. It is just these large self-energies that are the reason for using a relativistic treatment of nuclear structures.

The nucleon–nucleon potential obtained from (A.88) and (A.89) may be represented as the sum of a central part and a tensor part. For the π-exchange potential, we have

$$V_\pi(\boldsymbol{q}) = -\frac{1}{3}\left[\frac{f_\pi}{m_\pi}\right]^2 \frac{1}{m_\pi^2 + q^2}\Big[(3\boldsymbol{\sigma}_1\cdot\boldsymbol{q}\,\boldsymbol{\sigma}_2\cdot\boldsymbol{q} - \boldsymbol{\sigma}_1\cdot\boldsymbol{\sigma}_2 q^2) \\ + (\boldsymbol{\sigma}_1\cdot\boldsymbol{\sigma}_2)q^2\Big]\cdot(\boldsymbol{\tau}_1\cdot\boldsymbol{\tau}_2) = V_\pi^\mathrm{T}(\boldsymbol{q}) + V_\pi^\mathrm{c}(\boldsymbol{q}). \tag{A.91}$$

The central part is given by

$$V_\pi^\mathrm{c}(\boldsymbol{q}) = -\frac{1}{3}\left[\frac{f_\pi}{m_\pi}\right]^2 \boldsymbol{\sigma}_1\cdot\boldsymbol{\sigma}_2\boldsymbol{\tau}_1\cdot\boldsymbol{\tau}_2\left[1 - \frac{m_\pi^2}{m_\pi^2+q^2}\right]. \tag{A.92}$$

By performing a Fourier transformation, we obtain a repulsive contact term and an attractive Yukawa interaction:

$$V_\pi^\mathrm{c}(\boldsymbol{r}) = -\frac{m_\pi^3}{12\pi}\left[\frac{f_\pi}{m_\pi}\right]^2 \boldsymbol{\sigma}_1\cdot\boldsymbol{\sigma}_2\boldsymbol{\tau}_1\cdot\boldsymbol{\tau}_2\left[\frac{4\pi}{m_\pi^3}\delta(\boldsymbol{r}) - \frac{\exp(-m_\pi r)}{m_\pi r}\right], \tag{A.93}$$

A similar expression is obtained for the central potential for ρ exchange, with a strength two times larger. In a realistic many-body theory, such $\delta(\boldsymbol{r})$ terms are suppressed by short-range correlations due to ω exchange repulsion in the nucleon–nucleon potential at short distances. For this reason, the resulting contribution, i.e. the Yukawa term, is attractive.

In a nonrelativistic theory, there is a possibility of simulating short-range correlations in the π and ρ contributions by removing spurious δ components from the potential part of the nuclear Hamiltonian. In [335] this is realized by subtracting the zero-rank tensor part of the nucleon–nucleon potential originating from pion (pv coupling) and ρ (tensor coupling) exchanges, represented by

$$\delta\left[\Gamma_{\pi,\rho}(1,2)\right] = \frac{1}{4\pi q^2}\int \mathrm{d}\Omega_q\,\Gamma_{\pi,\rho}(1,2). \tag{A.94}$$

Here Γ_π and Γ_ρ are taken from (8.24)–(8.28).

A.10 Globally Chirally Invariant Lagrangian for the Model with an Axial Meson

A globally chirally invariant Lagrangian for the model with an axial meson was considered in [344] and is given by

$$\mathcal{L} = \bar{\psi}\Big[i\gamma^\mu\partial_\mu - g_\sigma\sigma - g_\omega\gamma^\mu\omega_\mu - g_v\boldsymbol{\tau}(\boldsymbol{\rho}_\mu + \gamma_5\boldsymbol{a}_\mu)\gamma^\mu - ig_\sigma\gamma_5\boldsymbol{\tau}\cdot\boldsymbol{\pi}\Big]\psi$$

$$+ \frac{C}{2}\bar{\psi}\gamma^\mu\Big[\gamma_5\boldsymbol{\tau}(\sigma\partial_\mu\boldsymbol{\pi} - \boldsymbol{\pi}\cdot\partial_\mu\sigma) + \boldsymbol{\tau}\cdot\boldsymbol{\pi}\times\partial_\mu\boldsymbol{\pi}\Big]\psi$$

$$+ \frac{1}{2}(\partial_\mu\sigma\cdot\partial^\mu\sigma + \partial_\mu\boldsymbol{\pi}\cdot\partial^\mu\boldsymbol{\pi}) - \frac{1}{4}(\omega_{\mu\nu}\omega^{\mu\nu} + \boldsymbol{\rho}_{\mu\nu}\cdot\boldsymbol{\rho}^{\mu\nu} + \boldsymbol{a}_{\mu\nu}\boldsymbol{a}^{\mu\nu})$$

$$- \frac{m_\sigma^2 - m_\pi^2}{8f_\pi^2}\left(\sigma^2 + \boldsymbol{\pi}^2 - f_\pi^2\frac{m_\sigma^2 - 3m_\pi^2}{m_\sigma^2 - m_\pi^2}\right)^2$$

$$+ \frac{1}{2}g_\omega^2(\sigma^2 + \boldsymbol{\pi}^2)\omega^2 + \frac{1}{2}m_V^2(\boldsymbol{\rho}_\mu\cdot\boldsymbol{\rho}^\mu + \boldsymbol{a}_\mu\boldsymbol{a}^\mu) + \mathcal{L}_{\text{tens}} + f_\pi m_\pi^2\sigma \ , \quad (A.95)$$

where the tensor fields $\omega_{\mu\nu}$, $\boldsymbol{\rho}_{\mu\nu}$, and $\boldsymbol{a}_{\mu\nu}$ are

$$\omega_{\mu\nu} = \partial_\mu\omega_\nu - \partial_\nu\omega_\mu \ , \quad \boldsymbol{\rho}_{\mu\nu} = \partial_\mu\boldsymbol{\rho}_\nu - \partial_\nu\boldsymbol{\rho}_\mu \ , \quad \boldsymbol{a}_{\mu\nu} = \partial_\mu\boldsymbol{a}_\nu - \partial_\nu\boldsymbol{a}_\mu \ . \quad (A.96)$$

The bold variables entering $\mathcal{L}$ refer to vectors in isotopic space; $g_{i,v}$ and $m_{i,v}$ ($i = (\sigma, \omega, \pi)$ and $v = (\rho, a)$) designate the effective meson-nucleon coupling constants and masses, respectively.

The term proportional to $C = (g_A - 1)f_\pi^{-2}$ is introduced to ensure the right value of the axial-vector form factor g_A. The constant f_π is the vacuum value of the scalar field, which determines the weak-interaction pion decay rate.

If retardation effects are not considered, the addend proportional to $\partial_\mu\sigma$ in this term can be neglected in the case of nuclear matter. The first contribution, proportional to $\partial_\mu\boldsymbol{\pi}$, has the form of a pseudovector coupling. In the nuclear medium, it is well known that this coupling is weaker than the pseudoscalar coupling. The calculations in [344] were restricted to the one-boson exchange approximation. Thus, we shall neglect the term proportional to C and the other terms in the Lagrangian involving the square (or a higher power) of the pion field.

The ω meson mass is introduced via a σ–ω interaction (see Chap. 7); this procedure guarantees saturation in nuclear matter.

The $\mathcal{L}_{\text{tens}}$ term represents the tensor interaction corresponding to the ρ–nucleon and a–nucleon vertices. One possible chirally invariant choice is

$$\mathcal{L}_{\text{tens}} = -\frac{if_v}{8M^2}\left(\bar{\psi}\boldsymbol{\tau}\{\gamma^\lambda\partial_\lambda, \sigma^{\mu\nu}\}\psi\boldsymbol{\rho}_{\mu\nu} + \bar{\psi}\boldsymbol{\tau}\gamma_5\{\gamma^\lambda\partial_\lambda, \sigma^{\mu\nu}\}\psi\boldsymbol{a}_{\mu\nu}\right) \ , \quad (A.97)$$

where f_v is the coupling constant, M is the nucleon mass, and the braces denote the anticommutator necessary to ensure hermiticity of the corresponding interaction in the Dirac equation.

Note that a Lagrangian containing the ρ meson also needs the presence of an axial meson a with a mass (m_V) and the same coupling constants (g_V, f_V) as the ρ to guarantee global chiral invariance. This follows from the transformation rules for the fields under infinitesimal chiral rotations (see (7.72)–(7.74)):

$$\delta\psi = \frac{i}{2}\boldsymbol{\varepsilon}\cdot\boldsymbol{\tau}\gamma_5\psi\,,\quad \delta\boldsymbol{\rho}_\mu = \boldsymbol{\varepsilon}\times\boldsymbol{a}_\mu\,,\quad \delta\boldsymbol{a}_\mu = \boldsymbol{\varepsilon}\times\boldsymbol{\rho}_\mu\,,\quad \delta\sigma = \boldsymbol{\varepsilon}\cdot\boldsymbol{\pi}\,,\quad (A.98)$$

$$\delta\boldsymbol{\pi} = -\boldsymbol{\varepsilon}\sigma\,,\quad \delta\omega_\mu = 0\,. \quad (A.99)$$

If the nucleons are on-shell in (A.97), the free Dirac equation yields the following tensor interaction term:

$$L_{\text{tens}}^{(\text{on shell})} = -\frac{f_v}{4M}\bar\psi(\boldsymbol{\tau}\sigma^{\mu\nu}\boldsymbol{\rho}_{\mu\nu} + \boldsymbol{\tau}\gamma_5\sigma^{\mu\nu}\boldsymbol{a}_{\mu\nu})\psi\,, \quad (A.100)$$

which is just the conventional form of the tensor interaction. By applying the transformation (A.98), (A.99) to (A.100) one can see that, to be chirally invariant, the Lagrangian should contain an odd number of Dirac matrices between the spinors $\bar\psi$ and ψ. This is just the case with (A.97), whereas in (A.100) this number is even. Previous considerations suggest the choice of (A.97) for the tensor interaction.

Equation (A.100) enables one to extract the value of the tensor coupling constant in the same way as the conventional tensor interaction does.

For global chiral symmetry, we have $m_a = m_\rho = m_V$ and $g_a = g_\rho$. It is shown in [344] that in this case the axial-meson contribution to the total exchange energy is greater than the corresponding contribution for the ρ meson. For local chiral symmetry $m_a \simeq \sqrt{2}m_\rho$, whereas $g_a = g_\rho g_A(m_a^2/m_\rho^2)$, where $g_A \simeq 1.25$, so the role of the axial meson is still increased.

A.11 Direct and Exchange Matrix Elements for the Two-Body Spin–Orbit Force

A scalar–isoscalar meson generates a two-body spin–orbit force of the Thomas form [59]:

$$-\frac{1}{4M^2}\frac{1}{r}\frac{dJ^S}{dr}\left\{\left[\boldsymbol{r}\times\boldsymbol{p}_1\right]\boldsymbol{\sigma}_1 - \left[\boldsymbol{r}\times\boldsymbol{p}_2\right]\boldsymbol{\sigma}_2\right\}\,, \quad (A.101)$$

while $\boldsymbol{r} \equiv \boldsymbol{r}_1 - \boldsymbol{r}_2$.

For a system consisting of only two particles, it is easy to obtain the following potential instead of the operator (A.101):

$$-\frac{1}{4M^2}\frac{1}{r}\frac{dJ^S}{dr}\left[(\boldsymbol{r}_1-\boldsymbol{r}_2)\times(\boldsymbol{p}_1-\boldsymbol{p}_2)\right](\boldsymbol{s}_1+\boldsymbol{s}_2)\,. \quad (A.102)$$

This can be obtained using the identity

$$\boldsymbol{\sigma}_{\{\genfrac{}{}{0pt}{}{1}{2}\}} = \frac{1}{2}(\boldsymbol{\sigma}_1+\boldsymbol{\sigma}_2) + \frac{1}{2}\left(\boldsymbol{\sigma}_{\{\genfrac{}{}{0pt}{}{1}{2}\}} - \boldsymbol{\sigma}_{\{\genfrac{}{}{0pt}{}{2}{1}\}}\right) \quad (A.103)$$

and the fact that $\boldsymbol{p}_1+\boldsymbol{p}_2 = 0$ in the center-of-mass system for two particles.

A vector–isoscalar object generates two forms of the two-body spin–orbit force [59]:

1. The Thomas operator,
$$-\frac{1}{4M^2}\frac{1}{r}\frac{\mathrm{d}J^\mathrm{V}}{\mathrm{d}r}\left\{\left[\boldsymbol{r}\times\boldsymbol{p}_1\right]\boldsymbol{\sigma}_1-\left[\boldsymbol{r}\times\boldsymbol{p}_2\right]\boldsymbol{\sigma}_2\right\}. \quad (A.104)$$

2. The Larmor operator,
$$2\frac{1}{4M^2}\frac{1}{r}\frac{\mathrm{d}J^\mathrm{V}}{\mathrm{d}r}\left\{\left[\boldsymbol{r}\times\boldsymbol{p}_2\right]\boldsymbol{\sigma}_1-\left[\boldsymbol{r}\times\boldsymbol{p}_1\right]\boldsymbol{\sigma}_2\right\}. \quad (A.105)$$

The total operator may be written in the following form:
$$-\frac{1}{4M^2}\frac{1}{r}\frac{\mathrm{d}J^\mathrm{V}}{\mathrm{d}r}\left\{\left[\boldsymbol{r}\times\boldsymbol{p}_1\right](\boldsymbol{\sigma}_1+2\boldsymbol{\sigma}_2)-\left[\boldsymbol{r}\times\boldsymbol{p}_2\right](\boldsymbol{\sigma}_2+2\boldsymbol{\sigma}_1)\right\}. \quad (A.106)$$

In the center-of-mass system, $\boldsymbol{p}_1 = -\boldsymbol{p}_2 = +\boldsymbol{p}$ for two particles, and instead of the operator (A.106), one obtains the following potential:
$$-\frac{3}{2}\frac{1}{M^2}\frac{1}{r}\frac{\mathrm{d}J^\mathrm{V}}{\mathrm{d}r}\left[\boldsymbol{r}_1-\boldsymbol{r}_2\times\boldsymbol{p}\right]\cdot\frac{1}{2}(\boldsymbol{\sigma}_1+\boldsymbol{\sigma}_2). \quad (A.107)$$

We should mention that for a many-body system one should start from the operator (A.101) for scalar–isoscalar mesons and from (A.104), (A.105) for vector–isoscalar mesons. Note also that in the Hartree approximation, the Larmor operator for the two-body spin–orbit interaction makes no contribution to the single-particle spin–orbit force for nuclei with an extra nucleon above closed shells.

Equation (A.106) can be reduced, utilizing (A.103) introduced earlier for $\boldsymbol{\sigma}$ matrices, to the following form:
$$-\frac{1}{8M^2}\frac{1}{r}\frac{\mathrm{d}J^\mathrm{V}}{\mathrm{d}r}\left\{3\left[\boldsymbol{r}_1-\boldsymbol{r}_2\times\boldsymbol{p}_1-\boldsymbol{p}_2\right](\boldsymbol{\sigma}_1+\boldsymbol{\sigma}_2)\right.$$
$$\left.-\left[\boldsymbol{r}_1-\boldsymbol{r}_2\times\boldsymbol{p}_1+\boldsymbol{p}_2\right](\boldsymbol{\sigma}_1-\boldsymbol{\sigma}_2)\right\}. \quad (A.108)$$

In the case of the ρ meson, the following replacement should be carried out in (A.108) $3 \to 3 + 4(f_\mathrm{V}/g_\mathrm{V})$ where f_V and g_V are the constants for the tensor and vector couplings. In the case of the isovector (scalar and vector) mesons, all the operators should be multiplied also by the factor $\boldsymbol{\tau}_1\cdot\boldsymbol{\tau}_2$. In the OBE scheme, the pion makes no contribution to the two-body spin–orbit force.

The spin–orbit splitting of the ith particle in the nucleus is given by
$$\sum_{k\leq A}[\langle ik|\,V^{\mathrm{so}}\,|ik\rangle - \langle ik|\,V^{\mathrm{so}}\,|ki\rangle], \quad (A.109)$$

where V^{so} is the two-body spin–orbit force, the sum in (A.109) being carried out over the particles with quantum numbers k. Equation (A.109) determines the value of the spin–orbit splitting in the Hartree–Fock approximation, however, the matrix elements are calculated with nonsymmetrized wave functions. We assume that V^{so} is a short-ranged force.

A.11 Matrix Elements for the Two-Body Spin–Orbit Force

First we consider the two-body spin–orbit force generated by scalar–isoscalar mesons. It has the following general structure:

$$f(r_{ik})\left\{\left[\bm{r}_i - \bm{r}_k \times \bm{p}_i\right]\bm{\sigma}_i - \left[\bm{r}_i - \bm{r}_k \times \bm{p}_k\right]\bm{\sigma}_k\right\}, \tag{A.110}$$

which can be reduced to the following form in the many-body case, using (A.103):

$$\frac{f(r_{ik})}{2}\left\{\left[\bm{r}_i - \bm{r}_k \times \bm{p}_i - \bm{p}_k\right](\bm{\sigma}_i + \bm{\sigma}_k) + \left[\bm{r}_i - \bm{r}_k \times \bm{p}_i + \bm{p}_k\right](\bm{\sigma}_i - \bm{\sigma}_k)\right\}. \tag{A.111}$$

Note that the operator (A.110) has a radial dependence $f(r_{ik})$ common to both terms, which is not written out here (see (A.102) and (A.108)).

We should mention also that the second term in (A.111) depends on the total momentum of the nucleon pair $\bm{p}_i + \bm{p}_k$ and on the difference between their spin operators $\bm{\sigma}_i - \bm{\sigma}_k$. In the two-body system, the second term in (A.111) does not operate. This can be seen by performing a transformation to the center-of-mass system, where $\bm{p}_i + \bm{p}_k = 0$. In the many-body case (A.111) is not reduced only to the first term (with a conventional structure), but is determined by both terms of this equation. The two-body spin–orbit force is a relativistic operator (of order v^2/c^2) and in the general case it does not satisfy the requirements of Galilean invariance. So, in what follows we consider the general form of the two-body spin–orbit operator, its structure being represented by (A.111).

Now we show that if for a nucleus with an extra particle (or hole) above doubly closed shells the value of the spin–orbit splitting in the Hartree approximation is given by $\sum_{k<A}\langle ik|\,V^{so}\,|ik\rangle$ (where V^{so} is given by (A.111)), one can easily calculate [60] the spin–orbit splitting in the Hartree–Fock approximation. To perform this calculation, one needs to assume only the short-range character of the two-body spin–orbit force. Let us follow this program. The nonsymmetrized two-particle wave function is given by

$$|ik\rangle = \varphi_{r_i}(\bm{r}_i)\varphi_{r_k}(\bm{r}_k)\varphi_{s_i}(\sigma_i)\varphi_{s_k}(\sigma_k)\varphi_{t_i}(\tau_i)\varphi_{t_k}(\tau_k), \tag{A.112}$$

where r_i, s_i, t_i are the quantum numbers, and $\bm{r}_i$, σ_i, τ_i are the space, spin, and isospin coordinates of the ith particle (a similar statement relates to the kth particle). We express the nonsymmetrized components of the wave function in (A.112) via the symmetrized wave functions determined by the following equations:

$$\varphi^{(+)}_{r_i r_k}(\bm{r}_i, \bm{r}_k) = \frac{1}{\sqrt{2}}\left[\varphi_{r_i}(\bm{r}_i)\varphi_{r_k}(\bm{r}_k) + \varphi_{r_i}(\bm{r}_k)\varphi_{r_k}(\bm{r}_i)\right], \tag{A.113}$$

$$\varphi^{(-)}_{r_i r_k}(\bm{r}_i, \bm{r}_k) = \frac{1}{\sqrt{2}}\left[\varphi_{r_i}(\bm{r}_i)\varphi_{r_k}(\bm{r}_k) - \varphi_{r_i}(\bm{r}_k)\varphi_{r_k}(\bm{r}_i)\right]. \tag{A.114}$$

It is easy to verify that

$$\varphi_{r_i}(\boldsymbol{r}_i)\varphi_{r_k}(\boldsymbol{r}_k) = \frac{1}{\sqrt{2}} \left(\varphi^{(+)}_{r_i r_k}(\boldsymbol{r}_i, \boldsymbol{r}_k) + \varphi^{(-)}_{r_i r_k}(\boldsymbol{r}_i, \boldsymbol{r}_k) \right) . \tag{A.115}$$

Now we consider the spin wave functions.

If the two-body system has a total spin $S = 0$, its wave function is antisymmetric and is given by

$$\varphi_{S=0}(\sigma_i, \sigma_k) = \frac{1}{\sqrt{2}} \left[\varphi_{s_i}(\sigma_i)\varphi_{s_k}(\sigma_k) - \varphi_{s_i}(\sigma_k)\varphi_{s_k}(\sigma_i) \right] . \tag{A.116}$$

If the total spin of the two-nucleon system is $S = 1$, three values of its projection are allowed, $S_z = 0, \pm 1$, and the three corresponding symmetric wave functions are given by

$$\varphi^{S=1}_{S_z=0}(\sigma_i, \sigma_k) = \frac{1}{\sqrt{2}} \left(\varphi_{s_i}(\sigma_i)\varphi_{s_k}(\sigma_k) + \varphi_{s_i}(\sigma_k)\varphi_{s_k}(\sigma_i) \right) , \tag{A.117}$$

$$\varphi^{S=1}_{S_z=\pm 1}(\sigma_i, \sigma_k) = \varphi_{s_i}(\sigma_i)\varphi_{s_i}(\sigma_k) . \tag{A.118}$$

We consider (A.117) and (A.118) separately as follows:

1. $S = 1$, $s_i + s_k = \pm 1 = S_z$, where s_i and s_k denote the z-projection of the spin of the ith and kth particles, respectively. We replace the product of the spin wave functions $\varphi_{s_i}(\sigma_i)\varphi_{s_k}(\sigma_k)$ appearing in (A.112) by the product $\varphi_{s_i}(\sigma_i)\varphi_{s_i}(\sigma_k)$, which is a symmetric function with respect to the permutation $\sigma_i \longleftrightarrow \sigma_k$.
2. $S = 1$, $s_i + s_k = 0 = S_z$, i.e. in this case s_i and s_k are different. Using (A.116) and (A.117), one can easily show that the product $\varphi_{s_i}(\sigma_i)\varphi_{s_k}(\sigma_k)$ appearing in (A.112) may be represented by calculating the sum of (A.116)–(A.118) in the following form:

$$\varphi_{s_i}(\sigma_i)\varphi_{s_k}(\sigma_k) = \frac{1}{\sqrt{2}} \left(\varphi^{S=0}(\sigma_i, \sigma_k) + (\varphi^{S=1}_{S_z=0}(\sigma_i, \sigma_k) \right) , \tag{A.119}$$

where the first term corresponds to an antisymmetric function and the second term to a symmetric one.

The wave function for the case $S = 0$, $S_z = 0$ may be written down in a similar way. All of what is said above relates also to the isospin wave function.

We consider the first term in (A.111). The case $S = 0$ is not interesting for the first term, since the spin–orbit interaction of the "conventional type" does not operate in a state with a spin equal to zero.

Taking into account all of what is said above, we split the sum over k in (A.109) into four parts. The first part is formed by those particles with quantum numbers k (i.e. summation is carried out over those particles) which have the same isospin and spin z-projection as the ith particle, i.e. $S_z = \pm 1$ and $T_z = \pm 1$. In the second part, summation is carried out over particles with quantum numbers k which have the same spin z-projection as the ith particle,

A.11 Matrix Elements for the Two-Body Spin–Orbit Force

and have an isospin z-projection opposite to the ith particle, i.e. $S_z = \pm 1$ and $T_z = 0$. In the third part, summation is carried out over particles with quantum numbers k which have the same isospin z-projection as the ith particle, and have a spin z-projection opposite to the ith particle, i.e. $S_z = 0$ and $T_z = \pm 1$. In the fourth part, summation is performed over particles with quantum numbers k which have spin and isospin z-projections opposite to the ith particle, i.e. $S_z = 0$ amd $T_z = 0$. So we consider independently the following four sums:

1. $S_z = \pm 1, T_z = \pm 1$.
2. $S_z = \pm 1, T_z = 0$.
3. $S_z = 0, T_z = \pm 1$.
4. $S_z = 0, T_z = 0$.

Consider the first case, $S_z = \pm 1, T_z = \pm 1$. Here,

$$|ik\rangle = \varphi_{r_i}(\mathbf{r}_i)\varphi_{r_k}(\mathbf{r}_k)\varphi_{s_i}(\sigma_i)\varphi_{s_k}(\sigma_k)\varphi_{t_i}(\tau_i)\varphi_{t_k}(\tau_k)$$
$$= \frac{1}{\sqrt{2}}\left(\varphi_{r_i r_k}^{(+)}(\mathbf{r}_i, \mathbf{r}_k) + \varphi_{r_i r_k}^{(-)}(\mathbf{r}_i, \mathbf{r}_k)\right)\varphi_{s_i}(\sigma_i)\varphi_{s_i}(\sigma_k)\varphi_{t_i}(\tau_i)\varphi_{t_i}(\tau_k) . \tag{A.120}$$

In the short-range approximation, the conventional part of the spin–orbit operator (the first term in (A.111)) operates only on a relative P state. In this case the space part of $|ik\rangle$ should be odd function in the "relative" coordinates, i.e. an antisymmetric function in the "absolute" coordinates. For this reason, $\varphi_{r_i r_k}^{(+)}(\mathbf{r}_i, \mathbf{r}_k)$ makes no contribution to the matrix element (A.109)

So, in the first group we have

$$|ik\rangle = \frac{1}{\sqrt{2}}\varphi_{r_i r_k}^{(-)}(\mathbf{r}_i, \mathbf{r}_k)\varphi_{s_i}(\sigma_i)\varphi_{s_i}(\sigma_k)\varphi_{t_i}(\tau_i)\varphi_{t_i}(\tau_k) , \tag{A.121}$$

$$|ki\rangle = -\frac{1}{\sqrt{2}}\varphi_{r_i r_k}^{(-)}(\mathbf{r}_i, \mathbf{r}_k)\varphi_{s_i}(\sigma_i)\varphi_{s_i}(\sigma_k)\varphi_{t_i}(\tau_i)\varphi_{t_i}(\tau_k) , \tag{A.122}$$

and for the first term of the operator (A.111), V_1^{so}, it is possible to obtain

$$\langle ik | V_1^{so} |ki\rangle = -\langle ik | V_1^{so} |ik\rangle . \tag{A.123}$$

If we take into account the negative sign of the second component in (A.109), we find that for the first group of particles ($S_z = \pm 1, T_z = \pm 1$), the matrix elements in (A.109) appear to be doubled in the Hartree–Fock approximation[3] in comparison with the corresponding group of matrix elements in the Hartree approximation.

---

[3] Note that in the present appendix we consider only the proper symmetry of the wave function when speaking about the Hartree–Fock approximation.

Consider the second case, $S_z = \pm 1$, $T_z = 0$. The nonsymmetrized wave function $|ik\rangle$ is given by

$$|ik\rangle = \frac{1}{\sqrt{2}}\varphi^{(-)}_{r_i r_k}(\boldsymbol{r}_i, \boldsymbol{r}_k)\varphi_{s_i}(\sigma_i)\varphi_{s_i}(\sigma_k)\frac{1}{\sqrt{2}}\left(\varphi^{T=0}(\tau_i, \tau_k) + \varphi^{T=1}_{T_z=0}(\tau_i, \tau_k)\right), \quad (A.124)$$

where the term $\varphi^{T=0}$ corresponds to the antisymmetric wave function, while the second term $\varphi^{T=1}$ corresponds to the symmetric wave function. For the wave function $|ki\rangle$, we have

$$|ki\rangle = -\frac{1}{\sqrt{2}}\varphi^{(-)}_{r_i r_k}(\boldsymbol{r}_i, \boldsymbol{r}_k)\varphi_{s_i}(\sigma_i)\varphi_{s_i}(\sigma_k)\frac{1}{\sqrt{2}}\left(\varphi^{T=1}(\tau_i, \tau_k) - \varphi^{T=0}(\tau_i, \tau_k)\right). \quad (A.125)$$

The matrix element

$$-\langle ik|\, V\, |ki\rangle \sim \left(\varphi^{T=1}(\tau_i, \tau_k)\varphi^{T=1}(\tau_i, \tau_k) - \varphi^{T=1}(\tau_i, \tau_k)\varphi^{T=0}(\tau_i, \tau_k)\right.$$
$$\left. + \varphi^{T=0}(\tau_i, \tau_k)\varphi^{T=1}(\tau_i, \tau_k) - \varphi^{T=0}(\tau_i, \tau_k)\varphi^{T=0}(\tau_i, \tau_k)\right) \quad (A.126)$$

equals zero for this group owing to the orthogonality and normalization conditions.

So one may conclude that the second group of particles makes no contribution to the exchange matrix element and contributes only to the direct matrix element. The same is valid for the fourth group of particles also.

Consider the third case, $S_z = 0$, $T_z = \pm 1$. For the wave function $|ik\rangle$, we have

$$|ik\rangle = \frac{1}{\sqrt{2}}\varphi^{(-)}_{r_i r_k}(\boldsymbol{r}_i, \boldsymbol{r}_k)\frac{1}{\sqrt{2}}\left(\varphi^{S=1}_{S_z=0}(\sigma_i, \sigma_k) + \varphi^{S=0}(\sigma_i, \sigma_k)\right)\varphi_{t_i}(\tau_i)\varphi_{t_i}(\tau_k), \quad (A.127)$$

where the term $\varphi^{S=1}$ corresponds to the symmetric wave function in spin variables, while the term $\varphi^{S=0}$ corresponds to the antisymmetric wave function. For the wave function $|ki\rangle$, we obtain

$$|ki\rangle = -\frac{1}{\sqrt{2}}\varphi^{(-)}_{r_i r_k}(\boldsymbol{r}_i, \boldsymbol{r}_k)\frac{1}{\sqrt{2}}\left(\varphi^{S=1}_{S_z=0}(\sigma_i, \sigma_k) - \varphi^{S=0}(\sigma_i, \sigma_k)\right)\varphi_{t_i}(\tau_i)\varphi_{t_i}(\tau_k). \quad (A.128)$$

Calculating the exchange matrix element with the wave functions given above, we conclude that a contribution to this matrix element is made only by the term containing $\varphi^{S=1}_{S_z=0}(\sigma_i, \sigma_k)$, i.e. for the third group of particles, the Hartree–Fock result appears to be doubled in comparison with the Hartree result (for the same group of particles).

Consider now the contribution of the second term of the two-body spin–orbit operator V_2^{so} (see (A.111)) to the spin–orbit splitting. This term makes no contribution to the exchange matrix elements. If the particles i and k have $S_z = \pm 1$, the exchange matrix element is diagonal in the spin variables and the diagonal matrix elements of the operator $\boldsymbol{\sigma}_i - \boldsymbol{\sigma}_k$ are equal to zero. If $S_z = 0$, the wave function has the form

A.11 Matrix Elements for the Two-Body Spin–Orbit Force

$$|ik\rangle = \frac{\left(\varphi_{r_i r_k}^{(+)}(\boldsymbol{r}_i, \boldsymbol{r}_k) + \varphi_{r_i r_k}^{(-)}(\boldsymbol{r}_i, \boldsymbol{r}_k)\right)}{\sqrt{2}} \times \frac{\left(\varphi^{S=1}(\sigma_i, \sigma_k) + \varphi^{S=0}(\sigma_i, \sigma_k)\right)}{\sqrt{2}}$$

× an isospin wave function depending on T_z, where the term $\varphi^{S=1}$ corresponds to the symmetric wave function and the term $\varphi^{S=0}$ corresponds to the antisymmetric wave function.

The spin part of the exchange matrix element is equal to

$$\frac{1}{2}\left\langle \varphi^{S=1}(\sigma_i, \sigma_k) + \varphi^{S=0}(\sigma_i, \sigma_k) \middle| \boldsymbol{\sigma}_i - \boldsymbol{\sigma}_k \middle| \varphi^{S=1}(\sigma_i, \sigma_k) - \varphi^{S=0}(\sigma_i, \sigma_k) \right\rangle . \tag{A.129}$$

Diagonal matrix elements of an odd operator are equal to zero. For this reason, we obtain

$$\frac{1}{2}\Big[\left\langle \varphi^{S=0}(\sigma_i, \sigma_k) \middle| \boldsymbol{\sigma}_i - \boldsymbol{\sigma}_k \middle| \varphi^{S=1}(\sigma_i, \sigma_k) \right\rangle \\ - \left\langle \varphi^{S=1}(\sigma_i, \sigma_k) \middle| \boldsymbol{\sigma}_i - \boldsymbol{\sigma}_k \middle| \varphi^{S=0}(\sigma_i, \sigma_k) \right\rangle\Big] = 0 , \tag{A.130}$$

since the matrix elements are calculated for a Hermitian operator, which has real matrix elements.

We should mention that in treating the second term in (A.111), we did not make use of the short-range character of this component (the space part of the wave function includes both $\varphi^{(+)}$ and $\varphi^{(-)}$).

Let us consider now the result of calculation of the spin–orbit splitting in the Hartree–Fock approximation $\Delta E(\text{HF})$ utilizing the operator (A.111) as a whole, and compare it with the corresponding value $\Delta E(\text{H})$ obtained by use of the Hartree approximation (for the same operator (A.111)). We designate the first term in (A.111) as V_1^{so} and the second term as V_2^{so}. We obtain

$$\Delta E(\text{H}) = \sum_{k \leq A} \langle ik | V_1^{\text{so}} + V_2^{\text{so}} | ik \rangle$$
$$= \sum_{k \leq A} \langle ik | V_1^{\text{so}} | ik \rangle + \sum_{k \leq A} \langle ik | V_2^{\text{so}} | ik \rangle$$
$$= \Delta E_1(\text{H}) + \Delta E_2(\text{H}) = 2\,\Delta E_1(\text{H}) , \tag{A.131}$$

$$\Delta E(\text{HF}) = \sum_{k \leq A} [\langle ik | V_1^{\text{so}} + V_2^{\text{so}} | ik \rangle - \langle ik | V_1^{\text{so}} + V_2^{\text{so}} | ki \rangle]$$
$$= \sum_{k \leq A} [\langle ik | V_1^{\text{so}} | ik \rangle - \langle ik | V_1^{\text{so}} | ki \rangle]$$
$$+ \sum_{k \leq A} [\langle ik | V_2^{\text{so}} | ik \rangle - \langle ik | V_2^{\text{so}} | ki \rangle] . \tag{A.132}$$

Consider the spin–orbit splitting of a neutron state. We now have

$$\Delta E(\text{HF}) = 2\,\Delta E_{1n}(\text{H}) + \Delta E_{1p}(\text{H}) + \Delta E_{2n}(\text{H}) + \Delta E_{2p}(\text{H}). \quad (A.133)$$

The subscripts 1 and 2 indicate the contributions from V_1^{so} and V_2^{so}, respectively, and n and p indicate the contributions from the interactions of the ith nucleon (a neutron) with neutrons and protons, respectively. So we have

$$\Delta E(\text{HF}) = 3\,\Delta E_{1n}(\text{H}) + 2\,\Delta E_{1p}(\text{H})\,. \quad (A.134)$$

Taking into account the following equations,

$$\Delta E_{1n}(\text{H}) = \Delta E_{2n}(\text{H})\,, \quad (A.135)$$

$$\Delta E_{1p}(\text{H}) = \Delta E_{2p}(\text{H})\,, \quad (A.136)$$

$$\Delta E_{1n}(\text{H}) = \frac{1}{2}\Delta E_{n}(\text{H})\,, \quad (A.137)$$

$$\Delta E_{2p}(\text{H}) = \frac{1}{2}\Delta E_{p}(\text{H})\,, \quad (A.138)$$

$$\Delta E_{n,p}(\text{H}) = \frac{N,Z}{A}\Delta E(\text{H})\,, \quad (A.139)$$

we obtain

$$\Delta E(\text{HF}) = \frac{1}{2}\left(3\,\Delta E_n(\text{H}) + 2\,\Delta E_p(\text{H})\right) = \frac{1}{2}\Delta E(\text{H})\left(2 + \frac{N}{A}\right). \quad (A.140)$$

So, for scalar–isoscalar mesons, we have finally

$$\Delta E(\text{HF}) = \frac{1}{2}\left(2 + \frac{N}{A}\right)\Delta E(\text{H}) \quad (A.141)$$

for a neutron state and

$$\Delta E(\text{HF}) = \frac{1}{2}\left(2 + \frac{Z}{A}\right)\Delta E(\text{H}) \quad (A.142)$$

for a proton state.

We emphasize once more that $\Delta E(\text{H})$ is a *partial* (i.e. caused by a given scalar–isoscalar meson) spin–orbit splitting calculated by the Hartree method and $\Delta E(\text{HF})$ is the corresponding Hartree–Fock splitting.

Consider now the partial spin–orbit splitting produced by the vector–isoscalar meson. Calculating this value by the Hartree–Fock method, we obtain the contribution to the spin–orbit splitting from the Larmor operator (A.105) also. The general structure of the two-body spin–orbit operator in this case is given by

$$\frac{f(r_{ik})}{2}\left\{3\left[\boldsymbol{r}_i - \boldsymbol{r}_k \times \boldsymbol{p}_i - \boldsymbol{p}_k\right](\boldsymbol{\sigma}_i + \boldsymbol{\sigma}_k) - \left[\boldsymbol{r}_i - \boldsymbol{r}_k \times \boldsymbol{p}_i + \boldsymbol{p}_k\right](\boldsymbol{\sigma}_i - \boldsymbol{\sigma}_k)\right\}. \quad (A.143)$$

A.11 Matrix Elements for the Two-Body Spin–Orbit Force

Again this operator contains two terms: the first term is Galilean-invariant, and is determined by $\boldsymbol{\sigma}_i + \boldsymbol{\sigma}_k$, while the second term does not satisfy the requirements of Galilean invariance and is determined by $\boldsymbol{\sigma}_i - \boldsymbol{\sigma}_k$.

The partial Hartree spin–orbit splitting in this case may be written down in the following form:

$$\Delta E(\mathrm{H}) = \Delta E_1(\mathrm{H}) + \Delta E_2(\mathrm{H}) = \frac{2}{3}\Delta E_1(\mathrm{H}) , \qquad (\mathrm{A}.144)$$

since from (A.143) it can be seen that

$$\Delta E_2(\mathrm{H}) = -\frac{1}{3}\Delta E_1(\mathrm{H}) . \qquad (\mathrm{A}.145)$$

Using

$$\Delta E(\mathrm{HF}) = 2\,\Delta E_{1\mathrm{n}}(\mathrm{H}) + \Delta E_{1\mathrm{p}}(\mathrm{H}) + \Delta E_{2\mathrm{n}}(\mathrm{H}) + \Delta E_{2\mathrm{p}}(\mathrm{H}) , \qquad (\mathrm{A}.146)$$

and (A.144) and (A.145), we obtain

$$\Delta E_{1\mathrm{n}}(\mathrm{H}) = \frac{3}{2}\Delta E_\mathrm{n}(\mathrm{H}) , \qquad (\mathrm{A}.147)$$

$$\Delta E_{1\mathrm{p}}(\mathrm{H}) = \frac{3}{2}\Delta E_\mathrm{p}(\mathrm{H}) . \qquad (\mathrm{A}.148)$$

Finally, for the partial splittings produced by vector–isoscalar mesons, we have

$$\Delta E(\mathrm{HF}) = \left(1 + \frac{3}{2}\frac{N}{A}\right)\Delta E(\mathrm{H}) \qquad (\mathrm{A}.149)$$

for a neutron state and

$$\Delta E(\mathrm{HF}) = \left(1 + \frac{3}{2}\frac{Z}{A}\right)\Delta E(\mathrm{H}) \qquad (\mathrm{A}.150)$$

for a proton state.

Consider now the case of scalar–isovector mesons. The two-body spin–orbit force is given in this case by the operator (A.111) multiplied by $\boldsymbol{\tau}_i \cdot \boldsymbol{\tau}_k$. Using the technique described above, we obtain the following for $N \neq Z$ nuclei:

$$\Delta E(\mathrm{HF}) = \frac{3}{2}\frac{N}{N-Z}\Delta E(\mathrm{H}) \qquad (\mathrm{A}.151)$$

for a neutron state and

$$\Delta E(\mathrm{HF}) = -\frac{3}{2}\frac{Z}{N-Z}\Delta E(\mathrm{H}) \qquad (\mathrm{A}.152)$$

for a proton state. For $N = Z$ nuclei,

$$\Delta E(\mathrm{HF}) = \frac{3}{2}\Delta E_\mathrm{n}(\mathrm{H}) \qquad (\mathrm{A}.153)$$

for a neutron state and

$$\Delta E(\mathrm{HF}) = \frac{3}{2}\Delta E_\mathrm{p}(\mathrm{H}) \tag{A.154}$$

for a proton state, where $\Delta E_\mathrm{n,p}(\mathrm{H})$ is the contribution to the partial (scalar–isovector) Hartree splitting produced by all neutron or proton states, respectively. We should mention that $\Delta E_\mathrm{n}(\mathrm{H}) > 0$, $\Delta E_\mathrm{p}(\mathrm{H}) > 0$.

Now, consider the case of vector–isovector mesons, with the tensor coupling included. In this case, we have, for $N \neq Z$ nuclei,

$$\Delta E(\mathrm{HF}) = \frac{1}{2}\frac{(5+8f/g)N + (4+4f/g)Z}{(1+2f/g)(N-Z)}\Delta E(\mathrm{H}) \tag{A.155}$$

for neutron states and

$$\Delta E(\mathrm{HF}) = \frac{1}{2}\frac{(5+8f/g)Z + (4+4f/g)N}{(1+2f/g)(Z-N)}\Delta E(\mathrm{H}) \tag{A.156}$$

for proton states. For $Z = N$ nuclei, we have

$$\Delta E(\mathrm{HF}) = \frac{9+12f/g}{2+4f/g}\Delta E_\mathrm{n}(\mathrm{H}) \tag{A.157}$$

for neutron states and

$$\Delta E(\mathrm{HF}) = \frac{9+12f/g}{2+4f/g}\Delta E_\mathrm{p}(\mathrm{H}) \tag{A.158}$$

for proton states.

A.12 Hartree–Fock Procedure for the Point-Coupling Model

Consider the most general form of a Lagrangian including all possible types of four-fermion couplings [536, 537, 539]:

$$\mathcal{L}(x) = \bar{\psi}(x)\left\{i\gamma_\nu\partial^\nu - M\right\}\psi(x) + \sum_i \mathcal{L}^{(i)}(x), \tag{A.159}$$

$$\mathcal{L}^{(i)}(x) = -\frac{1}{2}\alpha_i(\bar{\psi}(x)O^{(i)}\psi(x))(\bar{\psi}(x)O^{(i)}\psi(x)). \tag{A.160}$$

In (A.159) and (A.160), $O^{(i)} \otimes O^{(i)}$ include all invariant combinations of Dirac matrices:

$$\text{S, scalar,} \quad 1 \otimes 1; \tag{A.161}$$
$$\text{V, vector,} \quad \gamma^\mu \otimes \gamma_\mu; \tag{A.162}$$
$$\text{P, pseudoscalar,} \quad \gamma_5 \otimes \gamma_5; \tag{A.163}$$
$$\text{A, axial,} \quad \gamma^\mu\gamma_5 \otimes \gamma_\mu\gamma_5; \tag{A.164}$$
$$\text{T, tensor,} \quad \sigma_{\mu\nu} \otimes \sigma^{\mu\nu}. \tag{A.165}$$

They also include the combinations obtained from the combinations listed above by multiplying by $\boldsymbol{\tau} \otimes \boldsymbol{\tau}$ in isotopic space.

The exchange (Fock) terms can be easily calculated in this model. In the space of two-fermion states, the four-fermion coupling Lagrangian produces an interaction without retardation, which is of the Skyrme type because of relativistic invariance.

The exchange terms, due to the Skyrme-type interaction in the Lagrangian considered here, may be taken into account in the Hartree terms using a Fierz transformation (FT). This can be easily done by introducing new coupling parameters $\widetilde{\alpha}_i$ instead of α_i, the values of $\widetilde{\alpha}_i$ being determined by the condition

$$\sum_i \widetilde{\alpha}_i O^{(i)} \otimes O^{(i)} = \sum_i \alpha_i O^{(i)} \otimes O^{(i)} (1 - P_\tau \cdot P_\mathrm{D}), \tag{A.166}$$

where

$$P_\tau = \frac{1}{2}(1 \otimes 1 + \boldsymbol{\tau} \otimes \boldsymbol{\tau}) \tag{A.167}$$

is the isospin exchange operator and

$$P_\mathrm{D} = \frac{1}{4}\bigg(1 \otimes 1 + \gamma_5 \otimes \gamma_5 + \gamma_\mu \otimes \gamma^\mu \\ + \frac{1}{2}\sigma_{\mu\nu} \otimes \sigma^{\mu\nu} - \gamma_5 \gamma_\mu \otimes \gamma_5 \gamma^\mu \bigg) \tag{A.168}$$

is the exchange operator for Dirac spinors.

For this reason the four-fermion coupling model considered here is a perfect tool for investigating the effect of the Fock terms. Consider the case of nuclear matter. To follow the line of the σ–ω model, we take the coupling constants α_i equal to zero apart from α_S and α_V. The Hartree–Fock result can be obtained at once by changing the constants α_S and α_V, obtained in the Hartree approximation, to the respective values $\widetilde{\alpha}_\mathrm{S}$ and $\widetilde{\alpha}_\mathrm{V}$ obtained from the Fierz relation (A.166). In [539] it is shown that the Hartree–Fock approximation provides results which differ greatly from those of the Hartree approximation.

In [537], a consistent procedure for including the Dirac sea effects in the framework of the RPCM was suggested for the model Lagrangian given by (A.159) and (A.160). If the negative-energy states are to be included in the Hartree–Fock scheme, one must use some regularization to handle the divergences, a cutoff or form factors may be used, for example. The authors of [537] discuss a procedure that uses appropriate subtractions (introducing counterterms). Usually, if one considers a model with a renormalizable Lagrangian, an infinite number of counterterms are needed to eliminate the divergences of the theory. Nevertheless, one can obtain a renormalized Hartree–Fock theory with a finite number of counterterms, although the Lagrangian of the model (A.159), (A.160) is not renormalizable. However, this procedure was developed in [537] for a model without the $\mathcal{L}_{\text{hot}}$ and $\mathcal{L}_{\text{der}}$ terms (see (12.21) and (12.22)), which are essential for a description of the properties of finite nuclei. For this reason, we shall not consider this procedure here, a detailed description being given in [537].

The Hartree–Fock method corresponds to minimizing the expectation value $\langle \Phi | H | \Phi \rangle$ of the RPCM Hamiltonian H using Slater determinants

$$\Phi = \prod_{k=1}^{A} a_k^+ |0\rangle , \qquad (A.169)$$

where a_k^+ is the creation operator for a nucleon in state k, and A indicates the nuclear mass number. We can use Wick's theorem [7] to calculate the expectation value of H. Utilizing this theorem enables us to express the expectation values of products of many field operators via the expectation values of only two field operators. Consider the following example (see [66]):

$$\langle \Phi | :(\overline{\psi}\psi)^2: | \Phi \rangle = \langle \Phi | \overline{\psi}_\alpha \overline{\psi}_\beta \psi_\beta \psi_\alpha | \Phi \rangle$$
$$= \langle \Phi | \overline{\psi}_\alpha \psi_\alpha | \Phi \rangle \langle \Phi | \overline{\psi}_\beta \psi_\beta | \Phi \rangle - \langle \Phi | \overline{\psi}_\alpha \psi_\beta | \Phi \rangle \langle \Phi | \overline{\psi}_\beta \psi_\alpha | \Phi \rangle ,$$
$$(A.170)$$

where the colon in (A.170) indicates a normal ordering [7] with respect to the vacuum state $|0\rangle$. As can be seen from (A.170), we obtain in this case the well-known Hartree (direct) and Fock (exchange) contributions.

There is a corresponding procedure to express the exchange contributions in a form identical to the direct terms using the Fierz transformation formula [7, 66, 753–756]. Let us consider the 16 Dirac matrices $\{1, \gamma_5, \gamma_\mu, \gamma_5\gamma_\mu, \sigma_{\mu\nu}\}$, designated as Γ_A ($A = 1, 2, \ldots, 16$), which are linearly independent and form the basis of the Γ-algebra normalized by

$$\Gamma_A^2 = 1 , \quad \text{Tr}\,\Gamma_A = 0 , \quad \text{Tr}[\Gamma_A \cdot \Gamma_B] = 4\delta_{AB}$$
$$\text{(no summation) (if } \Gamma_A \neq 1) , \quad A, B = 1, 2, \ldots, 16 . \quad (A.171)$$

The matrix elements of Γ_A will be written below in the following form:

$$(\overline{a}\,\Gamma_A\,b) \equiv \overline{\psi}_a \Gamma_A \psi_b . \qquad (A.172)$$

An expansion of a product of nondiagonal matrix elements of Γ-matrices into products of diagonal matrix elements, for example

$$(\overline{a}\,\Gamma_A\,b)(\overline{b}\,\Gamma_A\,a) = \sum_{C,D=1}^{16} x_{CD}(\overline{a}\,\Gamma_C\,a)(\overline{b}\,\Gamma_D\,b) , \qquad (A.173)$$

will be referred to in what follows as a Fierz transformation (FT).

A.12 Hartree–Fock Procedure for the Point-Coupling Model

For the scalar case ($\Gamma_A = \Gamma_B = 1$), we have [754–756]

$$(\bar{a}\,b)(\bar{b}\,a) = \frac{1}{4}(\bar{a}\,a)(\bar{b}\,b) + \frac{1}{4}(\bar{a}\,\gamma_5\,a)(\bar{b}\,\gamma_5\,b) + \frac{1}{4}(\bar{a}\,\gamma_\mu\,a)(\bar{b}\,\gamma^\mu\,b)$$
$$- \frac{1}{4}(\bar{a}\,\gamma_5\gamma_\mu\,a)(\bar{b}\,\gamma_5\gamma^\mu\,b) + \frac{1}{8}(\bar{a}\,\sigma_{\mu\nu}\,a)(\bar{b}\,\sigma^{\mu\nu}\,b)\,. \quad (A.174)$$

From (A.174) it follows that if, for example, we have an interaction of scalar type in the relativistic Hartree approximation, we obtain contributions from pseudoscalar, vector, axial-vector, and tensor terms in addition to a contribution of scalar type in the relativistic Hartree–Fock approximation. Similar statements can be made in the case of other types of interaction (however, some of the components could be absent because of symmetry considerations).

We define the order of the FT by the number of the Γ-matrices taken into consideration. For example, to treat $\mathcal{L}_{4f}$ in a Hartree–Fock calculation within the RPCM, we need the FT of the second order given by (A.173). To consider $\mathcal{L}_{hot}$, i.e. three- and four-nucleon contact interactions, in the relativistic Hartree–Fock framework, we need to use FTs of higher (third and fourth) orders:

$$(\bar{a}\,\Gamma_A\,b)(\bar{b}\,\Gamma_B\,c)(\bar{c}\,\Gamma_C\,a) = \sum_{D,E,F=1}^{16} x_{DEF}(\bar{a}\,\Gamma_D\,a)(\bar{b}\,\Gamma_E\,b)(\bar{c}\,\Gamma_F\,c)\,, \quad (A.175)$$

$$(\bar{a}\,\Gamma_A\,b)(\bar{b}\,\Gamma_B\,c)(\bar{c}\,\Gamma_C\,d)(\bar{d}\,\Gamma_D\,a)$$
$$= \sum_{E,F,G,H=1}^{16} x_{EFGH}(\bar{a}\,\Gamma_E\,a)(\bar{b}\,\Gamma_F\,b)(\bar{c}\,\Gamma_G\,c)(\bar{d}\,\Gamma_H\,d)\,. \quad (A.176)$$

The proper treatment of the third- and fourth-order terms in the FT is discussed in [755, 756]. Consideration of the isovector components for scalar, vector, and pseudoscalar interactions causes no complications in the use of the FT. The terms of the RPCM Lagrangian containing derivatives of spinors can be treated in the relativistic Hartree–Fock approximation using the results obtained in [754]. We may say in summary that the FT is a suitable tool for expressing the Fock terms of the RPCM via the Hartree terms (see also the applications of point couplings in the Nambu–Jona–Lasinio model described in [755]).

References

1. S.S. Schweber: *Relativistic Quantum Field Theory* (Row, Peterson and Co., Evanston, IL 1961)
2. J.D. Bjorken, S.D. Drell: *Relativistic Quantum Mechanics* (McGraw-Hill, New York 1964); J.D. Bjorken, S.D. Drell: *Relativistic Quantum Fields* (McGraw-Hill, New York 1965)
3. P. Roman: *Introduction to Quantum Field Theory* (Wiley, New York 1969)
4. R.D. Mattuck: *A Guide to Feynman Diagrams in the Many-Body Problem* (McGraw-Hill, New York 1976)
5. N.N. Bogoliubov, D.V. Shirkov: *Introduction to the Theory of Quantized Fields* (Nauka, Moscow 1976)
6. L.D. Faddeev, A.A. Slavnov: *Gauge Fields. Introduction to Quantum Theory* (Benjamin/Cummings, Reading Massachusetts 1980)
7. C. Itzykson, J. B. Zuber: *Quantum Field Theory* (McGraw-Hill, New York 1980)
8. F. Mandl, G. Shaw: *Quantum Field Theory* (Wiley Interscience, New York 1984)
9. W. Greiner, B. Müller, J. Rafelski: *Quantum Electrodynamics of Strong Fields: with an Introduction into Modern Relativistic Quantum Mechanics* (Springer, Berlin, Heidelberg 1985)
10. B.D. Serot, J.D. Walecka: Adv. Nucl. Phys. **16** (1986), Special Issue (Plenum Press, New York/London)
11. J. Negele, H. Orland: *Quantum Many-Particle Systems* (Addison-Wesley, Redwood City 1988)
12. R.K. Bhaduri: *Models of Nucleon. From Quarks to Soliton* (Lecture Notes and Supplements in Physics) (Physics Department, McMaster University, Hamilton, Ontario 1988)
13. S.J. Chang: *Introduction to Quantum Field Theory*, Lecture Notes in Physics. Vol. 29 (World Scientific, Singapore 1990)
14. F. Gross: *Relativistic Quantum Mechanics and Field Theory* (Wiley Interscience, New York 1993)
15. M. Kaku: *Quantum Field Theory. A Modern Introduction* (Oxford University Press, New York 1993)
16. H. Umezawa: *Advanced Field Theory: Micro, Macro and Thermal Physics* (American Institute of Physics, New York 1993)
17. F.J. Yndurain: *Relativistic Quantum Mechanics and Introduction to Field Theory*, Texts and Monographs in Physics (Springer, Berlin, Heidelberg 1996)
18. S. Weinberg: *The Quantum Theory of Fields, Vol. 1, Foundations* (Cambridge University Press, Cambridge 1996); S. Weinberg: *The Quantum Theory of*

Fields, Vol. 2, Modern Applications (Cambridge University Press, Cambridge 1996); S. Weinberg: *The Quantum Theory of Fields, Vol. 3, Supersymmetry* (Cambridge University Press, Cambridge 2000).
19. B.H. Bransden, C.J. Joachain: *Quantum Mechanics*, 2nd edn. (Pearson Education, Harlow 2000)
20. D.H. Perkins: *Introduction to High Energy Physics*, 4th edn. (Cambridge University Press, Cambridge, New York 2000)
21. S.G. Bondarenko, V.V. Burov, A.V. Molochkov, G.I. Smirnov, H. Toki: Prog. Part. Nucl. Phys. **48**, 449 (2002)
22. N. Isgur, G. Karl: Phys. Rev. D **18**, 4187 (1978)
23. A. Valcarce, A. Buchmann, F. Fernandez, A. Faessler: Phys. Rev. C **51**, 1480 (1995)
24. K. Yazaki: Prog. Part. Nucl. Phys. **24**, 353 (1990)
25. K. Brauer, A. Faessler, F. Fernardez, K. Shimizu: Nucl. Phys. A **507**, 599 (1990)
26. W. Weise (ed.) *Quarks and Nuclei*, International Reviews of Nuclear Physics, Vol. 1. (World Scientific, Singapore 1984)
27. F. Myhrer, J. Wrolsden: Rev. Mod. Phys. **60**, 629 (1988)
28. G.T. Hooft: Nucl. Phys. B **75**, 461 (1974)
29. E. Witten: Nucl. Phys. B **160**, 57 (1979)
30. H. Müther, A. Polls: nucl-th/0001007
31. W. Greiner, A. Schäfer: *Quantum Chromodynamics* (Springer, Berlin 1995); J. Dobaczewski: nucl-th/0301069; K. Kubodera, in: *"Physics of Hadrons and Nuclei"*, Proc. International Symp. *Physics of Hadrons and Nuclei*, ed. by Y. Akaishi, Nucl.Phys. A **670** (2000), p.103c
32. C. Ordóñez, L. Ray, U. van Kolck: Phys. Rev. C **53**, 2086 (1996); V.G.J. Stocks, T.A. Rijken: Nucl. Phys. A **613**, 311 (1997); T.D. Cohen: Phys. Rev. C **55**, 67 (1997); U. van Kolck, J.L. Friar, T. Goldman: Phys. Lett. B **371**, 169 (1996); M. R. Robilotta: in *Proc. 20th Brazilian Workshop on Nuclear Physics*, ed. by S.R. Souza et al. (World Scientific, Singapore, 1997) p. 171; U. van Kolck: Prog. Part. Nucl. Phys. **43**, 337 (1999)
33. R.M. Barnett et al. (Particle Data Group): Phys. Rev. D **54**, 1 (1996); Eur. Phys. J. Vol. **15 C**, Nos. 1–4, 2000. (Rev. Part. Phys.)
34. R. Machleidt: Nucl. Phys. A **689**, 11c (2001); R. Machleidt, G.Q. Li: Phys. Rep. **242**, 5 (1994); R. Machleidt: Adv. Nucl. Phys. **19**, 189 (1989); R. Machleidt, K. Holinde, C. Elster: Phys. Rep. **149**, 1 (1987); K. Holinde: Phys. Rep. C **68**, 121 (1981); K. Erkelenz: Phys. Rep. C **13**, 191 (1974);
35. B. Serot: in Nucl. Phys. B Proc. Suppl. **39**, 281 (1995)
36. M.M. Nagels et al.: Nucl. Phys. B **147**, 189 (1979)
37. M. Kirchbach, L. Tiator: Nucl. Phys. A **604**, 385 (1996)
38. A. Stadler, F. Gross: Phys. Rev. Lett. **78**, 26 (1997)
39. B.L. Birbrair, V.N. Fomenko, L.N. Savushkin: J. Phys. G **8**, 1517 (1982); Preprint LNPI-N615, Leningrad, 1980
40. M. Lacombe et al.: Phys. Rev. D **12**, 1495 (1975); R. Vinh Mau: in *Mesons in Nuclei, Vol. 1*, ed. by M. Rho, D. H. Wilkinson (North-Holland, Amsterdam 1979) p. 151; M. Lacombe et al.: Phys. Rev. C **21**, 861 (1980); B. Loiseau: in *Proc. 20th Brazilian Workshop on Nuclear Physics*, ed. by S.R. Souza et al. (World Scientific, Singapore 1997), p. 139
41. A.D. Jackson, D.O. Riska, B. Verwest: Nucl. Phys. A **249**, 397 (1975)

42. G.E. Brown, A.D. Jackson: *The Nucleon–Nucleon Interaction* (North-Holland, Amsterdam 1976)
43. J.W. Durso, M. Saarela, G.E. Brown, A.D. Jackson: Nucl. Phys. A **278**, 445 (1977)
44. J.W. Durso, G.E. Brown, M. Saarela: Nucl. Phys. A **430**, 653 (1984)
45. S. Weinberg: Phys. Rev. Lett. **18**, 188 (1967)
46. J. Schwinger: Phys. Lett. B **24**, 473 (1967)
47. J. Wess, B. Zumino: Phys. Rev. **163**, 1727 (1967)
48. E.C.G. Sudarshan: Proc. Roy. Soc. A **305**, 319 (1968)
49. J. Binstock, R. Bryan, A. Gersten: Ann. Phys. **139**, 355 (1981)
50. A.E.S. Green, T. Sawada: Rev. Mod. Phys. **39**, 594 (1967); T. Ueda, A.E.S. Green: Phys. Rev. **174**, 1304 (1968); Nucl. Phys. B **10**, 289 (1969); K. Tominaga, T. Ueda, M. Yamaguchi, N. Kijima, D. Okamoto, K. Miyagawa, T. Yamada: Nucl. Phys. A **642**, 483 (1998)
51. L.N. Savushkin, V.N. Fomenko: Sov. J. Part. Nucl. **8**, 371 (1977)
52. V.G.J. Stoks, R.A.M. Klomp, C.P.F. Terheggen, J.J. de Swart: Phys. Rev. C **49**, 2950 (1994); R.B. Wiringa, V.G.J. Stoks, R. Schiawilla: Phys. Rev. C **51**, 38 (1995)
53. R. Brockmann, R. Machleidt: Phys. Rev. C **42**, 1965 (1990)
54. R. Machleidt, F. Sammarruca, Y. Song: Phys. Rev. C **53**, 1483 (1996)
55. J.L. Forest: nucl-th/99-05063
56. L.D. Miller: Phys. Rev. Lett. **28**, 1281 (1972); L.D. Miller, A.E.S. Green: Phys. Rev. C **5**, 241 (1972)
57. L.D. Miller: Phys. Rev. C **9** 537 (1974); Phys. Rev. C **12** 710 (1975); Phys. Rev. C **14** 706 (1976)
58. L.D. Miller: Ann. Phys. (NY) **91**, 1 (1975)
59. V.A. Krutov, L.N. Savushkin: J. Phys. A **6**, 93 (1973)
60. V.A. Krutov, V.N. Fomenko, L.N. Savushkin: J. Phys. A **7**, 372 (1974)
61. L.N. Savushkin: Izv. Akad. Nauk SSSR, Ser. Fiz. **39**, 167 (1975)
62. L.N. Savushkin, V.N. Fomenko: Yad. Fiz. **28**, 58 (1978); Sov. J. Nucl. Phys. **28**, 29 (1978)
63. R. Brockmann, W. Weise: Phys. Rev. C **16**, 1282 (1977)
64. R. Brockmann: Phys. Rev. C **18**, 1510 (1978)
65. R. Brockmann, R. Machleidt: Lecture Notes Phys. Springer, Berlin, **243**, 459 (1985)
66. B.A. Nikolaus, T. Hoch, D. Madland: Phys. Rev. C **46**, 1757 (1992)
67. T.D. Cohen, R.J. Furnstahl, D.K. Griegel: Phys. Rev. Lett. **67**, 961 (1991)
68. W.H. Furry: Phys. Rev. **50**, 784 (1936)
69. A.E.S. Green: Phys. Rev. **76**, A 460, L 870 (1949)
70. M.H. Johnson, E. Teller: Phys. Rev. **98**, 783 (1955)
71. H.P. Duerr, E. Teller: Phys. Rev. **101**, 494 (1956)
72. H.P. Duerr: Phys. Rev. **103**, 469 (1956)
73. H.J. Lipkin, A.N. Tavkelidze: Phys. Lett. **17**, 331 (1965)
74. J.D. Walecka: Ann. Phys. (NY) **83**, 491 (1974); Phys. Lett. B **59**, 109 (1975)
75. S.A. Chin: Ann. Phys. **108**, 301 (1977)
76. B.D. Serot, J.D. Walecka: Adv. Nucl. Phys. **16**, 1 (1986)
77. B.D. Serot: Rep. Prog. Phys. **55**, 1855 (1992)
78. L.N. Savushkin, V.N. Fomenko: *Lecture Notes for Young Scientists. Introduction to Meson Theory of Nuclear Interactions and Nuclear Systems*, R4-83-369 (JINR, Dubna 1983)

79. L.S. Celenza, C.M. Shakin: *Lecture Notes in Physics, Vol. 2* (World Scientific, Singapore 1986)
80. R. Malfliet: Prog. Part. Nucl. Phys. **21**, 207 (1988); B. ter Haar, R. Malfliet: Phys. Rep. C **149**, 208 (1987)
81. P.-G. Reinhard: Rep. Prog. Phys. **52**, 439 (1989); M. Bender, P.-H. Heenen, P.-G. Reinhard: Rev. Mod. Phys. **75**, 607 (2003)
82. N.V. Giai, L.N. Savushkin: Sov. J. Part. Nucl. **23**, 373 (1992)
83. J.D. Walecka: *Theoretical Nuclear and Subnuclear Physics.* (Oxford University Press, Oxford 1995)
84. P. Ring: Prog. Part. Nucl. Phys. **37**, 193 (1996)
85. B.D. Serot, J.D. Walecka: Int. J. Mod. Phys. E **6**, 515 (1997)
86. R.J. Furnstahl, R.J. Perry, B.D. Serot: Phys. Rev. C **40**, 321 (1989)
87. C.J. Horowitz, B.D. Serot: Phys. Lett. B **140**, 181 (1984)
88. R.J. Perry: Phys. Lett. B **182**, 269 (1986); Phys. Lett. B **199**, 489 (1987); Nucl. Phys. A **467**, 717 (1987); D.A. Wasson: Phys. Lett. B **210**, 41 (1988); T.D. Cohen, M.K. Banerjee, C.Y. Ren: Phys. Rev. C **36**, 1653 (1987)
89. K. Tanaka, W. Bentz, A. Arima: Nucl. Phys. A **518**, 229 (1990); K. Tanaka, W. Bentz, A. Arima, F. Beck: Nucl. Phys. A **528**, 676 (1991); K. Tanaka, W. Bentz: Nucl. Phys. A **540**, 383 (1992)
90. J. Caro, E. Ruiz Arriola, L.L. Salcedo: Phys. Lett. B **383**, 9 (1996)
91. C. Nash: *Relativistic Quantum Fields* (Academic Press, New York 1978)
92. Y.K. Gambhir, P. Ring, A. Thimet: Ann. Phys. **198**, 132 (1990); Y.K. Gambhir, P. Ring: Mod. Phys. Lett. **8**, 787 (1993); J.P. Maharana, L.S. Warrier, Y.K. Gambhir: Ann. Phys. **250**, 237 (1996)
93. Qi-Ren Zhang, W. Greiner: Mod. Phys. Lett. A **10**, 2809 (1995)
94. K. Sumiyoshi, H. Toki: Astrophys. J. **422**, 700 (1994)
95. H. Toki, Y. Sugahara, D. Hirata, B.V. Carlson, I. Tanihata: Nucl. Phys. A **524**, 633 (1991)
96. D. Hirata, H. Toki, T. Watabe, I. Tanihata, B.V. Carlson: Phys. Rev. C **44**, 1467 (1990)
97. J. Boguta, H. Stöcker: Phys. Lett. B **106**, 241, 245, 250, 255 (1981)
98. N.K. Glendenning: Astrophys. J. **293**, 470 (1985)
99. K. Sumiyoshi, H. Toki, R. Brockmann: Phys. Lett. B **276**, 393 (1992)
100. L.N. Savushkin: Sov. J. Nucl. Phys. **30**, 349 (1979)
101. C.J. Horowitz, B.D. Serot: Nucl. Phys. A **399**, 529 (1983)
102. J. Boguta, A.R. Bodmer: Nucl. Phys. A **292**, 413 (1977)
103. P.-G. Reinhard, M. Rufa, J. Maruhn, W. Greiner, J. Friedrich: Z. Phys. A **323**, 13 (1986); M. Del Estal, M. Centellas, X. Viñas, S.K. Patra: Phys. Rev. C **63**, 044321 (2001)
104. M. Rufa, P.-G. Reinhard, J.A. Maruhn, W. Greiner, M.R. Strayer: Phys. Rev. C **38**, 390 (1988)
105. L.I. Schiff: Phys. Rev. **84**, 1, 10 (1951)
106. S. Barshay, G.E. Brown: Phys. Rev. Lett. **34**, 1106 (1975)
107. B. Banerjee, N.K. Glendenning, M. Guylassy: Nucl. Phys. A **361**, 326 (1981)
108. A. Bouyssy, S. Marcos: Phys. Lett. B **124**, 139 (1983)
109. N.K. Glendenning, B. Banerjee, M. Guylassy: Ann. Phys. (NY) **149**, 1 (1983); N.K. Glendenning, P. Hecking, V. Ruck: Ann. Phys. (NY) **149**, 22 (1983)
110. J. Boguta, H. Stöcker: Phys. Lett. B **120**, 289 (1983)
111. J. Boguta, S.A. Moszkowski: Nucl. Phys. A **403**, 445 (1983)

112. A. Bouyssy, S. Marcos, Pham Van Thieu: Nucl. Phys. A **422**, 541 (1984)
113. P.-G. Reinhard: Z. Phys. A **329**, 257 (1988)
114. D. Hirata, K. Sumiyoshi, B.V. Carlson, H. Toki, I. Tanihata: Nucl. Phys. A **609**, 131 (1996)
115. A.R. Bodmer: Nucl. Phys. A **526**, 703 (1991)
116. Y. Sugahara, H. Toki: Nucl. Phys. A **579**, 557 (1994)
117. L.N. Savushkin, S. Marcos, M. López-Quelle, P. Bernardos, V.N. Fomenko, R. Niembro: Phys. Rev. C **55**, 167 (1997)
118. B. Bakamjian, L.H. Thomas: Phys. Rev. **92**, 1300 (1952)
119. L.L. Foldy: Phys. Rev. **122**, 275 (1961)
120. J. Carlson, V.R. Pandharipande, R. Schiavilla: Phys. Rev. C **47**, 484 (1993)
121. J.L. Forest, V.R. Pandharipande, J. Carlson, R. Shiavilla: Phys. Rev. C **52**, 576 (1995)
122. J.L. Forest, V.R. Pandharipande, J.L. Friar: Phys. Rev. C **52**, 568 (1995)
123. P. Amore, M.B. Barbaro, A. De Pace: Phys. Rev. C **53**, 2801 (1996)
124. J.L. Forest, V.R. Pandharipande, A. Arriaga: Phys. Rev. C **60**, 014002-1 (1999)
125. S.K. Patra, C.R. Praharaj: Phys. Rev. C **44**, 2552 (1991)
126. J. Blocki, M. Flocard: Nucl. Phys. A **273**, 45 (1976)
127. J.V. Noble: Nucl. Phys. A **329**, 354 (1979)
128. M. Jaminon, C. Mahaux, P. Rochus: Phys. Rev. C **22**, 2027 (1980)
129. R. Humphries: Nucl. Phys. A **182**, 580 (1972)
130. R.L. Mercer, L.G. Arnold, B.C. Clark: Phys. Lett. B **73**, 9 (1978)
131. L.G. Arnold, B.C. Clark: Phys. Lett. B **84**, 46 (1979)
132. B.L. Birbrair, L.N. Savushkin, V.N. Fomenko: Sov. J. Nucl. Phys. **35**, 664 (1982)
133. A. Bouyssy, S. Marcos: Phys. Lett. B **127**, 157 (1983)
134. T.H.R. Skyrme: Phil. Mag. **1**, 1043 (1956)
135. T.H.R. Skyrme: Nucl. Phys. **9**, 615 (1959)
136. D. Vautherin, D.M. Brink: Phys. Rev. C **5**, 626 (1972)
137. M. Beiner, H. Flocard, N.V. Giai, P. Quentin: Nucl. Phys. A **238**, 29 (1975)
138. A. Bohr, B. Mottelson: *Nuclear Structure Vol. 1* (Benjamin, New York 1969)
139. K. Sumiyoshi, D. Hirata, H. Toki, H. Sagawa: Nucl. Phys. A **552**, 437 (1993)
140. K. Pomorski, P. Ring, G.A. Lalazissis, A. Baran, Z. Loewski, B. Nerlo-Pomorska, M. Warda: Nucl. Phys. A **624**, 349 (1997)
141. M. Centelles, M. Del Estal, X. Viñas: nucl-th/9712002; M. Del Estal, M. Centelles, X. Viñas. S.K. Patra: Phys. Rev. C **63**, 024314 (2001)
142. A. Baran: Phys. Rev. C **61**, 024316 (2000)
143. B. Nerlo-Pomorska, K. Mazurek: nucl-th/0211049.
144. H. Kurasawa, T. Suzuki: nucl-th/0201035.
145. L.N. Savushkin: Vestn. Lening. Univ. **16**, 41 (1975); L.N. Savushkin: *The atomic nucleus as a relativistic system*, Thesis M. Sc. (Kiev 1982)
146. S.J. Lee, J. Fink, A.B. Balantekin, M.R. Strayer, A.S. Umar, P.G. Reinhard, J.A. Maruhn, W. Greiner: Phys. Rev. Lett. **57**, 2916 (1986)
147. M.M. Sharma, N.A. Nagarajan, P. Ring: Phys. Lett. B **312**, 377 (1993)
148. G.A. Lalazissis, J. König, P. Ring: Phys. Rev. C **55**, 540 (1997)
149. M. Rashdan: Phys. Lett. B **395** 141 (1997)
150. W. Koepf, Y.K. Gambhir, P. Ring, M.M. Sharma: Z. Phys. A **340**, 119 (1991)
151. M.M. Sharma, G.A. Lalazissis, W. Hillebrandt, P. Ring: Phys. Rev. Lett. **72**, 1431 (1994)

152. R. Brockmann, H. Toki: Phys. Rev. Lett. **68**, 3408 (1992)
153. C.J. Batty et al.: Adv. Nucl. Phys. **19**, 1 (1989)
154. P. Möller, J.R. Nix: nucl-th/9309028
155. I. Tanihata et al.: Phys. Rev. Lett. **55**, 2676 (1985)
156. I. Tanihata: in *Treatise on Heavy Ion Science. Vol. 8.*, ed. by D.A. Bromley (Plenum, New York 1989), p. 443; I. Tanihata: Prog. Part. Nucl. Phys. **35**, 505 (1995)
157. J.H. Hamilton: in *Treatise on Heavy Ion Science. Vol. 8*, ed. by D.A. Bromley (Plenum, New York 1989), p. 3
158. D. Hirata, H. Toki, I. Tanihata, P. Ring: Phys. Lett. B **314**, 168 (1993)
159. D.P. Menezes, C. Providencia: Phys. Rev. C **60**, 024313 (1999); Nucl. Phys. A **650**, 283 (1999); Phys. Rev. C **64**, 044306 (2002)
160. A.L. Espindola, D.P. Menezes: Phys. Rev. C **65**, 045803 (2002)
161. Y.K. Gambhir, J.P. Maharana, G.A. Lalazissis, C.P. Panos, P. Ring: Phys. Rev. C **62**, 054610 (2000)
162. J.L. Friar, J.W. Negele: Adv. Nucl. Phys. **8**, 219 (1975)
163. L.S. Celenza, A. Rosenthal, C.M. Shakin: Phys. Rev. C **31**, 232 (1985)
164. L.S. Celenza, A. Harindranath, A. Rosenthal, C.M. Shakin: Phys. Rev. C **31**, 946 (1985)
165. T.E.O. Erikson, A. Richter: Phys. Lett. B **183**, 249 (1987)
166. D.F. Geesaman, K. Saito, A.W. Thomas: Annu. Rev. Nucl. Part. Sci. **45**, 337 (1995); M. Arneodo: Phys. Rep. **240**, 301 (1994); E. Marco, E. Oset, P. Fernandez de Cordoba: Nucl. Phys. A **611**, 484 (1996); P. Fernandez de Cordoba, E. Marco, H. Müther, E. Oset, A. Faessler: Nucl. Phys. A **611**, 514 (1996); F. Gross, S. Liuti: Phys. Rev. C **45**, 1374 (1992)
167. M. Birse: Phys. Lett. B **299**, 186 (1993)
168. S.J. Wallace, F. Gross, J.A. Tjon: Phys. Rev. Lett. **74**, 228 (1995); Phys. Rev. C **53**, 860 (1996); J.A. Tjon: p.194.: in *Proc. 20th Brazilian Workshop on Nuclear Physics*, ed. by S.R. Souza et al. (World Scientific, Singapore 1997), p. 194
169. K. Saito, A.W. Thomas: Phys. Rev. C **52**, 2789 (1995)
170. B.L. Birbrair, L.N. Savushkin, V.N. Fomenko: Sov. J. Nucl. Phys. **38**, 25 (1983)
171. J. Sakurai: *Currents and Mesons* (University of Chicago Press, Chicago 1969); V. de Alfaro, S. Fubini, G. Furlan, C. Rossetti: *Currents in Hadron Physics* (North-Holland, Amsterdam 1973); H.B. O'Connell, B.C. Pearce, A.W. Thomas, A.G. Williams: Prog. Part. Nucl. Phys. **39**, 201 (1997)
172. M. Bawin, S.A. Coon: Phys. Rev. C **60**, 025207-1 (1999)
173. N. Isgur: hep-ph/9812243.
174. S. Marcos, N.V. Giai, L.N. Savushkin: Nucl. Phys. A **549**, 143 (1992)
175. G.E. Brown, R. Machleidt: Phys. Rev. C **50**, 1731 (1994)
176. C.W. De Jager, H. De Vries, C. De Vries: At. Data Nucl. Data Tables **14**, 479 (1974)
177. G.D. Alkhazov et al.: Phys. Lett. B **42**, 121 (1978)
178. L. Ray, P.E. Hodgson: Phys. Rev. C **20**, 2403 (1979)
179. G.D. Alkhazov et al.: Nucl. Phys. A **381**, 430 (1982)
180. K. Okamoto: Phys. Lett. **11**, 150 (1964); J.A. Nolen, J.P. Schiffer: Annu. Rev. Nucl. Sci. **19**, 471 (1969); S. Shlomo: Rep. Prog. Phys. **41**, 957 (1978); G.A. Miller, B.M.K. Nefkens, I. Šlaus: Phys. Rep. **194**, 1 (1990)

181. H. Margenau: Phys. Rev. **57**, 383 (1940)
182. H. Ohtsubo, M. Sano, M. Morita: J. Phys. Soc. Japan **34**, 509 (1973); Prog. Theor. Phys. **49**, 877 (1973)
183. M. Bawin, C.A. Hughes, G.L. Strobel: Phys. Rev. C **28**, 456 (1983)
184. A. Bouyssy, S. Marcos, J.-F. Mathiot: Nucl. Phys. A **415**, 497 (1984)
185. T. Matsui: Nucl. Phys. A **370**, 365 (1981)
186. H. Kurasawa, T. Suzuki: Phys. Lett. B **165**, 234 (1985)
187. S. Ichii, W. Bentz, A. Arima, T. Suzuki: Phys. Lett. B **192**, 11 (1987)
188. J.A. McNeil, R.D. Amado, C.J. Horowitz, M. Oka, J.R. Shepard, D.A. Sparrow: Phys. Rev. C **34**, 746 (1986)
189. P.G. Blunden: Nucl. Phys. A **464**, 525 (1987)
190. R.J. Furnstahl, B.D. Serot: Nucl. Phys. A **468**, 539 (1987)
191. U. Hofmann, P. Ring: Phys. Lett. B **214**, 307 (1988)
192. M. Gari, H. Hyuga: Z. Phys. A **277**, 291 (1976); Nucl. Phys. A **278**, 372 (1977)
193. M. Gari, H. Hyuga, B. Sommer: Phys. Rev. C **14**, 2196 (1976)
194. V.V. Burov, V.N. Dostovalov: Z. Phys. A **236** 254 (1987)
195. V.V. Burov, A.A. Goi, V.N. Dostovalov: Sov. J. Nucl. Phys. **45**, 616 (1987)
196. V.V. Burov, V.N. Dostovalov, S.Eh. Sus'kov: Czech. J. Phys. **41**, 1139 (1991); Sov. J. Part. Nucl. **23**, 317 (1992)
197. M. Chemtob, M. Rho: Nucl. Phys. A **163**, 1 (1971)
198. A. Arima: in *Progress in Particle and Nuclear Physics, Vol. 11*, ed. by D. Wilkinson (Pergamon, Oxford 1984), p. 53
199. W. Bentz, A. Arima, H. Hyuga, K. Shimizu, K. Yazaki: Nucl. Phys. A **436**, 593 (1985)
200. H. Kurasawa, T. Suzuki: Nucl. Phys. A **445**, 685 (1985)
201. G.E. Brown: Physica-Scripta **36** (2), 209 (1987)
202. J.R. Shepard, E. Rost, C.Y. Cheung, J.A. McNeil: Phys. Rev. C **37**, 1130 (1988)
203. K. Rutz, M. Bender, P.-G. Reinhard, J.A. Maruhn, W. Greiner: Nucl. Phys. A **634**, 67 (1998)
204. R.J. Glauber: in *Lectures in Theoretical Physics, Vol. 1*, ed. by W.E. Brittin et al. (Interscience, New York 1959), p. 315
205. A.K. Kerman, H. McManus, R. Thaler: Ann. Phys. (NY) **8**, 551 (1959)
206. M. Jaminon, C. Mahaux, P. Rochus: Phys. Rev. C **22**, 2027 (1980)
207. R.L. Mercer, L.G. Arnold, B.C. Clark: Phys. Lett. B **73**, 9 (1978)
208. L.G. Arnold, B.C. Clark: Phys. Lett. B **84**, 46 (1979)
209. J.V. Noble: Phys. Rev. C **17**, 2151 (1978)
210. J. Boguta: Phys. Lett. B **106**, 245 (1981)
211. A.T. Raikov, L.N. Savushkin: Bull. Leningrad State Univ. **4**, 148 (1976); P. Susan, C.S. Shastry, Y.K. Gambhir: Phys. Rev. C **50**, 2955 (1994)
212. J.V. Noble: Nucl. Phys. A **329**, 354 (1979)
213. B.L. Birbrair: in *Proc. 17th Winter School Leningrad Institute for Nuclear Physics*, (1982), p. 143; B.L. Birbrair: nucl-th/0110002; B.L. Birbrair, V.L. Ryazanov: Yad. Fiz. **64**, 471 (2001)
214. Y. Nogami, F.M. Toyama: Phys. Rev. C **42**, 2449 (1990)
215. Y. Nogami, A. Suzuki, F.M. Toyama: J. Phys. G: Nucl. Part. Phys. **18**, L173 (1992)
216. N.V. Giai, J. van de Wiele, L.N. Savushkin: Phys. Rev. C **52**, 2266 (1995)

217. J. Rapaport: Phys. Lett. B **92**, 233 (1980)
218. L.G. Arnold, B.C. Clark, R.L. Mercer: Phys. Rev. C **23**, 15 (1981)
219. J. Franz, H.P. Grotz, L. Lehmann, E. Rössle, H. Schmitt, L. Schmitt: Nucl. Phys. A **490**, 667 (1988)
220. R.W. Finlay, W.P. Abfalterer, G. Fink, E. Montei, T. Adami, P.W. Lisowski, G.L. Morgan, R.C. Haight: Phys. Rev. C **47**, 237 (1993)
221. R. Kozack, D.G. Madland: Phys. Rev. C **39**, 1461 (1989); Nucl. Phys. A **509**, 664 (1990)
222. Shen Qing-Biao, Feng Da-Chun, Zhuo Yi-Zhong: Phys. Rev. C **43**, 2773 (1991)
223. S. Hama, B.C. Clark, E.D. Cooper, H.S. Sherif, R.L. Mercer: Phys. Rev. C **41**, 2737 (1990)
224. J.A. McNeil, J.R. Shepard, S.J. Wallace: Phys. Rev. Lett. **50**, 1439 (1983)
225. J.R. Shepard, J.A. McNeil, S.J. Wallace: Phys. Rev. Lett. **50**, 1443 (1983)
226. B.C. Clark, S. Hama, R.L. Mercer, L. Ray, B.D. Serot: Phys. Rev. Lett. **50**, 1644 (1983)
227. G.W. Hoffman, L. Ray, M.L. Barlett, R. Fergerson, J. McGill, et al.: Phys. Rev. Lett. **47**, 1436 (1981); Phys. Rev. C **24**, 541 (1981)
228. L. Ray et al.: Phys. Rev. C **23**, 828 (1981); A. Rahbar, B. Aas, E. Bleszynski, M. Bleszynski, M. Haji-Saeid, et al.: Phys. Rev. Lett. **47**, 1811 (1981)
229. L. Ray, G.W. Hoffman: Phys. Rev. C **31**, 538 (1985)
230. S.J. Wallace: Annu. Rev. Nucl. Part. Sci. **37**, 267 (1987); Comments Nucl. Part. Phys. **13**, 27 (1984)
231. B.C. Clark et al.: in *10th Int. Conf. on Particles and Nuclei (PANIC)*, Heidelberg, Book of Abstracts, Vol. 1 (1984), p. c12
232. B.C. Clark et al.: Phys. Rev. Lett. **50**, 1644 (1983); Phys. Rev. Lett. **51**, 1809C (1983); Phys. Rev. C **28**, 1421 (1983)
233. R.A. Arndt et al.: Phys. Rev. D **8**, 97 (1983)
234. K. Kaki, H. Toki: Nucl. Phys. A **696**, 453 (2001)
235. B.C. Clark et al.: *10th Int. Conf. on Particles and Nuclei (PANIC)*, Heidelberg, Book of Abstracts, Vol. 1 (1984), p. c40
236. J. Coté et al.: Phys. Rev. Lett. **48**, 319 (1982); C.B. Dover et al.: Phys. Rev. C **28**, 2368 (1983)
237. C.J. Horowitz, B.D. Serot: Nucl. Phys. A **368**, 503 (1981); C.J. Horowitz: Phys. Rev. C **31**, 1340 (1985)
238. D. Garreta et al.: Phys. Lett. B **135**, 266 (1984)
239. M.B. Zhalov, L.N. Savushkin: Pis'ma Zh. Eksp. Teor. Fiz. **35**, 441 (1982)
240. J. Iwadare, S. Hatano: Prog. Theor. Phys. (Japan) **15**, 185 (1956)
241. J.S. Ball, G.F. Chew: Phys. Rev. **109**, 1385 (1958)
242. E.H. Auerbach, C.B. Dover, S.H. Kahana: Phys. Rev. Lett. **46**, 702 (1981)
243. G. Mao, H. Stöcker, W. Greiner: nucl-th/0112010
244. T. Bürvenich, I.N. Mishustin, L.M. Satarov, H. Stöcker, W. Greiner: nucl-th/0207011
245. H. Toki, H. Suganuma: in *Proceedings of the 14th RCNP Osaka International Symposium on Nuclear Reaction Dynamics of Nucleon–Hadron Many Body System. From Nucleon Spins and Mesons in Nuclei to Quark Lepton Nuclear Physics*, ed. by H. Ejiri, T. Noro, K. Takahisa, H. Toki (World Scientific, Singapore 1996), p. 99
246. A.B. Migdal: Rev. Mod. Phys. **50**, 107 (1978)

247. E. Oset, H. Toki, W. Weise: Phys. Rep. **83**, 281 (1982)
248. A.B. Migdal: in *Mesons in Nuclei, Vol. 3*, ed. by M. Rho and D.H. Wilkinson, chap. 25 (North-Holland, Amsterdam 1979), p. 941; R.F. Sawyer: in *Mesons in Nuclei, Vol. 3*, ed. by M. Rho and D.H. Wilkinson, chap. 26 (North-Holland, Amsterdam 1979), p. 991; G. Baym and D.K. Campbell, *Mesons in Nuclei, Vol. 3*, ed. by M. Rho and D.H. Wilkinson, chap. 27 (North-Holland, Amsterdam 1979), p. 1031; S.-O. Bäckman and W. Weise: in *Mesons in Nuclei, Vol. 3*, ed. by M. Rho and D.H. Wilkinson, chap. 28 (North-Holland, Amsterdam 1979), p. 1095
249. E. Stenlund et al. (eds.): *Proceedings of "Quark Matter"*, Nucl. Phys. A **566** (1994), Special Issue
250. H. Tezuka: Phys. Rev. C **24**, 288 (1981)
251. N.K. Glendenning, B. Banerjee, M. Guylassy: Ann. Phys. **149**, 1 (1983); N.K. Glendenning, P. Hecking, V. Ruck: Ann. Phys. **149**, 22 (1983)
252. J.F. Dawson, J. Piekarewicz: Phys. Rev. C **43**, 2631 (1991)
253. V.N. Fomenko: *Phenomenological Relativistic Theory of the Pion-Nucleus Interaction*. Thesis M. Sc. (Leningrad 1989)
254. K. Johnson, E.C.G. Sudarshan: Ann. Phys. **13**, 126 (1961)
255. R.D. Peccei: Phys. Rev. **176**, 1812 (1968)
256. A. Aurilia, H. Umezawa: Phys. Rev. **182**, 1682 (1969)
257. C.R. Hagen: Phys. Rev. D **4**, 2204 (1971)
258. L.M. Nath, B. Etemadi, J.D. Kimel: Phys. Rev. D **3**, 2153 (1971)
259. M.G. Olsson, E.T. Osypowski: Nucl. Phys. B **101**, 136 (1975)
260. M. Benmerrouche, R.M. Davidson, N.C. Mukhopadhyay: Phys. Rev. C **39**, 2339 (1989)
261. P. van Nieuwenhuizen: Phys. Rep. **68**, 89 (1981)
262. D.K. Griegel: Phys. Rev. C **43**, 904 (1991)
263. K. Wehrberger, F. Beck: Phys. Lett. B **270**, 1 (1991)
264. K. Wehrberger, R. Wittman: Nucl. Phys. A **513**, 603 (1990)
265. K. Wehrberger: Phys. Rep. **225**, 272 (1993)
266. F. de Jong, R. Malfliet: Phys. Rev. C **46**, 2567 (1992)
267. Hua-Bin Tang, P.J. Ellis: Phys. Lett. B **387**, 9 (1996)
268. Liang-gang Liu, Qi-fa Zhou: Z. Phys. A **357**, 27 (1997)
269. T. Matsui, B.D. Serot: Ann. Phys. **144**, 107 (1982)
270. T. Ericson, W. Weise: *Pions and Nuclei* (Clarendon, Oxford 1988)
271. F. Gross, J.W. Van Orden, K. Holinde: Phys. Rev. C **45**, 2094 (1992); Phys. Rev. C **41**, 1909 (1990)
272. P.F.A. Goudsmith, H.J. Leisi, E. Matsinos: Phys. Lett. B **271**, 290 (1991); Phys. Lett. B **299**, 6 (1993); P.F.A. Goudsmith, H.J. Leisi, E. Matsinos, B.L. Birbrair, A.B. Gridnev: Nucl. Phys. A **575**, 673 (1994)
273. S. Weinberg: Phys. Rev. Lett. **18**, 188 (1967); Phys. Rev. **166**, 1568 (1968)
274. R. Niembro, V.N. Fomenko, L.N. Savushkin, M. López-Quelle, S. Marcos, P. Bernardos: J. Phys. G: Nucl. Part. Phys. **24**, 1945 (1998)
275. A.M. Green: Rep. Prog. Phys. **39**, 1109 (1976)
276. F.J. Ynduráin: *Quantum Chromodynamics. An Introduction to the Theory of Quarks and Gluons*, Texts and Monographs in Physics (Springer, Berlin, Heidelberg 1983)
277. G. Höhler, H.P. Jakob, R. Strauss: Nucl. Phys. B **39**, 237 (1972)
278. Hua-Bin Tang, P.J. Ellis: Phys. Lett. B **387**, 9 (1996)

279. M.K. Banerjee, J. Milana: Phys. Rev. D **52**, 6451 (1995)
280. E. Jenkins, A.V. Manohar: Phys. Lett. B **259**, 353 (1991); E. Jenkins: Nucl. Phys. B **368**, 190 (1992)
281. C. Ordóñez, L. Ray, U. van Kolck: Phys. Rev. Lett. **72**, 1982 (1994); C. Ordóñez, U. van Kolck: Phys. Lett. B **291**, 459 (1992); U. van Kolck: Phys. Rev. C **49**, 2932 (1994)
282. V. Bernard, N. Kaiser, U. Meissner: Phys. Lett. B **378**, 337 (1996); Phys. Lett. B **382**, 19 (1996)
283. L.S. Ferreira, G. Cattapan: Phys. Rep. **362**, 303 (2002)
284. G. Chanfray, M. Ericson: Phys. Lett. B **141**, 167 (1984); P. Guichon, J. Delorme: Phys. Lett. B **263**, 157 (1991); T. Udagawa, S.W. Hong, F. Osterfeld: Phys. Lett. B **245**, 1 (1990); G. Chanfray, D. Davesne: nucl-th/9806086
285. S. Adler, R. Dashen: *Current Algebras* (Benjamin, New York 1968)
286. B.W. Lee: *Chiral Dynamics* (Gordon and Breach, New York 1972)
287. M. Rho, G.E. Brown: Comments Nucl. Part. Phys. **10**, 201 (1981)
288. U.-G. Meissner: Phys. Rep. **161**, 213 (1988)
289. K. Huang: *Quarks, Leptons, and Gauge Fields* (World Scientific, Singapore 1982)
290. A.W. Thomas: Adv. Nucl. Phys. **13**, 1, (1984)
291. G.E. Brown, M. Rho: Phys. Rep. **363**, 85 (2002)
292. S. Weinberg: Phys. Lett. B **251**, 288 (1990)
293. V. Koch: Int. J. Mod. Phys. E **6**, 203 (1997); K. Kubodera: Nucl. Phys. A **670**, 103c (2000)
294. T.D. Lee, G.C. Wick: Phys. Rev. D **9**, 2291 (1974)
295. T.D. Lee, M. Margulies: Phys. Rev. D **11**, 1591 (1975)
296. M. Gell-Mann, M. Lévy: Nuovo Cimento **16**, 705 (1960)
297. F. Gürsey: Nuovo Cimento **16**, 230 (1960)
298. F. Gürsey: Ann. Phys. **12**, 91 (1961)
299. J. Boguta: Phys. Lett. B **120**, 34 (1983)
300. J. Kunz, D. Masak, U. Post, J. Boguta: Phys. Lett. B **169**, 133 (1986)
301. A.K. Kerman, L.D. Miller: *2nd High Energy Ion Summer Study*, July 15–26, 1974, (Lawrence Berkeley Laboratory, Berkeley 1974)
302. S. Sarkar, S.K.Chowdhury: Phys. Lett. B **153**, 358 (1985)
303. W. Bentz, L.G. Liu, A. Arima: Ann. Phys. (NY) **188**, 61 (1988); L.G. Liu, W. Bentz, A. Arima: Ann. Phys. (NY) **194**, 387 (1989); W. Bentz, A. Arima, H. Baier: Ann. Phys. (NY) **200**, 127 (1990)
304. V.N. Fomenko, S. Marcos, L.N. Savushkin: J. Phys. G: Nucl. Part. Phys. **19**, 545 (1993)
305. V.N. Fomenko, L.N. Savushkin, S. Marcos, R. Niembro, M. López-Quelle: J. Phys. G: Nucl. Part. Phys. **21**, 53 (1995)
306. J. Kunz, D. Masak, U. Post: Phys. Lett. B **186**, 124 (1987)
307. P. Bernardos, V.N. Fomenko, M. López-Quelle, S. Marcos, R. Niembro, L.N. Savushkin: J. Phys. G: Nucl. Part. Phys. **22**, 361 (1996)
308. R.J. Furnstahl, B.D. Serot: Phys. Rev. C **47**, 2338 (1993); Phys. Lett. B **316**, 12 (1993)
309. R.J. Furnstahl, H.B. Tang, B.D. Serot: Phys. Rev. C **52**, 1368 (1995)
310. R.J. Furnstahl, B.D. Serot, H.B. Tang: Nucl. Phys. A **615**, 441 (1997); Nucl. Phys. A **598**, 539 (1996)

311. M. Bando, T. Kugo, S. Uehara, K. Yamawaki, T. Yanagida: Phys. Rev. Lett. **54**, 1215 (1985); M. Bando, T. Kugo, K. Yamawaki: Phys. Rep. **169**, 217 (1988)
312. G. Gelmini, B. Ritzi: Phys. Lett. B **357**, 431 (1995)
313. E. Heide, S. Rudaz, P.J. Ellis: Nucl. Phys. A **571**, 713 (1994)
314. G.W. Carter, P.J. Ellis. S. Rudaz: Nucl. Phys. A **603**, 367 (1996); G.W. Carter, P.J. Ellis: Nucl. Phys. A **628**, 325 (1998)
315. Pyongyan Ko, S. Rudaz: Phys. Rev. D **50**, 6877 (1994)
316. P.J. Ellis, E.K. Heide, S. Rudaz: Phys. Lett. B **282**, 271 (1992); Phys. Lett. B **287**, 413E (1992)
317. E.K. Heide, S. Rudaz, P.J. Ellis: Phys. Lett. B **293**, 259 (1992)
318. J. Ellis, J.I. Kapusta, K.A. Olive: Nucl. Phys. B **348**, 345 (1991)
319. I. Mishustin, J. Bondorf, M. Rho: Nucl. Phys. A **555**, 215 (1993)
320. G. Ripka, M. Jaminon: Ann. Phys. (NY) **218**, 51 (1992); Nucl. Phys. A **564**, 505 (1993)
321. B.D. Serot, J.D. Walecka: Acta Phys. Polonica **23B**, 655 (1992)
322. V.N. Fomenko, P. Ring, L.N. Savushkin: Nucl. Phys. A **579**, 438 (1994)
323. V.N. Fomenko, P. Ring, L.N. Savushkin: Phys. At. Nucl. (Engl. Transl.) **58**, 2156 (1995); Petersburg Nuclear Physics Institute, Preprint TH-5-1995, N2032, (Gatchina, St. Petersburg 1995)
324. W. Lin, B.D. Serot: Phys. Lett. B **233**, 23 (1989); Nucl. Phys. A **512**, 637 (1990)
325. V.N. Fomenko, S. Marcos, P. Ring, L.N. Savushkin: Phys. At. Nucl. (Engl. Transl.) **60**, 2149 (1997); V.N. Fomenko, S. Marcos, P. Ring, L.N. Savushkin: in *Proc. European Conf. on Advances in Nuclear Physics and Related Areas*, ed. by D.M. Brink et al. (Thessaloniki 1999), p. 734
326. P. Bernardos, V.N. Fomenko, S. Marcos, R. Niembro, M. López-Quelle, L.N. Savushkin: J. Phys. G **27**, 147 (2001)
327. S. Gasiorowicz, D.A. Geffen: Rev. Mod. Phys. **41**, 531 (1969)
328. W. Kluge: Rep. Prog. Phys. **54**, 1251 (1991); M.S. Birse: Z. Phys. A **355**, 231 (1996)
329. F. Klingl, N. Kaiser, W. Weise: Z. Phys. A **356**, 193 (1996)
330. P. Jain, R. Johnson, U.G. Meissner, N.W. Park, J. Schechter: Phys. Rev. D **37**, 3252 (1988)
331. S. Marcos, R. Niembro, M. López-Quelle, J. Navarro: Phys. Lett. **271**, 277 (1991)
332. R. Brockmann, W. Weise: Nucl. Phys. A **355**, 365 (1981)
333. M. Jaminon, C. Mahaux, P. Rochus: Nucl. Phys. A **365**, 371 (1981)
334. A. Bouyssy, S. Marcos, J.F. Mathiot, N.V. Giai: Phys. Rev. Lett. **55**, 1731 (1985)
335. A. Bouyssy, J.-F. Mathiot, N.V. Giai, S. Marcos: Phys. Rev. C **36**, 380 (1987)
336. P.G. Blunden, M.J. Iqbal: Phys. Lett. B **196**, 295 (1987)
337. J. Götz, J. Ramschütz, F. Weber, M.K. Weigel: Phys. Lett. B **226**, 213 (1989)
338. Jian-Kang Zhang, D.S. Onley: Phys. Rev. C **44**, 1915 (1991)
339. Jian-Kang Zhang, Yanhe Jin, D.S. Onley: Phys. Rev. C **48**, 2697 (1993)
340. R. Fritz, H. Müther, R. Machleidt: Phys. Rev. Lett. C **71**, 46 (1993); R. Fritz, H. Müther: Phys. Rev. C **49**, 633 (1994)
341. H.F. Boersma, R. Malfliet: Phys. Rev. C **49**, 1495 (1994)
342. Hua-lin Shi, Bao-qiu Chen, Zhong-yu Ma : Phys. Rev. C **52**, 144 (1995); Zhong-yu Ma, Hua-lin Shi, Bao-qiu Chen: Phys. Rev. C **50**, 3170 (1994)

343. P. Bernardos, V.N. Fomenko, N.V. Giai, M. López-Quelle, S. Marcos, R. Niembro, L.N. Savushkin: Phys. Rev. C **48**, 2665 (1993)
344. P. Bernardos, V.N. Fomenko, M. López-Quelle, S. Marcos, R. Niembro, L.N. Savushkin: J. Phys. G: Nucl. Part. Phys. **22**, 361 (1996)
345. S. Marcos, J.-F. Mathiot, M. López-Quelle, R. Niembro, P. Bernardos: Nucl. Phys. A **600**, 529 (1996)
346. H. Müller, B.D. Serot: Nucl. Phys. A **606**, 509 (1996)
347. J. Cohen, J.V. Noble: Phys. Rev. C **46**, 801 (1992)
348. J. Marež, B.K. Jennings: Phys. Rev. C **49**, 2472 (1994)
349. J. Marež, B.K. Jennings: Nucl. Phys. A **585**, 347c (1995)
350. M. Jaminon: Nucl. Phys. A **402**, 366 (1983)
351. M.R. Anastasio, L.S. Celenza, W.S. Pong, C.M. Shakin: Phys. Rep. **100**, 327 (1983)
352. R. Brockmann, R. Machleidt: Phys. Lett. B **149**, 283 (1984)
353. X. Campi, D.W.L. Sprung: Nucl. Phys. A **194**, 401 (1972)
354. S. Marcos, L.N. Savushkin, V.N. Fomenko, M. López-Quelle, R. Niembro: nucl-th/0307063
355. R. Niembro, P. Bernardos, M. López-Quelle, S. Marcos: Phys. Rev. C **64**, 055802 (2001)
356. O. Haxel, J.H.D. Jensen, H.E. Suess: Phys. Rev. **75**, 1766 (1949); Z. Phys. **128**, 295 (1950)
357. M. Goeppert-Mayer: Phys. Rev. **75**, 1969 (1949); Phys. Rev. **78**, 16 (1950)
358. F. de Jong, H. Lenske: Phys. Rev. C **57**, 3099 (1998); S. Kubis, M. Kutchera: Phys. Lett. B **399**, 191 (1997); V. Greco, M. Colonna, M. Di Toro, G. Fabbri, F. Matera: Phys. Rev. C **64**, 045203 (2001); J. Buervenich, D.G. Madland, J.A. Maruhn, P.G. Reinhard: Phys. Rev. C **65**, 044308 (2002)
359. H. Shen, Y. Sugahara, H. Toki: Phys. Rev. C **55**, 1211 (1997)
360. M. López-Quelle, N.V. Giai, S. Marcos, L.N. Savushkin: Phys. Rev. C **61**, 064321-1 (2000)
361. A. Arima, M. Harvey, K. Shimizu: Phys. Lett. B **30**, 517 (1969)
362. K.T. Hecht, A. Adler: Nucl. Phys. A **137**, 129 (1969)
363. A. Bohr, I. Hamamoto, B.R. Mottelson: Phys. Scr. **26**, 267 (1982)
364. B.R. Mottelson: Nucl. Phys. A **522**, 1 (1991)
365. O. Castanos, M. Moshinsky, C. Quesne: Phys. Lett. B **277**, 238 (1992)
366. A.L. Blokhin, C. Bahri, J.P. Draayer: J. Phys. A **29**, 2039 (1996)
367. A.L. Blokhin, C. Bahri, J.P. Draayer: Phys. Rev. Lett. **74**, 4149 (1995); T. Beuschel, A.L. Blokhin, J.P. Draayer: Nucl. Phys. A **619** (1997) 119; A.L. Blokhin, T. Beuschel, J.P. Draayer, C. Bahri: Nucl. Phys. A **612**, 163 (1997)
368. C. Bahri, J.P. Draayer, S.A. Moszkowski: Phys. Rev. Lett. **68**, 2133 (1992)
369. J.N. Ginocchio: Phys. Rev. Lett. **78**, 436 (1997)
370. J.N. Ginocchio, D.G. Madland: Phys. Rev. C **57**, 1167 (1998)
371. J.N. Ginocchio, A. Leviatan: nucl-th/9710019; Phys. Lett. B **425**, 1 (1998); Phys. Rev. Lett. **87**, 072502 (2001)
372. A. Leviatan, J.N. Ginocchio: nucl-th/0108016
373. J.N. Ginocchio: J. Phys. G: Nucl. Part. Phys. **25**, 617 (1999); Phys. Rep. **315**, 231 (1999); nucl-th/0012078
374. J. Meng, K. Sugawara-Tanabe, S. Yamaji, P. Ring, A. Arima: RIKEN-AF-NP-272, Riken, Wako, Saitama, October 1997; Phys. Rev. **58C**, R628 (1999); K. Sugawara-Tanabe, A. Arima: Phys. Rev. **58C**, R3065 (1998)

375. Y.K. Gambhir, J.P. Maharana, C.S. Warke: Eur. Phys. J. A **3**, 255 (1998); G.A. Lalazissis, Y.K. Gambhir, J.P. Maharana, C.S. Warke, P. Ring: nucl-th/9806009; Phys. Rev. **58C**, R45 (1998)
376. K. Sugawara-Tanabe, J. Meng, S. Yamaji, A. Arima: J. Phys. G: Nucl. Part. Phys. **25**, 811 (1999)
377. J. Meng, K. Sugawara-Tanabe, S. Yamaji, A. Arima: Phys. Rev. C **59**, 154 (1999)
378. J. Meng, I. Tanihata: Nucl. Phys. A **650**, 176 (1999)
379. J. Ginocchio: Phys. Rev. C **59**, 2487 (1999)
380. J. Ginocchio: Phys. Rev. Lett. **82**, 4599 (1999); nucl-th/9901022, Vol. 3, 12 April, 1999
381. S. Marcos, L.N. Savushkin, M. López-Quelle, P. Ring: Phys. Rev. C **62**, 054309 (2000); S. Marcos, M. López-Quelle, R. Niembro, L.N. Savushkin, P. Bernardos: Phys. Lett. B **513**, 30 (2001); S. Marcos, M. López-Quelle, R. Niembro, L.N. Savushkin, P. Bernardos: Eur. Phys. J. **17**, 173 (2003); P. Alberto, M. Fiolhais, M. Malheiro, A. Delfino, M. Chiapparini: Phys. Rev. Lett. **86**, 5015 (2001); P. Alberto, M. Fiolhais, M. Malheiro, A. Delfino, M. Chiapparini: Phys. Rev. C **65**, 034307 (2002)
382. M. López-Quelle, L.N. Savushkin, S. Marcos, R. Niembro: Eur. Phys. J. **17**, 173 (2003); M. López-Quelle, L.N. Savushkin, S. Marcos, R. Niembro: Nucl. Phys. A (2003), in press
383. H. Leeb, S. Wilmsen: Phys. Rev. C **62**, 024602 (2000)
384. J.S. Bell, H. Ruegg: Nucl. Phys. B **98**, 151 (1975)
385. R. Brockmann, R. Machleidt:, in *Nuclear Methods and the Nuclear Equation of State*, ed. M. Baldo. International Reviews of Nuclear Physics, Vol. 8 (World Scientific, Singapore 1999), p. 129; H. Müther: in *Nuclear Methods and the Nuclear Equation of State*, ed. M. Baldo. International Reviews of Nuclear Physics, Vol. 8 (World Scientific, Singapore 1999), p. 170
386. B.D. Day: Rev. Mod. Phys. **39**, 719 (1967); Rev. Mod. Phys. **50**, 495 (1978)
387. B. ter Haar, R. Malfliet: Phys. Rep. **149**, 207 (1987)
388. R. Malfliet: Prog. Part. Nucl. Phys. **21**, 207 (1988); F. de Jong, R. Malfliet: Phys. Rev. C **44**, 998 (1991)
389. C.J. Horowitz, B.D. Serot: Nucl. Phys. A **464**, 613 (1987)
390. P. Poschenrieder, M.K. Weigel: Phys. Rev. C **38**, 471 (1988); H. Huber, F. Weber, M.K. Weigel: Phys. Rev. C **51**, 1790 (1995)
391. A. Amorim, J.A. Tjon: Phys. Rev. Lett. **68**, 772 (1992); M.K. Banerjee, J.A. Tjon: nucl-th/9711029; Phys. Rev. C **58**, 2120 (1998); nucl-th/0101008; M.K. Banerjee: nucl-th/9806054
392. C. Fuchs, E. Lehmann, L. Sehn, F. Scholz, T. Kubo, J. Zipprich, A. Faessler: Nucl. Phys. A **603**, 471 (1996)
393. L. Sehn, H.H. Wolter: Nucl. Phys. A **601**, 473 (1996); C. Fuchs, L. Sehn, H.H. Wolter: Nucl. Phys. A **601**, 505 (1996); H. Elsenhans, L. Sehn, A. Faessler, H. Müther, N. Ohtzuka, H.H. Wolter: Nucl. Phys. A **536**, 750 (1992)
394. L. Sehn, C. Fuchs, A. Faessler: Phys. Rev. C **56**, 216 (1997); C. Fuchs, T. Waindzoch, A. Faessler, D.S. Kosov: Phys. Rev. C **58**, 2022 (1998)
395. T.S.H. Lee, F. Tabakin: Nucl. Phys. A **191**, 332 (1972)
396. G.E. Brown, A.D. Jackson, T.T.S. Kuo: Nucl. Phys. A **133**, 481 (1969)
397. A.A. Logunov, A.N. Tavkhelidze: Nuovo Cimento **29**, 380 (1963)
398. R. Blankenbecler, R. Sugar: Phys. Rev. **142**, 1051 (1966)

399. K. Holinde, R. Machleidt: Nucl. Phys. A **280**, 429 (1977)
400. V.R. Pandharipande, R.A. Smith: Nucl. Phys. A **237**, 507 (1975)
401. K. Holinde, R. Machleidt: Nucl. Phys. A **247**, 495 (1975)
402. K. Holinde, R. Machleidt: Nucl. Phys. A **256**, 479 (1976)
403. A.D. Jackson, D.O. Riska, B. Verwest: Nucl. Phys. A **249**, 397 (1975)
404. L.S. Celenza, C.M. Shakin: Phys. Rev. C **24**, 2704 (1981)
405. H. Müther, R. Machleidt, R. Brockmann: Phys. Rev. C **42**, 1981 (1990); A. Trasobares, A. Polls, A. Ramos, H. Müther: Nucl. Phys. A **640**, 471 (1998); H. Müther, S. Ulrych, H. Toki: Int. J. Mod. Phys. **8E**, 179 (1999)
406. R. Machleidt, R. Brockmann: Phys. Lett. B **160**, 364 (1985)
407. M.I. Haftel, F. Tabakin: Nucl. Phys. A **158**, 1 (1970)
408. B. Keister, R.B. Wiringa: Phys. Lett. B **173**, 5 (1986)
409. G.E. Brown, W. Weise, G. Baym, J. Speth: Comments Nucl. Phys. Part. **17**, 39 (1987)
410. J.L. Forest, V.R. Pandharipande, J.L. Friar: Phys. Rev. C **152**, 568 (1995)
411. L.S. Celenza, Shun-fu Gao, C.M. Shakin: Phys. Rev. C **41**, 1768 (1990); L.S. Celenza, H.B. Ai, Shun-fu Gao, C.M. Shakin: Phys. Rev. C **39**, 236 (1989); B. Ai, L.S. Celenza, A. Harindranath, C.M. Shakin: Phys. Rev. C **35**, 2299 (1987)
412. L. Engvik, E. Osnes, M. Hjorth-Jensen, G. Bao, E. Ostgaard: Astrophys. J. **469**, 794 (1996)
413. S. Ulrych, H. Müther: Phys. Rev. C **56**, 1788 (1997)
414. M.L. Cescato, P. Ring: Phys. Rev. C **57**, 134 (1998)
415. H. Lenske, C. Fuchs: Phys. Lett. B **345**, 355 (1995)
416. C. Fuchs, H. Lenske, H.H. Wolter: Phys. Rev. C **52**, 3043 (1995); S. Typel, H.H. Wolter: Nucl. Phys. A **656**, 331 (1999)
417. S. Gmuca: Nucl. Phys. A **547**, 447 (1992); J. Phys. G **17**, 1115 (1991)
418. R. Rapp, R. Machleidt, J.W. Durso, G. Brown: nucl-th/9706006; Phys. Rev. Lett. **82**, 1827 (1999)
419. C.-H. Lee, T.T.S. Kuo, G.Q. Li, G.E. Brown: Phys. Rev. C **57**, 3488 (1998)
420. D.P. Murdock, C.J. Horowitz: Phys. Rev. C **35**, 1442 (1987)
421. Y. Miyama, T. Suzuki: Phys. Lett. B **253**, 23 (1991); Y. Miyama: Phys. Lett. B **215**, 602 (1988); Y. Miyama, T. Suzuki: in *Relativistic Nuclear Many-Body Physics*, ed. by B.C. Clark, R.J. Perry, J.P. Vary (World Scientific, Singapore 1989), p. 371; T. Suzuki: in *Nuclear Physics at RCNP, from Nucleon Meson Nuclear Physics to Quark Lepton Nuclear Physics*, ed. by T. Yamazaki, Y. Mizuno, H. Ejiri (RCNP, Osaka 1996), p 75
422. A.F. Bielajew: Nucl. Phys. A **367**, 358 (1981)
423. W. Koepf, P. Ring: Nucl. Phys. A **493**, 61 (1989); Nucl. Phys. A **511**, 279 (1990)
424. H. Kurasawa, T. Suzuki: Phys. Lett. B **154**, 16 (1985)
425. M. Asakawa, C.M. Ko, P. Levai, X.J. Qiu: Phys. Rev. C **46**, R1159 (1992); G. Chanfray, P. Schuck: Nucl. Phys. A **545**, 2710 (1992); Nucl. Phys. A **555**, 329 (1993)
426. X. Ji: Phys. Lett. B **208**, 19 (1988)
427. C. Bedau, F. Beck: Nucl. Phys. A **560**, 518 (1993)
428. M. Asakawa: Nucl. Phys. A **629**, 344c (1998)
429. M. Herrmann, B.L. Friman, W. Noerenberg: Nucl. Phys. A **560**, 411 (1993)
430. G.E. Brown, M. Rho: Phys. Rev. Lett. **66**, 2720 (1991)

431. C. Adami, G.E. Brown: Phys. Rep. **234**, 1 (1993)
432. T.D. Cohen, R.J. Furnstahl, D.K. Griegel, X. Jin: Prog. Part. Nucl. Phys. **35**, 221 (1995)
433. K. Saito, T. Maruyama, K. Soutome: Phys. Rev. C **40**, 407 (1989)
434. S. Nishizaki, H. Kurasawa, T. Suzuki: Nucl. Phys. A **462**, 687 (1987); H. Kurasawa, T. Suzuki: Prog. Theor. Phys. **84**, 1030 (1990)
435. K. Tanaka, W. Bentz, A. Arima, F. Beck: Nucl. Phys. A **528**, 676 (1991)
436. J.C. Caillon, J. Labarsouque: Phys. Lett. B **311**, 19 (1993)
437. H.C. Jean, J. Piekarewicz, A.G. Williams: Phys. Rev. C **49**, 1981 (1994)
438. W. Weise: Nucl. Phys. A **574**, 347c (1994)
439. H. Shiomi, T. Hatsuda: Phys. Lett. B **334**, 281 (1994); H. Kuwabara, T. Hatsuda: Prog. Theor. Phys. **96**, 1163 (1995)
440. T. Hatsuda, H. Shiomi, H. Kuwabara: Prog. Theor. Phys. **95**, 1009 (1996)
441. T. Hatsuda, S.H. Lee: Phys. Rev. C **46**, R34 (1992); S.H. Lee: nucl-th/9705048.
442. V.L. Eletsky, B.L. Ioffe: Phys. Rev. Lett. **78**, 1010 (1997)
443. A.K. Dutt-Mazumder: Nucl. Phys. A **611**, 442 (1996)
444. J.-F. Mathiot: Nucl. Phys. A **601**, 380 (1996)
445. M. Nakano, N. Noda, T. Mitsumori, K. Koide, H. Kouno, A. Hasegawa, Liang-Gang Liu: Phys. Rev. C **56**, 3287 (1997)
446. R. Rapp, R. Machleidt, J.W. Durso, G.E. Brown: Phys. Rev. Lett. **82**, 1827 (1999)
447. R. Rapp, J.W. Durso, Z. Aouissart, G. Chanfray, O. Krehl, P. Schuck, J. Speth, J. Wambach: Phys. Rev. C **59**, R1237 (1999)
448. T. Hatsuda, T. Kunihiro, H. Shimizu: Phys. Rev. Lett. **82**, 2840 (1999)
449. T. Noro et al.: Nucl. Phys. A **663&664**, 517c (2000)
450. V. Bernard, U.G. Meissner: Nucl. Phys. A **489**, 647 (1988)
451. K. Wehrberger, F. Beck: Phys. Rev. C **35**, 298 (1988); K. Wehrberger, F. Beck: Phys. Rev. C **37**, 1148 (1988)
452. P.G. Blunden, P. McCorquodale: Phys. Rev. C **38**, 1861 (1988)
453. R.J. Furnstahl: Phys. Lett. B **152**, 313 (1985); Phys. Rev. C **38**, 370 (1988)
454. R.J. Furnstahl, C.E. Price: Phys. Rev. C **40**, 1398 (1989)
455. J.R. Shepard, E. Rost, J.A. McNeil: Phys. Rev. C **40**, 2320 (1989)
456. M. L'Huillier, N.V. Giai: Phys. Rev. C **39**, 2022 (1989)
457. M. L'Huillier, N.V. Giai, M. Auerbach: Nucl. Phys. A **519**, 83c (1990)
458. J.F. Dawson, R.J. Furnstahl: Phys. Rev. C **42**, 2009 (1990)
459. C.J. Horowitz, J. Piekarewicz: Nucl. Phys. A **511**, 461 (1990)
460. C.E. Price, E. Rost, J.R. Shepard, J.A. McNeil: Phys. Rev. C **45**, 1089 (1992)
461. D.S. Oakley, J.R.Shepard, N. Auerbach: Phys. Rev. C **45**, 2254 (1992)
462. Z.Y. Ma, H. Toki, N.V. Giai: Nucl. Phys. A **627**, 1 (1997)
463. Z.Y. Ma, N.V. Giai, H. Toki, M. L'Huillier: Phys. Rev. C **55**, 2385 (1997)
464. Z.Y. Ma, H. Toki, B. Chen, N.V. Giai: Prog. Theor. Phys. **98**, 917 (1997)
465. V.N. Fomenko, L.N. Savushkin, H. Toki: *Relativistic RPA Theory for Finite Nuclei* (1997, unpublished); see also RCNP Annual Report (Osaka University, Osaka 1997)
466. S. Shlomo, D.H. Youngblood: Phys. Rev. C **47**, 529 (1993) and references therein
467. Y. Fujita et al.: Phys. Rev. C **32**, 425 (1985)
468. A. Van der Woude: Prog. Part. Nucl. Phys. **18**, 217 (1987) and references therein

469. A. Erell, J. Alster, J. Lichtenstadt, M.A. Moinester, J.D. Bowman, M.D. Cooper, F. Irom, H.S. Matis, E. Plasetzky, U. Sennhauser, O. Ingram: Phys. Rev. Lett. **52**, 2134 (1984); H.S. Matis, E. Plasetzky, U. Sennhauser: Phys. Rev. C **34**, 1822 (1986)
470. J.D. Bowman: in *Nuclear Structure*, ed. by R. Broglia, G. Hagemann, B. Herskind (North-Holland, Amsterdam 1985), p. 549
471. B.L. Berman, S.C. Fultz: Rev. Mod. Phys. **47**, 713 (1975); R. Bengère: *Photonuclear Reactions*, Lecture Notes in Physics, Vol. 61 (Springer, Berlin, Heidelberg 1977)
472. P.G. Hansen, B. Jonson: Europhys. Lett. **4**, 409 (1987); K. Ikeda: Nucl. Phys. A **538**, 355c (1992)
473. D. Vretenar, H. Berghammer, P. Ring: Nucl. Phys. A **581**, 679 (1995); P. Ring, D. Vretenar, B. Podobnik: Nucl. Phys. A **598**, 107 (1996); B. Podobnik, D. Vretenar, P. Ring: Z. Phys. A **354**, 375 (1996); D. Vretenar, G.A. Lalazissis, R. Behnsch, W. Pöschl, P. Ring: Nucl. Phys. A **621**, 853 (1997)
474. R.J. Furnstahl, C.J. Horowitz: Preprint IU/NTC 87-6, Indiana University, 1987; Nucl. Phys. A **485**, 632 (1988); B.L. Friman, P. Henning: Preprint GSI-88-06, GSI, Darmstadt 1988
475. J. Milana: Phys. Rev. C **44**, 527 (1991)
476. M.P. Allendez, B.D. Serot: Phys. Rev. C **45**, 2975 (1992)
477. Zhong-yu Ma, Nguyen Van Giai, A. Wandelt, D. Vretenar, P. Ring: nucl-th/9910054, Vol. 4.
478. H. Kurasawa, T. Suzuki: Phys. Lett. B **474**, 262 (2000)
479. D. Vretenar, A. Wandelt, P. Ring: Phys. Lett. B **487**, 334 (2000); D. Vretenar, P. Ring, G. Lalazissis, N. Paar: Nucl. Phys. A **649**, 290 (1999)
480. J. Piekarewicz: Phys. Rev. C **62**, 051304 (2000); Phys. Rev. C **64**, 024307 (2001)
481. C.J. Horowitz, J. Piekarewicz: Phys. Rev. Lett. **62**, 391 (1989)
482. D. Vretenar, N. Paar, P. Ring, G.A. Lalazissis: Nucl. Phys. A **692**, 496 (2001)
483. D. Vretenar, N. Paar, P. Ring, T. Nikšić: Phys. Rev. C **65**, 021301R (2002)
484. R.J. Furnstahl, J. Piekarewicz, B.D. Serot: nucl-th/0205048
485. V. Greco, M. Colonna, M. Di Toro, F. Matera: nucl-th/0205046
486. J. König, P. Ring: Phys. Rev. Lett. **71**, 3079 (1993); A.V. Afanasjev, J. König, P. Ring: Phys. Lett. B **367**, 11 (1996); Nucl. Phys. A **608**, 107 (1996); J. König: Ph.D. Thesis, Technical University of Munich, Munich (1996); P. Ring, A. Afanasjev: Prog. Part. Nucl. Phys. **38**, 137 (1997); A.V. Afanasjev, J. König, P. Ring: Phys. Rev. C **60**, 051303 (1999)
487. K. Kaneko, M. Nakano, M. Matsuzaki: Phys. Lett. B **317**, 261 (1993); H. Madokoro, M. Matsuzaki: Phys. Rev. C **56**, R2934 (1997); H. Madokoro, M. Matsuzaki: nucl-th/9712063
488. P. Bonche, D. Vautherin: Nucl. Phys. A **372**, 496 (1981); R. Ogasawara, K. Sato: Prog. Theor. Phys. **70**, 1569 (1983)
489. W. Hillebrandt, R.G. Wolff: in *Nucleosynthesis-Challenges and New Developments*, ed. by W.D. Arnett and J.M. Truran (University of Chicago Press, Chicago 1985), p. 131; J.M. Lattimer, C.J. Pethick, D.G. Ravenhall, D.Q. Lamb: Nucl. Phys. A **432**, 646 (1985)
490. J.M. Lattimer, F.D. Swesty: Nucl. Phys. A **535**, 331 (1991)
491. H. Shen, H. Toki, K. Oyamatsu, K. Sumiyoshi: Nucl. Phys. A **637**, 435 (1998)
492. D. Hirata, K. Sumiyoshi, I. Tanihata, Y. Sugahara, T. Tachibana, H. Toki: Nucl. Phys. A **616**, 438c (1997)

493. K. Oyamatsu: Nucl. Phys. A **561**, 431 (1993)
494. M. Centelles, X. Viñas, M. Barranco, S. Marcos, R.J. Lombard: Nucl. Phys. A **537**, 486 (1992)
495. M. Centelles, X. Viñas, M. Barranco, P. Schuck: Ann. Phys. (NY) **221**, 165 (1993)
496. M. Centelles, X. Viñas, M. Barranco, N. Ohtsuka, A. Faessler, D.T. Khoa, H. Müther: Phys. Rev. C **47**, 1091 (1993)
497. M. Centelles, X. Viñas, P. Schuck: Nucl. Phys. A **567**, 611 (1994)
498. S. Haddad, M. Weigel: Nucl. Phys. A **578**, 471 (1994); K. Strobel, F. Weber, Ch. Shaab, M.K. Weigel: Int. J. Mod. Phys. E **6**, 669 (1997)
499. H. Müller, R.M. Drezler: Z. Phys. A **341**, 417 (1992); Nucl. Phys. A **563**, 649 (1993)
500. R.N. Schmid, E. Engel, R.M. Dreizler: Phys. Rev. C **52**, 164 (1995)
501. R.N. Schmid, E. Engel, R.M. Dreizler: Phys. Rev. C **52**, 2804 (1995)
502. C. Speicher, E. Engel, R.M. Dreizler: Nucl. Phys. A **562**, 569 (1993)
503. D. Von-Eiff, S. Haddad, M.K. Weigel: Phys. Rev. C **46**, 230 (1992)
504. D. Von-Eiff, W. Stocker, M.K. Weigel: Phys. Rev. C **50**, 1436 (1994)
505. D. Von-Eiff, J.M. Pearson, W. Stocker, M.K. Weigel: Phys. Lett. B **324**, 279 (1994); Phys. Rev. C **50**, 831 (1994)
506. K. Sumiyoshi, H. Kuwabara, H. Toki: Nucl. Phys. A **581**, 725 (1995)
507. H.A. Bethe, M.B. Johnson: Nucl. Phys. A **230**, 301 (1974)
508. K. Sumiyoshi, K. Oyamatsu, H. Toki: Nucl. Phys. A **595**, 327 (1995); K. Oyamatsu, I. Tanihata, Y. Sugahara, K. Sumiyoshi, H. Toki: Nucl. Phys. A **634**, 3 (1998)
509. S. Schaab, F. Weber, M.K. Weigel, N.K. Glendenning: Nucl. Phys. A **605**, 531 (1996)
510. C.J. Horowitz, K. Wehrberger: Phys. Lett. B **266**, 236 (1991); Nucl. Phys. A **531**, 665 (1991)
511. P.A.M. Guichon: Phys. Lett. B **200**, 235 (1988)
512. S. Fleck, W. Bentz, K. Shimizu, K. Yazaki: Nucl. Phys. A **510**, 731 (1990)
513. K. Saito, A.W. Thomas: Phys. Lett. B **327**, 9 (1994); K. Saito, A.W. Thomas: Phys. Rev. C **51**, 2757 (1995)
514. T. de Grand, R.L. Jaffe, K. Johson, J. Kiskis: Phys. Rev. D **12**, 2060 (1975)
515. V.K. Mishra, G. Fai, P.C. Tandy, M.R. Frank: Phys. Rev. C **46**, 1143 (1992); M.K. Banerjee: Phys. Rev. C **45**, 1359 (1992); E. Naar, M.C. Birse: J. Phys. G **19**, 555 (1993)
516. X. Jin, B.K. Jennings: Phys. Rev. C **54**, 1427 (1996); Phys. Lett. B **374**, 13 (1996); Phys. Rev. C **55**, 1567 (1997); H. Müller, B.K. Jennings: Nucl. Phys. A **640**, 55 (1998)
517. P.K. Panda, A. Mishra, J.M. Eisenberg, W. Greiner: Phys. Rev. C **56**, 3134 (1997)
518. G.E. Brown, M. Buballa, Z. Li, J. Wambach: Nucl. Phys. A **593**, 295 (1995); G.E. Brown, M. Rho: hep-ph/9504250
519. K. Saito, A.W. Thomas: Phys. Lett. B **335**, 17 (1994)
520. K. Saito, A. Michels, A.W. Thomas: Phys. Rev. C **46**, R2149 (1992); A.W. Thomas, K. Saito, A. Michels: Aust. J. Phys. **46**, 3 (1993); K. Saito, A.W. Thomas: Nucl. Phys. A **574**, 659 (1994)
521. P. Guichon, K. Saito, E. Rodionov, A.W. Thomas: Nucl. Phys. A **601**, 349 (1996)

522. P.G. Blunden, G.A. Miller: Phys. Rev. C **54**, 359 (1996)
523. K. Saito, K. Tsushima, A.W. Thomas: Nucl. Phys. A **609**, 339 (1996)
524. R. Tegen, R. Brockmann, W. Weise: Z. Phys. A **307**, 339 (1982); R. Tegen, M. Schedl, W. Weise: Phys. Lett. B **125**, 9 (1983)
525. K. Tsushima, K. Saito, A.W. Thomas: Phys. Lett. B **411**, 9 (1997)
526. K. Saito, K. Tsushima, A.W. Thomas: Phys. Rev. C **55**, 2637 (1997); A.W. Thomas: Nucl. Phys. A **629**, 20c (1998)
527. M.K. Banerjee: Phys. Rev. C **45**, 1359 (1992); M.K. Banerjee, J.A. Tjon: Phys. Rev. C **56**, 497 (1997)
528. H. Toki: Z. Phys. A **294**, 173 (1980)
529. M. Oka, K. Yazaki: Phys. Lett. B **90**, 41 (1980)
530. A. Faessler, F. Fernandez, G. Luebeck, K. Shimizu: Phys. Lett. B **112**, 201 (1980); Nucl. Phys. A **402**, 555 (1983)
531. Y. Fujiwara, C. Nakamoto, Y. Suzuki: Phys. Rev. Lett. **76**, 2242 (1996)
532. H. Toki, U. Meyer, A. Faessler, R. Brockmann: Phys. Rev. **58C**, 3749 (1998)
533. H. Shen, H. Toki: Phys. Rev. C **61**, 045205 (2000)
534. K. Saito: nucl-th/0207053
535. P. Papazoglou, S. Schramm, J. Schaffner-Bielich, H. Stöcker, W. Greiner: Phys. Rev. C **57**, 2576 (1998); P. Papazoglou, D. Zsgchiesche, S. Schramm, J. Schaffner-Bielich, H. Stöcker, W. Greiner: nucl-th/9806087
536. P. Manakos, T. Mannel: Z. Phys. A **330**, 223 (1988)
537. P. Manakos, T. Mannel: Z. Phys. A **334**, 481 (1989)
538. B.A. Nikolaus, D.G. Madland, P. Manakos: Bull. Am. Phys. Soc. **35**, 976 (1990); T. Hoch, B.A. Nikolaus, D.G. Madland: Bull. Am. Phys. Soc. **35**, 1651 (1990)
539. T. Hoch, D. Madland, P. Manakos, T. Mannel, B.A. Nikolaus, D. Strottman: Phys. Rep. **242**, 253 (1994)
540. J.L. Friar, D.G. Madland, B.W. Lynn: Phys. Rev. C **53**, 3085 (1996); J.L. Friar: nucl-th/0005076
541. J.J. Rusnak, R.J. Furnstahl: Nucl. Phys. A **627**, 495 (1997); R.J. Furnstahl, B.D. Serot: nucl-th/9907073; R.J. Furnstahl, J.V. Steele, N. Tirfessa: Nucl. Phys. A **671**, 396 (2000); R.J. Furnstahl, B.D. Serot: Nucl. Phys. A **671**, 447 (2000); R.J. Furnstahl, B.D. Serot: Comments Nucl. Part. Phys. A **23**, 2 (2000)
542. A. Manohar, H. Georgi: Nucl. Phys. B **234**, 189 (1984)
543. T. Mannel, T. Ohl, P. Manakos: Z. Phys. A **335**, 341 (1990)
544. B.W. Lynn: Nucl. Phys. B **402**, 281 (1993)
545. G. 't Hooft: in *Recent Developments in Gauge Theories*, ed. by G. 't Hooft et al. (Plenum, New York 1980), p. 135; J. Polchinski: in *Recent Directions in Particle Theory*, ed. by J. Harvey and J. Polchinski (World Scientific, Singapore 1993), p. 235
546. S. Weinberg: Physica. **96A**, 327 (1979)
547. T. Bürvenich, D.G. Madland, J.A. Maruhn, P.G. Reinhard: Phys. Rev. C **65**, 044308 (2002)
548. J. Zimányi, S.A. Moszkowski: Phys. Rev. C **42**, 1416 (1990)
549. M. Barranco, R.J. Lombard, S. Marcos, S.A. Moszkowski: Phys. Rev. C **44**, 178 (1991); P. Bernardos, R.J. Lombard, M. López-Quelle, S. Marcos, R. Niembro: Phys. Rev. C **62**, 024314 (2000)
550. M.M. Sharma, S.A. Moszkowski, P. Ring: Phys. Rev. C **44**, 2493 (1991)

551. H. Feldmeier, J. Lindner: Z. Phys. A **341**, 83 (1991)
552. W. Koepf, M.M. Sharma, P. Ring: Nucl. Phys. A **533**, 95 (1991)
553. N.K. Glendenning, F. Weber, S.A. Moszkowski: Phys. Rev. C **45**, 844 (1992); N.K. Glendenning: Phys. Rev. D **46**, 1274 (1992)
554. Zhi-Xin Qian, Hong-Qui Song, Ru-Keng Su: Phys. Rev. C **48**, 154 (1993)
555. S.K. Choudhury, R. Rakshit: Phys. Rev. C **48**, 598 (1993)
556. B. Datta, P.K. Sahu: Phys. Lett. B **318**, 277 (1993)
557. K. Miyazaki: Prog. Theor. Phys. **91**, 1271 (1994); Prog. Theor. Phys. **93**, 137 (1995)
558. A. Delfino, C.T. Coelho, M. Malheiro: Phys. Rev. C **51**, 2188 (1995); Phys. Lett. B **345**, 361 (1995)
559. Jian-Kang Zhang, D.S. Onley: Phys. Rev. C **44**, 2230 (1991)
560. S. Typel, T. von Chossy, H.H. Wolter: nucl-th/0210090
561. P. Mitra, G. Gangopadyay, B. Malakar: Phys. Rev. C **65**, 034329 (2002)
562. P.J. Ellis, E.K. Heide, S. Rudaz, M. Prakash: Phys. Rep. **242**, 379 (1994)
563. H. Kouno, K. Koide, T. Mitsumori, N. Noda, A. Hasegawa, M. Nakano: Phys. Rev. C **52**, 135 (1995)
564. R. Aguirre, O. Civitarese, A.L. de Paoili: Nucl. Phys. A **597**, 543 (1996); Nucl. Phys. A **579**, 573 (1994)
565. C. Greiner, P.-G. Reinhard: Z. Phys. A **342**, 379 (1992); R.J. Lombard, S. Marcos, J. Marež: Phys. Rev. C **51**, 1784 (1995); M. Chiapparini, A. Delfino, M. Malheiro, A. Gattone: Z. Phys. A **357**, 47 (1997); A. Bhattacharyya, S. Raha: Phys. Rev. C **53**, 522 (1996); T.S. Biro, J. Zimányi: Phys. Lett. B **391**, 1 (1997)
566. S.J. Brodsky: Comments Nucl. Part. Phys. **12**, 213 (1984)
567. M. Thies: Phys. Lett. B **162**, 255 (1985); Phys. Lett. B**166**, 23 (1986)
568. E.D. Cooper, B.K. Jennings: Nucl. Phys. A **458**, 717 (1986)
569. E. Bleszynski, M. Bleszynski, T. Jaroczewicz: Phys. Rev. Lett. **59**, 423 (1987)
570. J. Achtzehnter, L. Wilets: Phys. Rev. C **38**, 5 (1988)
571. T. Jaroszewicz, S.J. Brodsky: Phys. Rev. C **43**, 1946 (1991)
572. B. Blattel, V. Koch, W. Cassing, U. Mosel: Phys. Rev. C **38**, 1767 (1988)
573. H.-T. Elze, M. Gyulassy, D. Vasak, H. Heniz, H. Stöcker, W. Greiner: Mod. Phys. Lett. A **2**, 451 (1987)
574. Yu.B. Ivanov: Nucl. Phys. A **474**, 669 (1987)
575. C.M. Ko, Q. Li, R. Wang: Phys. Rev. Lett. **59**, 1084 (1987); C.M. Ko, Q. Li: Phys. Rev. C **37**, 2270 (1988)
576. B. Blattel, V. Koch, W. Cassing, U. Mosel: Phys. Rev. C **38**, 1767 (1988)
577. C.M. Ko: Nucl. Phys. A **495**, 321c (1989); Q. Li, J.Q. Wu, C.M. Ko: Phys. Rev. C **39**, 849 (1989)
578. M. Nielsen, J. da Providencia: Phys. Rev. C **40**, 2377 (1989)
579. X. Jin, Y. Zhuo, X. Znang, M. Sano: Nucl. Phys. A **506**, 655 (1990)
580. W. Botermans, R. Malfliet: Phys. Lett. B **215**, 617 (1988); W. Botermans, R. Malfliet: Phys. Rep. **198**, 115 (1990)
581. S.J. Wang, W. Cassing: Nucl. Phys. A **495**, 371c (1989); W. Cassing, S.J. Wang: Z. Phys. A **337**, 1 (1990)
582. A. Dellafiore, F. Matera: Phys. Rev. C **44**, 2456 (1991); F. Matera, V.Yu. Denisov: Phys. Rev. C **49**, 2816 (1994)
583. C. Fuchs, L. Sehn, H.H. Wolter: Nucl. Phys. A **545**, 151 (1992) Prog. Part. Nucl. Phys. **30**, 247 (1993); C. Fuchs, E. Lehmann, R.K. Puri, L. Sehn, A. Faessler, H.H. Wolter: J. Phys. G **22**, 131 (1996)

584. L. Sehn, H.H. Wolter: Nucl. Phys. A **601**, 473 (1996)
585. C. Fuchs, L. Sehn, H.H. Wolter: Nucl. Phys. A **601**, 505 (1996)
586. T. Gaitanos, C. Fuchs, H.H. Wolter: Nucl. Phys. A **650**, 97 (1999)
587. D. Bailin, A. Love: Phys. Rep. **107**, 325 (1984)
588. H. Kucharek, P. Ring: Z. Phys. A **339**, 23 (1991)
589. F.B. Guimarães, B.V. Carlson, T. Frederico: Phys. Rev. C **54**, 2385 (1996)
590. B.V. Carlson, T. Frederico, F.B. Guimarães: Phys. Rev. C **56**, 3097 (1997)
591. F. Matera, G. Fabbri, A. Dellafiore: Phys. Rev. C **56**, 228 (1997)
592. Ø. Elgarøy, L. Engvik, M. Hjorth-Jensen, E. Osnes: Phys. Rev. Lett. **77**, 1428 (1996); Ø. Elgarøy, L. Engvik, M. Hjorth-Jensen, E. Osnes: Phys. Rev. C **57**, R1069 (1998)
593. L.P. Gorkov: Sov. Phys. JETP **7**, 505 (1958)
594. J. Dobaczewski, H. Flocard, J. Treiner: Nucl. Phys. A **422**, 103 (1984)
595. T. Gonzalez-Llarena, J.L. Egido, G.A. Lalazissis, P. Ring: Phys. Lett. B **379**, 13 (1996)
596. W. Greiner: in *Proc. European Conf. on Advances in Nuclear Physics and Related Areas*, ed. by D.M. Brink, M.E. Grypeos, S.E. Massen (Thessaloniki 1999), p. 973
597. G.A. Lalazissis, D. Vretenar, W. Poschl, P. Ring: Phys. Lett. B **418**, 7 (1998); Nucl. Phys. A **632**, 363 (1998)
598. D. Vretenar, G. Lalazissis, P. Ring: Phys. Rev. Lett. **82**, 4595 (1999)
599. J. Meng, P. Ring: Phys. Rev. Lett. **80**, 460 (1998)
600. J. Meng: Phys. Rev. C **57**, 1229 (1998)
601. J. Meng: Nucl. Phys. A **635**, 3 (1998)
602. M.M. Sharma, A.R. Farhan, S. Mythili: Phys. Rev. C **61**, 054306 (2000)
603. B.V. Carlson, D. Hirata: Phys. Rev. C **62**, 054310 (2000)
604. S. Sugimoto, K. Sumiyoshi, H. Toki: Phys. Rev. C **64**, 054310 (2001)
605. M. Matsuzaki: Phys. Rev. C **58**, 3407 (1998)
606. M. Serra, A. Rummel, P. Ring: Phys. Rev. C **65**, 014304 (2002)
607. W. Pöschl, D. Vretenar, G.A. Lalazissis, P. Ring: Phys. Rev. Lett. **79**, 3841 (1997)
608. G.A. Lalazissis, D. Vretenar, P. Ring, M. Stoitsov, L.M. Robledo: Phys. Rev. C **60**, 014310 (1999)
609. G.A. Lalazissis, D. Vretenar, P. Ring: Nucl. Phys. A **650**, 133 (1999); Nucl. Phys. A **679**, 481 (2001)
610. T. Nikšić, D. Vretenar, P. Ring, G.A. Lalazissis: Phys. Rev. C **65**, 054320 (2002)
611. J. Meng, H. Toki, J.Y. Zeng, S.Q. Zhang, S.-G. Zhou: Phys. Rev. C **65**, 041302R (2002)
612. J. Meng, S.-G. Zhou, I. Tanihata: Phys. Lett. B **532**, 209 (2002)
613. D.C. Zheng, L. Zamick, H. Müther: Phys. Rev. C **45**, 275 (1992)
614. L. Zamick, D.C. Zheng, H. Müther: Phys. Rev. C **45**, 2763 (1992)
615. L. Zamick, D.C. Zheng: Phys. Rep. **242**, 233 (1994)
616. G.E. Brown, H. Müther, M. Prakash: Nucl. Phys. A **506**, 565 (1990)
617. M.M. Sharma, G.A. Lalazissis, P. Ring: Phys. Lett. B **317**, 9 (1993)
618. G.A. Lalazissis, M.M. Sharma, J. König, P. Ring: in *Proc. Int. Conf. Nuclear Shapes and Nuclear Structure at Low Excitation Energies*, ed. by M. Vergnes, D. Goutte, P.-H. Heenen, J. Sauvage (Antibes, Editions Frontieres, Gif-sur-Yvette 1994), p. 161

619. M.M. Sharma, G.A. Lalazissis, J. König, P. Ring: Phys. Rev. Lett. **74**, 3744 (1995)
620. P.-G. Reinhard, H. Flocard: Nucl. Phys. A **584**, 467 (1995)
621. S. Marcos, L.N. Savushkin, M. López-Quelle, R. Niembro, P. Bernardos: nucl-th/0009087; Phys. Lett. B **507**, 135 (2001)
622. G.A. Lalazissis, D. Vreternar, W. Poschl, P. Ring: Phys. Lett. B **418**, 7 (1998); Nucl. Phys. A **632**, 363 (1998)
623. S. Yoshida, H. Sagawa: Nucl. Phys. A **658**, 3 (1999)
624. I.S. Towner, J.C. Hardy: in *Symmetries and Fundamental Interactions in Nuclei*, ed. by W.C. Haxton, E.M. Henley (World Scientific, Singapore 1995), p. 183
625. Y. Nedjadi, J.R. Rook: Nucl. Phys. A **528**, 537 (1991)
626. A.O. Gattone, E.D. Izquierdo, M. Chiapparini: Phys. Rev. C **46**, 788 (1992)
627. A. Gil, M. Kleinmann, H. Müther, E. Oset: Nucl. Phys. A **584**, 621 (1995)
628. Tae-Sun Park, I.S. Towner, K. Kubodera: Nucl. Phys. A **579**, 381 (1994)
629. S.M. Ananyan, B.D. Serot, J.D. Walecka: nucl-th/0207019
630. S. Boffi, C. Giusti, F.D. Pacati: Phys. Rep. **226**, 1 (1993)
631. C.J. Horowitz: Phys. Rev. C **57**, 3430 (1998)
632. C.J. Horowitz, S.J. Pollock, P.A. Souder, R. Michaels: Phys. Rev. C **63**, 025501 (2001)
633. R.J. Furnstahl: nucl-th/0112085
634. J. Schaffner, C.B. Dover, A. Gal, C. Greiner, D.J. Millener, H. Stöcker: Ann. Phys. **235**, 35 (1994)
635. J. Ellis, J.I. Kapusta, K.A. Olive: Nucl. Phys. B **348**, 345 (1991)
636. A. Bouyssy, J. Hüfner: Phys. Lett. B **64**, 276 (1976)
637. R. Brockmann, W. Weise: Phys. Lett. B **69**, 167 (1977); Nucl. Phys. A **355**, 365 (1981)
638. J. Boguta, S. Bohrmann: Phys. Lett. B **102**, 93 (1981)
639. W. Brückner et al.: Phys. Lett. B **79**, 157 (1978)
640. H.J. Pirner: Phys. Lett. B **85**, 190 (1979)
641. J.V. Noble: Phys. Lett. B **89**, 325 (1980)
642. C.B. Dover, A. Gal: Prog. Part. Nucl. Phys. **12**, 171 (1984)
643. B.K. Jennings: Phys. Lett. B **246**, 325 (1990)
644. J. Cohen, J.V. Noble: Phys. Rev. C **46**, 801 (1992)
645. C. Greiner, J. Schäffner-Bielich: nucl-th/9801062
646. M. Rufa, H. Stöcker, P.-G. Reinhard, J. Maruhn, W. Greiner: J. Phys. G **13**, 143 (1987)
647. J. Mareš, J. Žofka: Z. Phys. A **333**, 209 (1989); Z. Phys. A **345**, 47 (1993)
648. M. Rufa, J. Schaffner, J. Maruhn, H. Stöcker, W. Greiner, P.-G. Reinhard: Phys. Rev. C **42**, 2469 (1990)
649. J. Cohen: Phys. Rev. C **48**, 1346 (1993)
650. E.D. Cooper, B.K. Jennings, J. Mareš: Nucl. Phys. A **585**, 157 (1995)
651. C.B. Dover, H. Feshbach, A. Gal: Phys. Rev. C **51**, 541 (1955)
652. N.K. Glendenning, D. Von-Eiff, M. Haft, H. Lenske, M.K. Weigel: Phys. Rev. C **48**, 889 (1993)
653. X. Jin, R.J. Furnstahl: Phys. Rev. C **49**, 1190 (1994)
654. X. Jin, M. Nielsen: Phys. Rev. C **51**, 347 (1995)
655. R.J. Lombard, S. Marcos, J. Mareš: Phys. Rev. C **50**, 2900 (1994); Phys. Rev. C **51**, 1784 (1995)

656. J. Marež, B.K. Jennings: Nucl. Phys. A **585**, 347c (1995)
657. K. Tsushima, K. Saito, J. Haidenbauer, A.W. Thomas: Nucl. Phys. A **630**, 691 (1998)
658. J.A. Caballero, T.W. Donnely, E. Moya de Guerra, J.M. Udías: Nucl. Phys. A **632**, 323 (1998)
659. S. Gardner, J. Piekarewicz: Phys. Rev. C **50**, 2822 (1994)
660. A. Picklesimer, J.W. van Orden, S.J. Wallace: Phys. Rev. C **32**, 1312 (1985); A. Picklesimer, J.W. van Orden: Phys. Rev. C **35**, 266 (1987); Phys. Rev. C **40**, 290 (1989)
661. J.P. McDermott: Phys. Rev. Lett. **65**, 1991 (1990); Y. Jin, D.S. Onley, L.E. Wright: Phys. Rev. C **45**, 1311 (1992)
662. Y. Jin, D.S. Onley: Phys. Rev. C **45**, 377 (1994); M. Hedayati-Poor, J.I. Johnson, H.S. Sherif: Phys. Rev. C **51**, 2044 (1995)
663. I. Bobeldijk et al.: Phys. Rev. Lett. **73**, 26 (1994)
664. J.M. Udías, P. Sarriguren, E. Moya de Guerra, E. Garrido, J.A. Caballero: Phys. Rev. C **48**, 2731 (1993)
665. J.M. Udías, P. Sarriguren, E. Moya de Guerra, E. Garrido, J.A. Caballero: Phys. Rev. C **51**, 3246 (1995)
666. J.M. Udías, P. Sarriguren, E. Moya de Guerra, J.A. Caballero: Phys. Rev. C **53**, R1488 (1996)
667. J. Gao et al.: Phys. Rev. Lett. **84**, 3265 (2000)
668. K. Kaki: Nucl. Phys. A **601**, 445 (1996)
669. S.J. Wallace: Nucl. Phys. A **631**, 137c (1998)
670. A. Stadler: Nucl. Phys. A **631**, 152c (1998)
671. M.C. Birse: Phys. Rev. C **51**, 1083 (1995)
672. F. Ritz, H. Göller, T. Wilbois, H. Arenhövel: Phys. Rev. C **55**, 2214 (1997)
673. U. van Kolck, G.A. Miller, D.O. Riska: Phys. Lett. B **388**, 679 (1996)
674. D.O. Riska: Nucl. Phys. A **606**, 251 (1996)
675. J. Adam, A. Stadler, M.T. Peña, F. Gross: nucl-th/9702055
676. F. Gross: Phys. Rev. **186**, 1448 (1969); Phys. Rev. C **26**, 2203, 2226 (1982)
677. M.J. Zuilhof, J.A. Tjon: Phys. Lett. B **84**, 31 (1979); Phys. Rev. C **24**, 736 (1981)
678. F. de Jong, K. Nakayama: Phys. Lett. B **385**, 33 (1996)
679. G.H. Martinus, O. Scholten, J.A. Tjon: Phys. Lett. B **402**, 7 (1997)
680. A.Yu. Korchin, O. Scholten: Nucl. Phys. A **581**, 493 (1995)
681. E. Hummel, J.A. Tjon: Phys. Rev. C **42**, 423 (1990)
682. E. Hummel, J.A. Tjon: Phys. Rev. C **49**, 21 (1994)
683. N.K. Devine, S.J. Wallace: Phys. Rev. C **48**, 973 (1993)
684. J.W. van Orden, N.K. Devine, F. Gross: Phys. Rev. Lett. **75**, 4369 (1995)
685. J.L. Forest, V.R. Pandharipande, A. Arriaga: nucl-th/9805033
686. A. Lépine-Szily: in *Proc. 20th Brazilian Workshop on Nuclear Physics*, ed. by S.R. Souza et al. (World Scientific, Singapore 1997), p. 261
687. D. Hirata, K. Sumiyoshi, I. Tanihata, Y. Sugahara, H. Toki. RIKEN-AF-NP-268, Riken, Wako, Saitama, October 1997
688. Y. Sugahara: Doctoral Thesis, Tokyo Metropolitan University (1995)
689. D. Hirata, K. Sumiyoshi, B.V. Carlson, H. Toki, I. Tanihata: Nucl. Phys. A **609**, 131 (1996);
690. K. Oyamatsu, I. Tanihata, Y. Sugahara, K. Sumiyoshi, H. Toki: Nucl. Phys. A **634**, 3 (1998)

691. R. Morlock, R. Kunz, A. Mayer, M. Jaeger, A. Mueller, J.W. Hammer, P. Mohr, H. Oberhummer, G. Staudt, V. Koelle: Phys. Rev. Lett. **79**, 3831 (1997)
692. Zhongzhou Ren, A. Faessler, A. Bobyk: Phys. Rev. C **57**, 2752 (1998)
693. G.A. Lalazissis, S. Raman: Phys. Rev. C **58**, 1467 (1998)
694. Zhongzhou Ren, W. Mittig, F. Sarazin: Nucl. Phys. A **652**, 250 (1999); Zhongzhou Ren, H. Toki: Nucl. Phys. A **689**, 691 (2001)
695. Y.K. Gambhir: Nucl. Phys. A **570**, 101c (1994)
696. J.A. Sheikh: Phys. Rev. C **48**, 476 (1993)
697. D. Vretenar, W. Pöschl, G.A. Lalazissis, P. Ring: Phys. Rev. C **57**, R1060 (1998)
698. W. Greiner: Int. J. Mod. Phys. E **5**, 1 (1996)
699. M. Bender, K. Rutz, P.-G. Reinhard, J.A. Maruhn, W. Greiner: Phys. Rev. C **58**, 2126 (1998)
700. S.K. Patra, Cheng-Li Wu, C.R. Praharaj, R.K. Gupta: Nucl. Phys. A **651**, 117 (1999)
701. S.K. Patra, W. Greiner, R.K. Gupta: J. Phys. G: Nucl. Part. Phys. **26**, L65 (2000)
702. K. Rutz, M. Bender, T. Bürvenich, T. Schilling, P.-G. Reinhard, J.A. Maruhn, W. Greiner: Phys. Rev. C **56**, 238 (1997)
703. M. Bender, K. Rutz, P.-G. Reinhard, J.A. Maruhn, W. Greiner: Phys. Rev. C **60**, 034304-1 (1999)
704. S.K. Patra, R.K. Gupta, W. Greiner: Mod. Phys. Lett. A **12**, 1727 (1997)
705. S.K. Patra, C.-L. Wu, C.R. Praharaj, R.K. Gupta: Nucl. Phys. A **651**, 117 (1999)
706. A.T. Kruppa, M. Bender, W. Nazarewicz, P.-G. Reinhard, T. Vertse, S. Cwiok: Phys. Rev. C **61**, 034313 (2000)
707. D. Vasak, H. Stöcker, B. Müller, W. Greiner: Phys. Lett. B **93**, 243 (1980); D. Vasak, B. Müller, W. Greiner: Phys. Scr. **22**, 25 (1980)
708. T. Lippert, U. Becker, N. Grün, W. Scheid, G. Soff: Phys. Lett. B **207**, 366 (1988)
709. I.N. Mishustin, L.M. Satarov, H. Stöcker, W. Greiner: Phys. Rev. C **52**, 3315 (1995)
710. I.N. Mishustin, L.M. Satarov, H. Stöcker: in *Proc. Int. Conf. on Nuclear Physics at the Turn of the Millenium*, Wilderness, South Africa, 1996, ed. by H. Stöcker, A. Gallman, J.H. Hamilton (World Scientific, Singapore 1997). p. 322
711. I.N. Mishustin, L.M. Satarov, H. Stöcker, W. Greiner: Phys. Rev. C **57**, 2552 (1998)
712. T. Wakasa et al.: Phys. Rev. C **55**, 2909 (1997)
713. K. Ikeda, S. Fujii, J.I. Fujita: Phys. Rev. Lett. **3**, 271 (1963)
714. C. De Conti, A.P. Galeão, F. Krmpotić: Phys. Lett. B **444**, 14 (1998); Phys. Lett. B **494**, 46 (2000); K. Kurasawa, T. Suzuki, N.V. Giai: nucl-th/0306080
715. K.J. Raywood et al.: Phys. Rev. C **41**, 2836 (1990)
716. T. Wakasa et al.: Phys. Lett. B **426**, 257 (1998)
717. H. Toki, W. Weise: Phys. Lett. B **97**, 12 (1980)
718. W.H. Dickhof, A. Faessler, J. Meyer-ter-Vehn, H. Müther: Phys. Rev. C **23**, 1154 (1981)
719. I. Towner: Phys. Rep. **155**, 263 (1987)

720. H. Toki, W. Weise: Phys. Rev. Lett. **42**, 1034 (1979)
721. E. Shiino, Y. Saito, M. Ichimura, H. Toki: Phys. Rev. C **34**, 1004 (1986)
722. H. Toki, W. Weise: Z. Phys. A **292**, 389 (1978)
723. Y. Futami, H. Toki, W. Weise: Phys. Lett. B **77**, 37 (1978)
724. S. Krewald, K. Nakayama, J. Speth: Phys. Rep. **161**, 103 (1988)
725. H. Toki, I. Tanihata: Phys. Rev. C **59**, 1196 (1999)
726. K. Yoshida, H. Toki: Nucl. Phys. A **648**, 75 (1999)
727. H. Sasaki et al.: in *Proceedings of the 14th RCNP International Symposium on Nuclear Reaction Dynamics of Nucleon–Hadron Many-Body System*, ed. by H. Ejiri, T. Noro, K. Takahashi, H. Toki (World Scientific, Singapore 1996), p. 21
728. J.B. McCelland et al.: Phys. Rev. Lett. **69**, 582 (1992)
729. T.N. Taddeucci et al.: Phys. Rev. Lett. **73**, 3516 (1994)
730. C.J. Horowitz, J. Piekarewicz: Phys. Rev. C **50**, 2540 (1994)
731. H. Toki, S. Sugimoto, K. Ikeda: nucl-th/0110017
732. L.S. Warrier, Y.K. Gambhir: Phys. Rev. C **49**, 871 (1994)
733. K. Rutz, M. Bender, P.-G. Reinhard, J.A. Maruhn: Phys. Lett. B **468**, 1 (1999)
734. V. Blum, J.A. Maruhn, P.-G. Reinhard, W. Greiner: Phys. Lett. B **323**, 262 (1994)
735. K. Rutz, J.A. Maruhn, P.-G. Reinhard, W. Greiner: Nucl. Phys. A **590**, 680 (1995)
736. M. Lutz: Nucl. Phys. A **677**, 241 (2000)
737. M. Lutz, B. Friman, C. Appel: Phys. Lett. B **474**, 7 (2000)
738. N. Kaiser, S. Fritsch, W. Weise: Nucl. Phys. A **697**, 255 (2002)
739. N. Kaiser, S. Fritsch, W. Weise: Nucl. Phys. A **700**, 343 (2002)
740. P. Finelli, N. Kaiser, D. Vretenar, W. Weise: nucl-th/0205016
741. N. Kaiser: nucl-th/0206056
742. L. Corragio, A. Covello, A. Gargano, N. Itaco, T.T.S. Kuo, D.R. Entem, R. Machleidt: nucl-th/0206025
743. D.R. Entem, R. Machleidt: Phys. Lett. B **524**, 93 (2002)
744. B.C. Clark, S. Hama, S.G. Kalberman, E.D. Cooper, R.L. Mercer: Phys. Rev. C **31**, 694 (1985)
745. R.J. Furnstahl, J.J. Rusnak, B.D. Serot: Nucl. Phys. A **632**, 607 (1998)
746. C.E. Price, G.E. Walker: Phys. Lett. **155**B, 17 (1985)
747. Y. Iwasaki, H. Kouno, A. Hasagawa, M. Nakano: nucl-th/0008018
748. R. Manka, I. Bednarek: nucl-th/0011084
749. M.M. Sharma, M.A. Nagarajan, P. Ring: Ann. Phys. (NY) **231**, 110 (1994)
750. D. Vretenar, T. Nikšić, P. Ring: Phys. Rev. C **65**, 024321 (2002)
751. R. Pezer, A. Ventura, D. Vretenar: nucl-th/0205068
752. G. Mao: nucl-th/0211034
753. M. Fierz: Z. Phyzik **104**, 553 (1937)
754. Y. Takahashi: *An Introduction to Field Quantization* (Pergamon, Oxford 1968); Phys. Rev. D **26**, 2169 (1982); J. Math. Phys. **24**, 1783 (1983)
755. S.P. Klevansky: Rev. Mod. Phys. **64**, 649 (1992)
756. A. Maruhn, T. Bürvenich, D.G. Madland: nucl-th/0007010; J. Comput. Phys. **169**, 238 (2001)

Index

anomalous kink 267
antiproton–nucleus scattering 97
axial meson 29, 138, 307–309
axial-vector form factor 121, 308

Brueckner–Hartree–Fock approach 5, 173

chiral symmetry 2, 5, 29, 113–117, 122, 125, 131, 135, 245, 246, 248, 250–253, 274, 290, 292
  breaking 116, 119, 245, 251
  global 120, 125, 126, 309
  local 309
Coester band 188, 190, 191, 193, 197
Coulomb–nuclear interference 92
cranking approach 216

Δ-isobar 99–103, 107, 109, 111, 112, 174, 182, 183, 185–187, 190, 272–274
Dirac matrices 7, 30, 45, 106, 251, 293, 294, 298, 309, 318, 320

effective-field theory 251, 286, 290
electroweak interactions 268, 269
EMC effect 75, 241
EMC-effect 239
equation of state of nuclear matter 3, 5, 40, 45, 47, 48, 68, 70, 197, 200, 219, 223, 257, 258, 278, 286, 291

Fierz transformation 159, 319, 320
fission barrier 285
four-dimensional notation 27, 63, 293
Fubini terms 100

Gamow–Teller resonance 282
gauge symmetry 125, 292

giant resonances 5, 62, 201, 203, 204, 208, 210, 212, 220, 285, 301
Goldberger–Treiman relation 116, 117, 124
Goldstone boson 5, 115
Green function 63, 103, 105, 107, 108, 174, 204, 269

Hartree–Fock–Bogoliubov approximation 263
heavy-ion collisions 261, 262, 281
Higgs mechanism 125

incompressibility 47, 70, 73, 124, 133, 135, 192, 208, 209, 212, 239, 242, 249, 250, 253, 256–259, 278, 286, 302
isotopic symmetry 99, 104, 113, 117, 128, 290, 292, 296
  local 125

Levenberg–Marquardt algorithm 248
Lorentz covariance 290

meson effective mass 130, 141, 142, 154–156, 200, 302
meson self-coupling terms 48, 67, 68, 154, 265, 266

Nambu–Jona-Lasinio model 321
naturalness 251–253
neutrino 160, 236, 237
neutron halo 74, 212, 265, 277, 279, 291
neutron star 3, 6, 45, 47, 68, 160, 200, 219, 232–236, 278, 291
  cooling 291
  matter 232–234
  merging 291
  profile 68, 232, 234, 236

NL-SH parameterization 67, 69–73, 203, 208–213
NL1 parameterization 67–74, 86, 215, 281
NL2 parameterization 67
NL3 parameterization 67
NLC parameterization 67
NLRa parameterization 67
no-sea approximation 41, 98, 205, 247
nonrelativistic Hartree–Fock approximation 4, 52, 165, 197, 248
nuclear magnetic moment 28, 62, 75, 80, 82–84, 86–88, 214, 268, 285, 301
nuclear magnetism 85, 87, 217, 285
nuclei
  deformed 1, 2, 48, 64–68, 164–166, 220, 279
  exotic 74, 250, 268, 276, 277, 279, 280, 291
  hypernuclei 149, 244, 270, 271
  spherical 49, 65, 133, 165, 166, 204, 220, 221, 226, 249, 302
  spin-saturated 164
  spin-unsaturated 5, 164
  superheavy 279, 280, 301
  unstable 2, 68, 70, 74, 203, 212, 219, 235, 236, 278, 290
nucleon effective mass 29, 39, 45, 48, 54, 60, 61, 83, 130, 131, 163, 181, 188, 191, 201, 208, 240, 245, 280, 283–285, 289, 300–302

Okamoto–Nolen–Schiffer anomaly 80, 241
one-boson exchange potentials 25, 29, 32, 36, 164, 173, 174, 180, 190, 199
optical potential 1, 5, 40, 89, 93, 94, 97, 98, 138, 174, 200, 262, 273, 300

(p,n) spin experiments 283, 284
parity conservation 290
pion condensation 99, 102, 104, 110, 190, 202, 203, 282, 283, 289, 290
  surface 284
pionic polarization 5, 104
point-coupling model 159, 259, 267, 318
  relativistic 246, 248–253

proton halo 279
pseudoscalar coupling 28, 111, 124, 139, 190, 305
pseudospin symmetry 1, 165, 166, 168, 252, 291
pseudovector coupling 28, 36, 111, 139, 274, 308

quantum kinetic equations 262
quark–meson coupling models 239, 245

radioactive nuclear beams 212, 236, 290
relativistic effective Lagrangians 3, 41, 125, 135, 138, 141, 199, 203, 205–209, 212, 215, 248
relativistic Hartree–Fock approximation 2, 5, 28, 48, 124, 137, 138, 141, 146, 154, 171, 199, 201, 246, 291, 321
relativistic mean-field approximation 2–5, 39, 41, 44, 48, 49, 60, 61, 64, 67, 68, 137, 159, 201, 215–217, 220, 233, 239, 267, 268, 270, 271, 277, 278, 281, 285, 301
  cranked 2
relativistic parity 20, 295
relativistic RPA 2, 5, 84, 85, 201–203, 206, 214, 237, 291
relativistic transport theory 263
renormalizability 290
rotation and reflection invariance 298

saturation 1, 3–5, 36, 40, 41, 44, 53, 56, 57, 60, 62–64, 68, 70, 73, 98, 119–121, 133, 145, 149–151, 157, 159, 174, 181, 182, 186, 188–195, 198–200, 235, 241–243, 257, 261, 263, 264, 277, 278, 286, 289, 291, 308
scalar derivative coupling models 253
self-interactions 1–3, 14, 42, 47, 48, 61, 63, 67, 80, 81, 126, 131, 138, 140, 141, 153, 154, 157–159, 194, 199–201, 203, 206, 207, 214, 216, 259, 271, 290, 291, 296
Skyrme–Hartree–Fock model 52, 219, 268, 278

spatial components of the vector fields 44, 214, 284
spin–orbit interaction 40, 42, 97, 159, 160, 170, 268, 279, 310, 312
supernova 3, 219, 220, 232, 235, 236, 291

tensor couplings 42, 76, 139, 143, 150, 153, 164, 270, 301–303, 307
Thomas–Fermi method 219, 220
time-even fields 284
time-odd fields 284, 285
time-reversal invariance 85, 87, 144, 298, 299

TM1 parameterization 51, 69–74, 203, 208–213, 219, 220, 278
TM2 parameterization 51, 69–71, 73

vector dominance model 2, 5, 75, 76, 81, 292

Walecka model 40, 41, 47, 53, 58–60, 62, 85, 91, 125, 132, 175, 216, 239, 253, 254, 256, 258, 296
Weinberg transformation 100, 305
Wigner–Seitz cell 221, 222, 228

Printing: Saladruck Berlin
Binding Lüderitz&Bauer, Berlin